AF581913

Power Electronics Applications in Renewable Energy Systems

Power Electronics Applications in Renewable Energy Systems

Special Issue Editor

Gilsoo Jang

MDPI • Basel • Beijing • Wuhan • Barcelona • Belgrade

Special Issue Editor
Gilsoo Jang
Korea University
Korea

Editorial Office
MDPI
St. Alban-Anlage 66
4052 Basel, Switzerland

This is a reprint of articles from the Special Issue published online in the open access journal *Energies* (ISSN 1996-1073) from 2019 to 2020 (available at: https://www.mdpi.com/journal/energies/special_issues/electronics_applications).

For citation purposes, cite each article independently as indicated on the article page online and as indicated below:

LastName, A.A.; LastName, B.B.; LastName, C.C. Article Title. *Journal Name* **Year**, *Article Number*, Page Range.

ISBN 978-3-03928-700-0 (Pbk)
ISBN 978-3-03928-701-7 (PDF)

© 2020 by the authors. Articles in this book are Open Access and distributed under the Creative Commons Attribution (CC BY) license, which allows users to download, copy and build upon published articles, as long as the author and publisher are properly credited, which ensures maximum dissemination and a wider impact of our publications.
The book as a whole is distributed by MDPI under the terms and conditions of the Creative Commons license CC BY-NC-ND.

Contents

About the Special Issue Editor

Gilsoo Jang (Professor) received his BE from the Department of Electrical Engineering, Korea University, Korea, in 1991, where he also received his MS degree in 1994. He received his Ph.D. from Electrical Engineering, Iowa State University, USA, in 1997. He was a visiting scientist at the Department of Electrical and Computer Engineering, Iowa State University, USA, from 1997 to 1998, a senior researcher at Power Systems Lab, Korea Electric Power Research Institute, Korea, from 1998 to 2000, and a visiting professor at the Department of Electrical Engineering, Cornell University, USA, from 2006 to 2007.

Dr. Jang has been a professor at the School of Electrical Engineering, Korea University since 2000. He was the Associate Dean of the College of Engineering from 2011 to 2013, and the Dean of the School of Electrical Engineering from 2015 to 2017. Since 2017, he has been the director of the Grid Connected Power Electronics Research Center in Korea, with the support of the National Research Foundation of Korea.

He has participated in academic societies as a journal editor of IEEE Transactions on Smart Grid, where he was a technical reviewer for international journals and conferences, a conference organizer, and so on. He is a member of CIGRE, and a senior member of IEEE and KIEE.

Preface to "Power Electronics Applications in Renewable Energy Systems"

This Special Issue of *Energies*, "Power electronics applications in renewable energy systems", aims to publish promising and novel methods and techniques to maintain the stable operation of power grids with power electronics-based renewable resources. This issue can be categorized into the following three parts.

○ PCS control and structure in renewable energy systems
   - A study to reduce the adverse effects of renewable energy sources in power systems through a control strategy and a novel structure of power electronic devices in renewable energy systems

No.	Title	Authors	Remarks
1	Improved Deadbeat FC-MPC Based on the Discrete Space Vector Modulation Method with Efficient Computation for a Grid-Connected Three-Level Inverter System	Alsofyani, I.M.; Lee, K.-B.	*Energies* **2019**, *12*(16), 3111;
2	A Modified DSC-Based Grid Synchronization Method for a High Renewable Penetrated Power System Under Distorted Voltage Conditions	Li, T.; Li, Y.; Yang, J.; Ge, W.; Hu, B.	*Energies* **2019**, *12*(21), 4040
3	A Novel Hybrid Converter Proposed for Multi-MW Wind Generator for Offshore Applications	Luqman, M.; Yao, G.; Zhou, L.; Zhang, T.; Lamichhane, A.	*Energies* **2019**, *12*(21), 4167
4	Reduction of DC Current Ripples by Virtual Space Vector Modulation for Three-Phase AC–DC Matrix Converters	Dang, H.-L.; Jun, E.-S.; Kwak, S.	*Energies* **2019**, *12*(22), 4319
5	Temporary Fault Ride-Through Method in Power Distribution Systems with Distributed Generations Based on PCS	Lee, J.-H.; Jeon, S.-G.; Kim, D.-K.; Oh, J.-S.; Kim, J.-E	*Energies* **2020**, *13*(5), 1123;

○ Application of power electronics in a power system with renewable energy

- Applied technologies of power electronic equipment in a power system, considering the variability of renewable energy resources

No.	Title	Authors	Remarks
1	Coordinated Frequency and State-of-Charge Control with Multi-Battery Energy Storage Systems and Diesel Generators in an Isolated Microgrid	Chang, J.-W.; Lee, G.-S.; Moon, H.-J.; Glick, M.B.; Moon, S.-I.	*Energies* **2019**, *12*(9), 1614
2	A Control Strategy of Modular Multilevel Converter with Integrated Battery Energy Storage System Based on Battery Side Capacitor Voltage Control	Wang, Z.; Lin, H.; Ma, Y.	*Energies* **2019**, *12*(11), 2151;
3	Accuracy Improvement Method of Energy Storage Utilization with DC Voltage Estimation in Large-Scale Photovoltaic Power Plants	Yoo, Y.; Jang, G.; Kim, J.-H.; Nam, I.; Yoon, M.; Jung, S.	*Energies* **2019**, *12*(20), 3907;
4	Practical Application Study for Precision Improvement Plan for Energy Storage Devices Based on Iterative Methods	Suh, J.; Yoon, M.; Jung, S.	*Energies* **2020**, *13*(3), 656;

○ Other applications related to renewable energy resources

- Technologies to address the adverse effects of the increasing penetration of renewable energy resources in a power system

No.	Title	Authors	Remarks
1	Development of Floquet Multiplier Estimator to Determine Nonlinear Oscillatory Behavior in Power System Data Measurement	Choi, N.; Cho, H.; Lee, B.	*Energies* **2019**, *12*(10), 1824
2	Preliminary Design of Multistage Radial Turbines Based on Rotor Loss Characteristics under Variable Operating Conditions	Tong, Z.; Cheng, Z.; Tong, S.	*Energies* **2019**, *12*(13), 2550;
3	Microgeneration of Electricity Using a Solar Photovoltaic System in Ireland	Virupaksha, V.; Harty, M.; McDonnell, K.	*Energies* **2019**, *12*(23), 4600;
4	Design and Testing of a Low Voltage Solid-State Circuit Breaker for a DC Distribution System	Tracy, L.; Sekhar, P.K.	*Energies* **2020**, *13*(2), 338;

5	A Two-Stage Industrial Load Forecasting Scheme for Day-Ahead Combined Cooling, Heating and Power Scheduling	Park, S.; Moon, J.; Jung, S.; Rho, S.; Baik, S.W.; Hwang, E.	*Energies* **2020**, *13*(2), 443;

Gilsoo Jang
Special Issue Editor

Article

Improved Deadbeat FC-MPC Based on the Discrete Space Vector Modulation Method with Efficient Computation for a Grid-Connected Three-Level Inverter System

Ibrahim Mohd Alsofyani and Kyo-Beum Lee *

Department of Electrical and Computer Engineering, Ajou University, 206, World cup-ro, Yeongtong-gu, Suwon 16499, Korea

* Correspondence: kyl@ajou.ac.kr; Tel.: +82-31-219-2376

Received: 5 July 2019; Accepted: 8 August 2019; Published: 13 August 2019

Abstract: The utilization of three-level T-type (3L T-type) inverters in finite set-model predictive control (FS-MPC) of grid-connected systems yielded good performance in terms of current ripples and total harmonic distortions. To further improve the system's performance, discrete space vector modulation (DSVM) was utilized to synthesize a higher number of virtual voltage vectors. A deadbeat control (DBC) method was used to alleviate the computational burden and provide the optimum voltage vector selection. However, 3L inverters are known to suffer from voltage deviation, owing to the imbalance of the neutral-point voltage. We have proposed a simplified control strategy for balancing the neutral point in the FS-MPC with DSVM and DBC of grid-connected systems, not requiring a weighting factor or additional cost function calculation. The effectiveness of the proposed method was validated using simulation and experiment results. Our experimental results show that the execution time of the proposed algorithm was significantly reduced, while its current quality performance was not affected.

Keywords: deadbeat control; discrete space vector modulation; computation efficiency; model predictive control; grid connected system; three-level system; T-type inverter

---

1. Introduction

Model predictive control (MPC) has become an attractive alternative for controlling power electronic applications, such as motor drives and power converters [1]. There are two main categories of MPC: (1) continuous MPC (CMPC), in which output is generated and delivered to a modulator, and (2) finite-set MPC (FS-MPC), which can control a finite number of feasible switching states using a predefined cost function [2–5]. Among the two types, FS-MPC is preferable, owing to its many advantages, such as the fast dynamic response, intuitive appeal, inclusion of constraints and nonlinearities, and easy implementation. However, an important drawback of the original method is its variable switching frequency and large current ripples, which requires the use of large passive filter components [2,6].

Numerous studies aiming to improve the performance of classical FS-MPC for both power converters and motor drives have been performed. To reduce current ripples and alleviate harmonic distortion, an attempt was made in [7,8] to increase the prediction horizon of FS-MPC. Although good performance was achieved, intensive experimentation is still necessary for determination of correct weighting factors and control horizons [8], which is computationally demanding [9]. A deadbeat solution was suggested in Reference [10] for a two-level voltage-source inverter, which allows the computational load of FS-MPC to be reduced by reducing the complete enumeration for the whole voltage vectors. Although this solution helped to address the problem of computational intensity, the large torque ripples could not be eliminated.

FS-MPC based on discrete space vector modulation (DSVM) was proposed in References [11,12] to reduce current ripples and guarantee a constant switching frequency. The main advantage of DSVM is that it allows the number of degrees of freedom to be increased by synthesizing various virtual voltage vectors in the space vector diagram [12]. Similarly to classical FS-PTC, the optimal voltage vector is selected to minimize the objective error in the respective cost function, and is applied to the inverter using space vector modulation (SVM). Nevertheless, the main issue associated with the DSVM approach is its high computational burden, owing to a large lookup table that holds the initialized virtual voltage vectors. To solve this problem, deadbeat control was utilized to consider a limited number of virtual voltage vectors, regardless of their number [13]. In this way, the calculation time was significantly reduced, making the method suitable for realistic applications.

Although two-level inverters (2L inverters) are extensively used for power converters and motor drives for generation of voltage vectors applied to terminals [14], they suffer from some issues. Two-level inverters require a very high switching frequency; hence, a higher harmonic current distortion is generated, owing to the limitation of voltage levels. In addition, the maximal DC link voltage is constrained due to the rating of the semiconductors. Therefore, multilevel inverters (ML inverters) have been considered an attractive solution capable of solving the above-mentioned problems and synthesizing output voltages with several discrete levels. Three-level inverters (3L inverters), such as neutral-point clamped (NPC) and T-type inverters, are the most prominent topologies of ML inverters. Compared with 2L inverters, the number of degrees of freedom for obtaining the voltage vectors is higher, which yields better current quality and better control. Despite the advantages of 3L inverters, neutral-point voltage balancing seriously affects their control performance [15], causing higher ripples and distortion of stator currents. Hence, 3L inverters require high-rated capacitors, owing to their unequal voltage distribution, which, in turn, results in a higher voltage stress on the semiconductor switches.

It is worth mentioning it is complicated to include a NPC voltage balance variable in the cost function when implementing DBC. Thus, an algorithm for the DC link capacitor voltage balance should be separately applied for proper 3L inverter operation [16–20]. For example, in Reference [16], a calculated zero-voltage sequence was used for neutral-point balancing, while in Reference [18], the time-offset injection method was used for the same purpose. In Reference [20], a deadbeat model of predictive control combined with the discrete space vector modulation method was used for grid-connected systems using T-type 3L inverters. Two cost functions were used: one for selecting the optimal voltage, and another for the compensated voltage offset, because the neutral-point voltage problem of 3L inverters cannot be included as a variable in the cost function, due to the use of DBC method. The optimal voltage vectors were then synthesized using the SVM method for the entire sampling duration. Nevertheless, the use of two cost functions increased the computational burden of the control system.

This paper proposes a simplified control method for balancing the neutral point in the FS-MPC with the DSVM and DBC of grid-connected systems. Therefore, unlike the approach in Reference [20], the proposed method does not require additional cost functions for balancing the capacitance voltage. The proposed method led to a significant reduction in computation time while maintaining the current quality performance. This method was simulated and experimentally verified on a grid-connected, three-level T-type voltage source inverter.

2. System Modeling

Essentially, there are three switching states for three-level topologies such as neutral-point clamped (NPC) and T-type inverters. Figure 1 shows the topology of a grid-connected, three-level T-type voltage source inverter. The output poles of the T-type inverter can be connected to three different levels of the source voltage, namely the positive bus bar "*P*," the negative bus bar "*N*," and the neutral point

"0" [21]. With three-phase to two-phase transformation, the model of the inverter in the stationary d–q frame is given by:

$$u = R \cdot \vec{i} + L \cdot \frac{di}{dt} + e. \quad (1)$$

In Equation (1), R, L, u, i, and e, are the load resistance, filter inductance, inverter voltage vector, output current vector, and grid voltage vector, respectively. Because the top capacitance voltage (V^{top}) and the bottom capacitance voltage (V^{bottom}) can become unequal in the three-level voltage source inverter (VSI) (and hence will produce poor-quality output current and distorted output voltage), the capacitor voltages should be observed and taken into account at every time step, to ensure that they become balanced. The dynamic equations of the two capacitor voltages are given by:

$$V^{top} = V^{top} + I_{NP} \cdot (T_s/C) \quad (2)$$

$$V^{bottom} = V^{bottom} + I_{NP} \cdot (T_s/C) \quad (3)$$

where C is the capacitance of each capacitor, T_s is the sampling time, and I_{NP} is the neutral-point current.

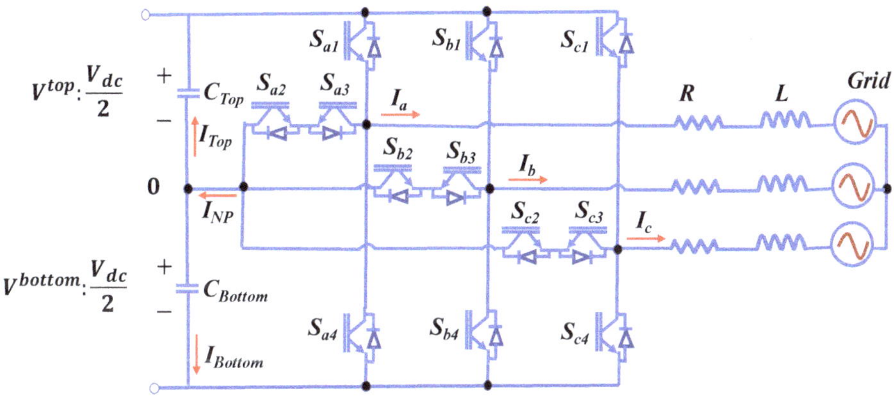

Figure 1. Circuit diagram of three-level T-type inverter.

3. FS-MPC Based DSVM with Deadbeat Control

3.1. Classical FS-MPC

The classical three-level FS-MPC uses only 19 voltage vectors as shown in Figure 2, defined by switching states, for prediction, and uses three stages: (1) estimation, (2) prediction, and (3) cost function optimization. An optimal control action is selected by minimizing a predefined cost function. All of the controlled objectives are included in the cost function in terms of errors; the errors are calculated by their respective references. The performance and the required computational burden of the model have been analyzed for the 3L T-type VSI.

To predict the future behavior of the inverter, the continuous-time model of Equation (1) should be approximated by a discrete-time model, using the normal forward Euler approximation with the sampling period T_s, as:

$$dy/dt = (y(k+1) - y(k))/T_s. \quad (4)$$

Thus, the discretized models of current are given as:

$$i_{dq}(k+1) = i_{dq}(k) + (T_s/L)\big(u_{dq}(k) - R \cdot i_{dq}(k) - e_{dq}(k)\big). \quad (5)$$

To predict $i(k+1)$ in Equation (5), $i(k)$, $e(k)$, and $v(k)$ are required. The current $i(k)$ is measured using the hardware sensors on the stationary d–q axis. Assuming that $e(k)$ does not change much during T_s, $e(k)$ can be estimated by shifting Equation (5) one step backward, as

$$e_{dq}(k) \approx e_{dq}(k-1) \tag{6}$$

$$e_{dq}(k) = u_{dq}(k-1) - (L/T_s)\left[(1-(R{\cdot}T_s/L)){\cdot}i_{dq}(k) - i_{dq}(k-1)\right]. \tag{7}$$

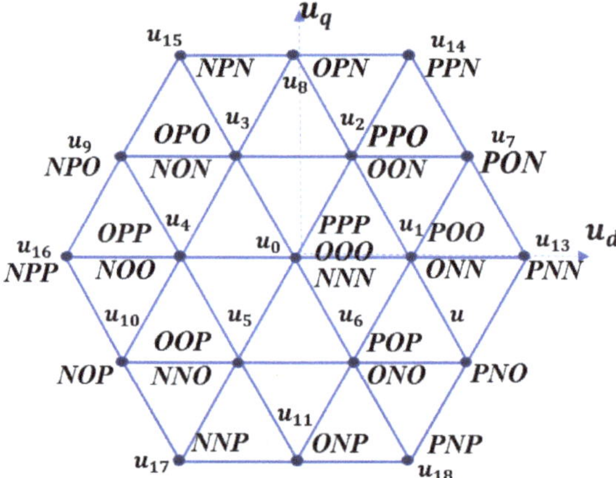

Figure 2. Space vector diagram of conventional finite-set model predictive control (FS-MPC).

As it is evident that there is a one-step delay in the digital control system, the voltage vector at time k will not be applied until time $k+1$. Therefore, to obtain the optimal voltage vector among the 19 voltage vectors, the cost function (*CF*) used to measure the errors between the references and the predictions in the stationary d–q frame was defined, as in Equation (8). To remove the delay, the voltage vector at time $k+2$ should be used in the cost function of Equation (8), instead of $k+1$:

$$CF = \left|i^*_d(k+2) - i_d(k+2)\right| + \left|i^*_q(k+2) - i_q(k+2)\right|. \tag{8}$$

The voltage vector, which yields the minimal *CF*, will be selected as the optimal vector u^{opt} and will be applied to the grid terminal by the inverter during the next sampling time.

3.2. FS-MPC Based on DSVM

The approach that uses FS-MPC based on the DSVM strategy follows the same route for predicting the state variables that is used in the classical FS-MPC approach, as described in the previous subsection, except that the selected voltage vectors are obtained from various virtual voltage vectors for prefix time intervals [20]. These virtual voltage vectors are obtained by subdividing the space vector diagram (SVD) into M equal parts.

For example, Figure 3 shows the virtual voltage vectors where the space vector diagram is subdivided into three equal parts. The overall number of voltage vectors (T) in the space vector diagram is:

$$T(M) = 3{\cdot}M{\cdot}(M+1) + 1. \tag{9}$$

According to Equation (9), the virtual voltage vectors, which are normally much higher, are calculated to obtain the current predictions and cost function. To enforce the actual output current vector to approach the reference current vector in the next step [20], the optimal voltage vector is applied to the inverter using the DSVM strategy during the entire time T_s. However, when taking into account all of the candidate voltage vectors for the current prediction, the computational burden increases dramatically, and the method becomes unsuitable for actual control systems.

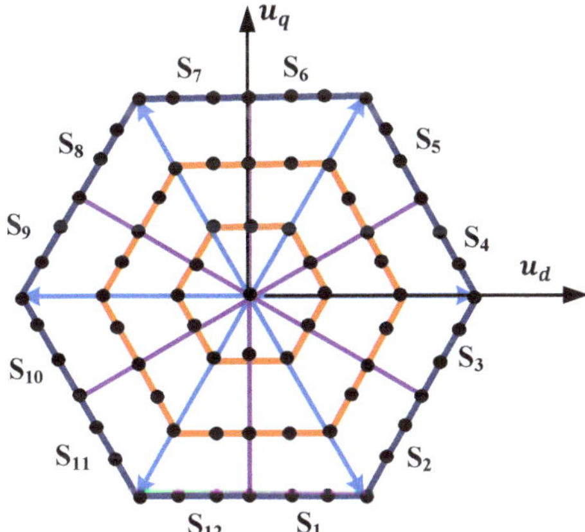

Figure 3. Space vector diagram of conventional model predictive control with discrete space vector modulation (DSVM-MPC).

3.3. Deadbeat Control Strategy

As mentioned earlier, considering all virtual voltage vectors significantly increases the computational burden on the prediction process. Therefore, a control method, namely deadbeat control (DBC), for alleviating the computational burden on the digital signal processor (DSP) is required. In addition, all of the voltage vector values in the stationary *d–q* frame should be predefined in a lookup table [16]. As a result, more complex lookup tables are required with increasing voltage vectors [16]. Thus, the deadbeat control method is utilized for reducing the computational burden associated with instantaneous computation of candidate voltage vectors.

The virtual voltage vectors in the stationary *d–q* frame are defined as [20]:

$$u_d(x,y) = (V_{dc}/6{\cdot}M)[(a+2{\cdot}e){\cdot}x+3b{\cdot}y] \tag{10}$$

$$u_q(x,y) = (\sqrt{3}{\cdot}V_{dc}/6{\cdot}M)[c{\cdot}x+(d+2{\cdot}f){\cdot}y]. \tag{11}$$

In Equations (10) and (11), x and y are the coordinates of the different sectors, and the coefficients $(a, b, \ldots f)$ are obtained from Table 1 [20]. Figure 4 schematically shows the voltage vectors for sectors S_4, S_5, and S_6 for $M = 3$.

Table 1. Coefficients for Each Sector.

	a	b	c	d	e	f
S_1	1	1	−1	1	0	0
S_2	−1	1	1	1	0	0
S_3	0	0	0	0	1	1
S_4	0	0	0	0	−1	1
S_5	1	−1	1	1	0	0
S_6	−1	−1	−1	1	0	0
S_7	−1	−1	1	−1	0	0
S_8	1	−1	−1	−1	0	0
S_9	0	0	0	0	−1	−1
S_{10}	0	0	0	0	1	−1
S_{11}	−1	1	−1	−1	0	0
S_{12}	1	1	1	−1	0	0

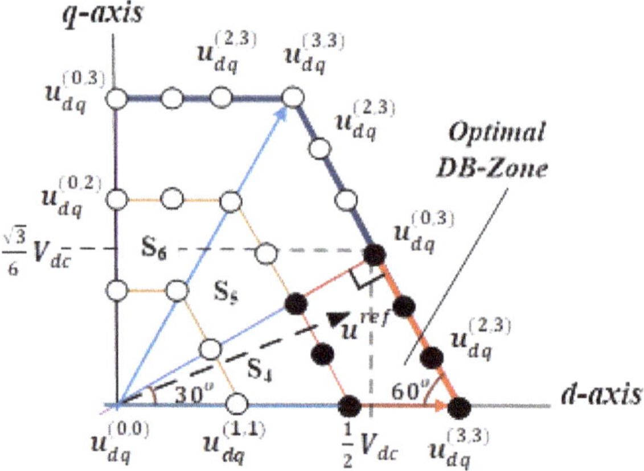

Figure 4. Space vector diagram of conventional DSVM-MPC.

To determine the candidate voltage vectors, the deadbeat method uses the reference inverter voltage vector phase (θ^*) and magnitude ($|u^*|$). Assuming that the control works correctly, it is possible to assume:

$$i(k+2) = i^*(k+2). \tag{12}$$

Under the assumption of Equation (12), it is possible to calculate the reference voltage vector as:

$$u^*_{dq}(k+1) = R \cdot i^*_{dq}(k+1) + (T_s/L)\left(i^*_{dq}(k+2) - i_{dq}(k+1)\right) + e_{dq}(k+1). \tag{13}$$

Hence, θ^* and u^* can be expressed as:

$$\theta^* = tan^{-1}\left(u^*_d(k+1)/u^*_q(k+1)\right) \tag{14}$$

$$\left|u^*(k+1)\right| = \sqrt{\left(u^*_d(k+1)\right)^2 + \left(u^*_q(k+1)\right)^2}. \tag{15}$$

Among the twelve sectors on the space vector diagram, a single sector is selected by θ^*. Since u^{opt} exists in the vicinity of the circle, this method determines two concentric hexagonal diagrams (SVD_1 and SVD_2), as given below:

$$SVD_1 = |u^*| \cdot (M/V_{dc}) \cdot (3/2) \tag{16}$$

$$SVD_2 = [|u^*| \cdot (M/V_{dc}) \cdot (3/2)] + 1. \tag{17}$$

Hence, only the voltage vectors within SVD_1 and SVD_2 are taken into account during the calculation and prediction processes. In this way, the candidate voltage vectors are restricted, which significantly reduces the computational burden on the DSP. Finally, the optimal voltage vector u^{opt} is selected at $k + 2$, using the cost function in Equation (8), before it is sent in the next sampling instant to the space vector modulator. To simplify our discussion in the next section, deadbeat DSVM-MPC was used to identify the FS-MPC with DSVM and deadbeat control.

4. Deadbeat DSVM-MPC with Proposed Neural Point Balancing Method

In this paper, the deadbeat DSVM-MPC used virtual voltage vectors obtained using the DSVM strategy, and their values were calculated instantly for the current prediction [20]. Although good performance can be obtained with deadbeat DSVM-MPC, the 3L T-type VSI topology can lead to an unbalanced neutral-point voltage, which increases the voltage stress on the switching device. It also increases the total harmonic distortion (THD) of the output current, because a low-order harmonic will appear in the output voltage. A large deviation of the DC link capacitance voltage is caused by the inconsistency in switching or imbalance of DC capacitors, owing to the manufacturing tolerance [22].

It is worth mentioning that there are various modulation strategies to synthesize output voltages, which can be categorized into two common types: continuous-based modulation (CPWM), such as sine pulse-width modulation (SPWM), and discontinuous-based modulation, namely discontinuous pulse-width modulation (DPWM). To optimize the performance of the 3L T-type VSI system, the voltages of the in-series connected DC link capacitors should be balanced. Unlike our previous work [20], wherein the problem of balancing the capacitor voltages was treated using a separate cost function to modify the offset voltage in SPWM (increasing the computational burden), the proposed deadbeat DSVM-MPC implements a modification in DPWM using a hysteresis capacitance voltage control. The main advantage of this proposed method is that it is straightforward and easily implemented, without additional hardware or extensive computation. Furthermore, it is known that by using DPWM, the switching losses are reduced, and better harmonic characteristics can be obtained for high modulation indices, compared with inverters that use continuous pulse-width modulation [23–26]. Although there are several different DPWM methods, conventional 60° DPWM is most commonly used for systems with the unity power factor. The idea behind the 60° DPWM method is schematically shown in Figure 5. Therefore, the pole reference voltages to be applied to the VSI are described by:

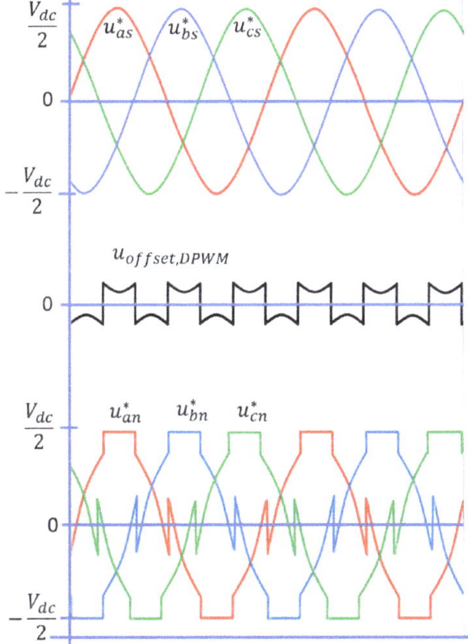

Figure 5. Reference voltage, offset voltage, and pole reference voltages of the conventional 60° discontinuous pulse-width modulation (DPWM).

$$\begin{aligned} u_{an}^* &= u_{as}^* + u_{offset,DPWM} \\ u_{bn}^* &= u_{bs}^* + u_{offset,DPWM} \\ u_{cn}^* &= u_{cs}^* + u_{offset,DPWM} \end{aligned} \tag{18}$$

where u^*_{an}, u^*_{bn}, and u^*_{cn} are the pole reference voltages to be applied to the VSI, whereas u^*_{as}, u^*_{bs}, and u^*_{cs} are the optimal reference voltages by deadbeat DSVM-MPC of each phase, respectively. The voltage $u_{offset,DPWM}$ is the offset voltage used in the DPWM, which is calculated as follows:

$$\begin{cases} u_{offset,DPWM} = \frac{V_{dc}}{2} - u_{\max}, & (u_{\max} + u_{\min} > 0) \\ u_{offset,DPWM} = -\frac{V_{dc}}{2} - u_{\min}, & (u_{\max} + u_{\min} < 0) \end{cases} \tag{19}$$

where u_{max} and u_{min} are, respectively, the maximum and the minimum values among the phase reference voltages.

Figure 6a depicts the imbalance of the DC link capacitor voltage for the 3L T-type VSI. Note that when the switch of the either phase is locked in the P state, the top DC link capacitor voltage V^{top} is decreased and the bottom DC link capacitor voltage V^{bottom} is increased. Conversely, if the switch of the same switch is locked in the N state, the top and the bottom DC link capacitor voltages are increased and decreased, respectively. Thus, clamping plays a major role in decreasing or increasing the top and bottom capacitance voltages. Figure 6b shows the proposed DPWM method using the hysteresis capacitance voltage band (ΔHB_{cv}). The proposed neutral-point voltage balancing method uses a compensated voltage offset ($u_{offset,cv}$) depending on the top and bottom capacitor voltages in the linear modulation range. The $u_{offset,cv}$ has an opposite influence from $u_{offset,DPWM}$ on changing the direction of the top and bottom voltages, which is given as:

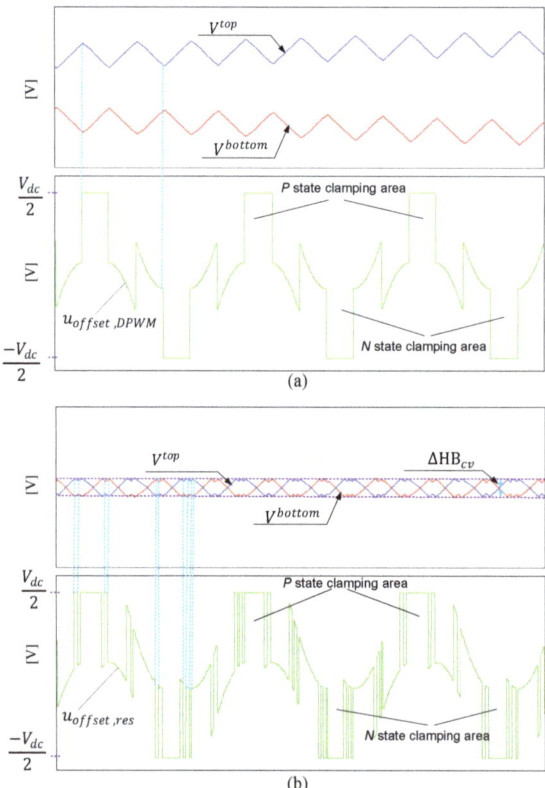

Figure 6. Behavior of top and bottom capacitance voltages with respect to: (**a**) conventional DPWM; (**b**) proposed DPWM.

$$\begin{cases} u_{offset,cv} = \frac{V_{dc}}{2} - u_{\max}, & (V^{top} > V^{bottom}) \\ u_{offset,cv} = -\frac{V_{dc}}{2} - u_{\min}, & (V^{top} < V^{bottom}) \end{cases}. \quad (20)$$

The proposed method seeks to maintain the advantage of diminishing the stress on transistors and minimizing the power loss, while simultaneously achieving a balanced DC link capacitance voltage with a stable and acceptable capacitance voltage error $|E_{cv}|$, which is defined as

$$|E_{cv}| = \left|V^{top} - V^{bottom}\right|. \quad (21)$$

The capacitance voltage error $|E_{cv}|$ is inevitable; thus, it should be limited to an acceptable error band ΔHB_{cv}, to avoid large deviations of the DC link capacitance voltage and high switching. The limited error band is defined as $|E_{limit}|$. The resultant voltage offset ($u_{offset,res}$) can then be designed depending on the following condition:

$$\begin{cases} u_{offset,res} = u_{offset,DPWM} + u_{offset,cv}, & (|E_{cv}| > |E_{\text{limit}}|) \\ u_{offset,res} = u_{offset,DPWM}, & (|E_{cv}| \le |E_{\text{limit}}|) \end{cases}. \quad (22)$$

According to Equation (22), if the capacitance voltage error $|E_{cv}|$ exceeds the limited error band, the $u_{offset,cv}$ will be injected into the $u_{offset,DPWM}$ to have an opposite effect on the clamped voltage, otherwise, the $u_{offset,DPWM}$ will continue with its normal operation.

The effect of the resultant voltage offset $u_{offset,res}$ on one of the pole reference voltages to be applied to the VSI is seen in Figure 6b. Note that the clamping areas are almost similar to the conventional DPWM; at the same time, the error of the DC link capacitance voltage is stable. In addition, there is a short clamping area injected by the $u_{offset,cv}$ to maintain the capacitance error range. It is noteworthy that the switching frequency of $u_{offset,res}$ and the non-switching area depend mainly on the setting of $|E_{\text{limit}}|$; increasing the limited error band will result in a lower switching of $u_{offset,res}$ and will deteriorate the quality of the current, whereas reducing the limited error band will cause undesirable switching frequency of $u_{offset,res}$ and a small clamping area. Thus, it is recommended that the tradeoff error band limit for achieving desirable current control performance be determined.

5. Simulation Results

Simulations using the PSIM software tool were conducted to validate the proposed method. The system configuration was similar to that shown in Figure 1. The DC link voltage (V_{dc}) was 300 V, and was distributed equally between the top capacitance voltage (V^{top}) and the bottom capacitance voltage (V^{bottom}). The total DC link capacitance was 2200 µF and the switching frequency was 10 kHz. The value of $|E_{limit}|$ was set empirically at 2.6 V. The simulation parameters are given in Table 2. After synthesizing the reference voltage vectors with the minimized cost function using the deadbeat DSVM-MPC system, the reference voltage vector was applied to the grid-connected 3L T-type VSI.

Table 2. System parameters.

Variable	Description	Value	Unit
P_r	Rated power	2.1	kW
V_{dc}	DC link voltage	300	V
f	Fundamental frequency	60	Hz
e	Grid voltage	100	Vrms
T_s	Sampling time	100	μs
R	Load resistance	1	ohm
L	Filter inductance	2	mH
M	Modulation index	0.954	

Figure 7 shows the simulation results for the optimal voltage vectors in the *d–q* frame for the deadbeat DSVM-MPC, which followed the reference current using only four candidate voltage vectors [20]. Figure 8 shows the simulation results for the deadbeat DSVM-MPC using the balancing method in Reference [20] and using the proposed DPWM method. It can be seen that the current waveforms became highly distorted before implementation of the neutral-point balancing of capacitance voltage, owing to large voltage deviations. Obviously, the total harmonic distortion (THD was is significantly reduced using either one of the balancing methods. Nevertheless, the proposed balancing method performed better in terms of THD, as shown in Figure 8b.

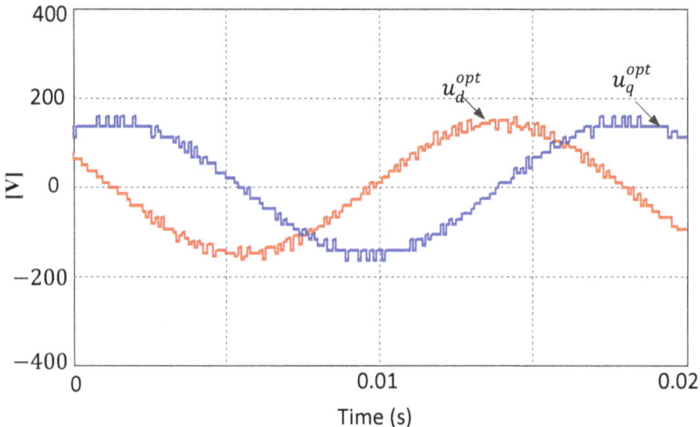

Figure 7. Simulation results of the optimal voltage vectors for the deadbeat FS-MPC with DSVM.

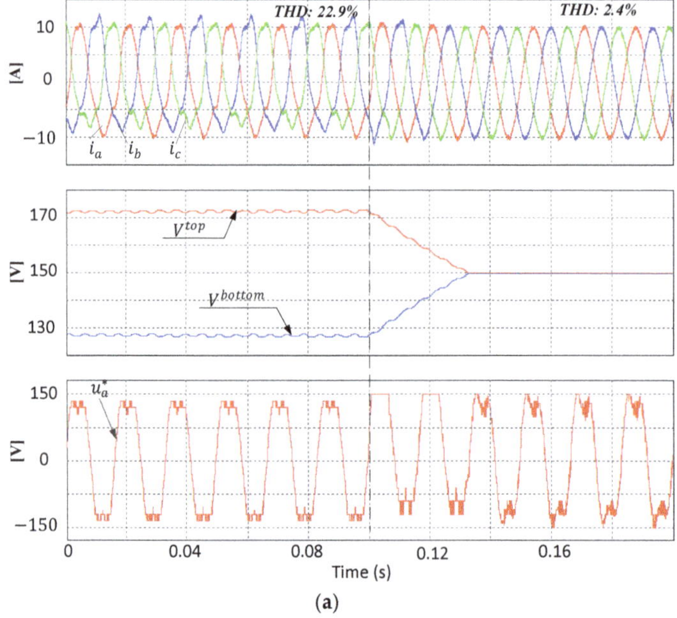

(a)

Figure 8. *Cont.*

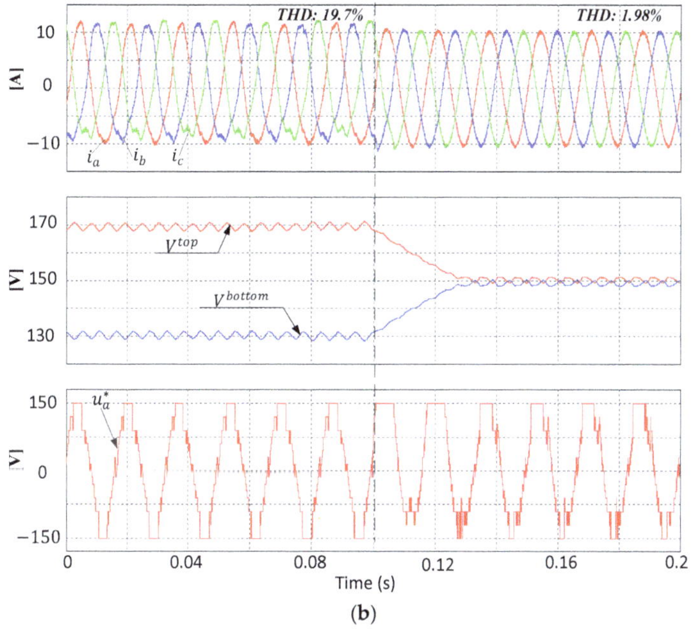

(b)

Figure 8. Simulation results of capacitance voltage balancing for deadbeat DSVM-MPC. (**a**) Conventional balancing method [20]. (**b**) Proposed balancing method.

Figure 9 shows the simulation results for dynamic response for the deadbeat DSVM-MPC system, using the two balancing methods. As can be seen, both methods exhibited a fast dynamic current response, because the MPC method was used. On the other hand, the proposed method exhibited smaller THD compared with the method in Reference [20].

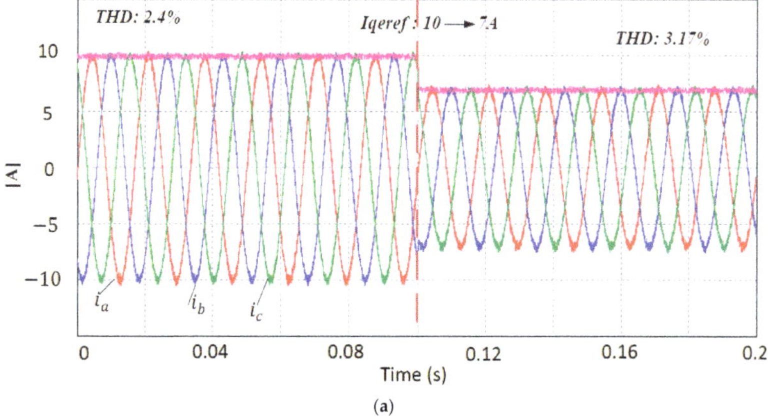

(a)

Figure 9. *Cont.*

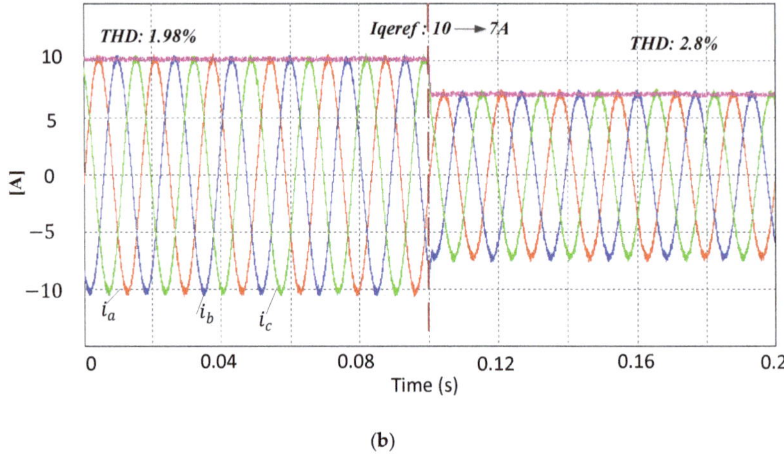

(**b**)

Figure 9. Simulation results of current dynamic response for deadbeat DSVM-MPC with (**a**) conventional balancing method [20] and (**b**) proposed balancing method.

The robustness of the proposed method was shown by varying the impedances of the grid connected system including the inductance (L) and resistance (R) (Figure 10). The values of resistance and inductance were increased (at t = 0.1 s) by 100% and 50% of the nominal values, respectively. It can be obviously seen that there was a negligible increase in the THD. Therefore, it was confirmed that the proposed balancing method in FS-MPC with DSVM and DBC is robust to the variations of impedances.

Figure 11a,b depicts the frequency spectrum of the phase voltage for the conventional method and proposed method, respectively. Apparently, the first harmonic component (i.e., switching frequency (f_{sw})) was greatly reduced with the proposed algorithm, compared to the conventional method. This indicates the proposed method has less switching frequency owing to the use of the DPWM modulator, as mentioned previously.

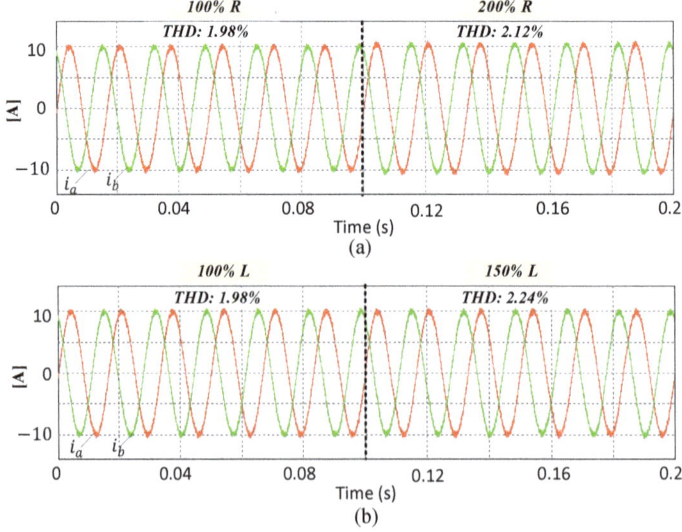

Figure 10. Simulation results of grid impedance variation for the proposed method.

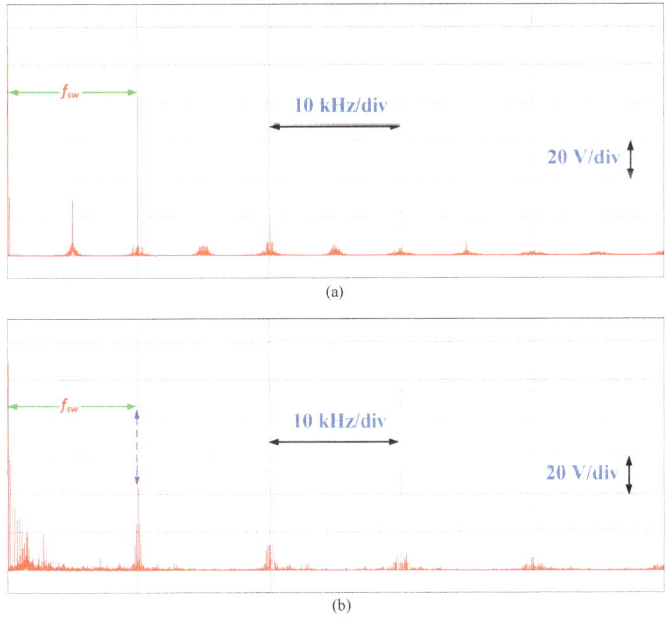

Figure 11. Simulation results of frequency spectrum for the phase voltage with (**a**) conventional balancing method [20] and (**b**) proposed balancing method.

6. Experimental Results

The proposed control system was further validated on a prototype grid-connected VSI that was used in a laboratory setup, as shown in Figure 12. The experimental parameters of the prototype were similar to those that were assumed in the simulation study, as shown in Table 2. The control system was configured using a DSP named TMS320F28377. In addition, 10-FZ12NMA080SH01-M260F-3 from Vincotech was employed to configure the three-level inverter system. To ensure a fair comparison of these methods, the experimental conditions were the same as those that were assumed in the simulation study. Thus, the limited error band $|E_{limit}|$ was also set at 2.6 V, which was similar to that in the simulation. To verify the effectiveness of the proposed method in the deadbeat DSVM-MPC, it was compared against the conventional neutral-point balancing method [20].

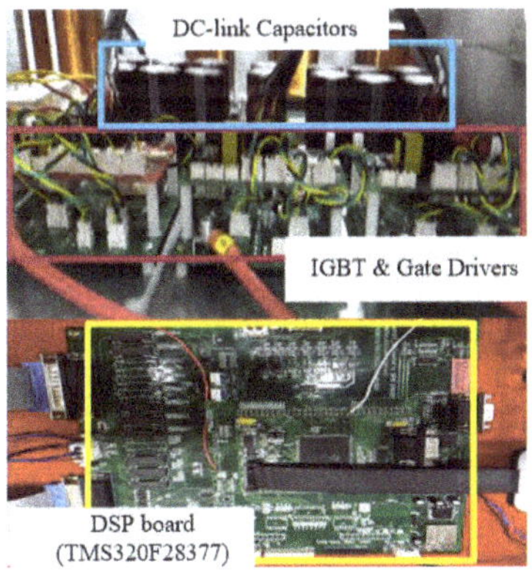

Figure 12. Experimental setup.

Figure 13 shows the experimental results for the grid current of the deadbeat DSVM-MPC, for both balancing methods. Similarly to the simulation results, it was observed that the proposed balancing method demonstrated a good capability of balancing the deviation of the top and bottom capacitance voltages. As can be seen, before applying either of the balancing methods, the difference between the top and bottom capacitor voltages was very high, and the output current became distorted owing to the neutral-point voltage imbalance. However, when the neutral-point voltage was balanced, the distortion of the output current disappeared. Figure 14 shows the experimental results for the dynamic current performance from 10 A to 7 A for the deadbeat DSVM-MPC system, using both the conventional and the proposed balancing methods. As can be seen, the settling time was very short because the MPC method was used. However, the proposed balancing method exhibited lower THD by around 4.9% before and after the current reference change, when compared with the conventional neutral-point balancing method [20].

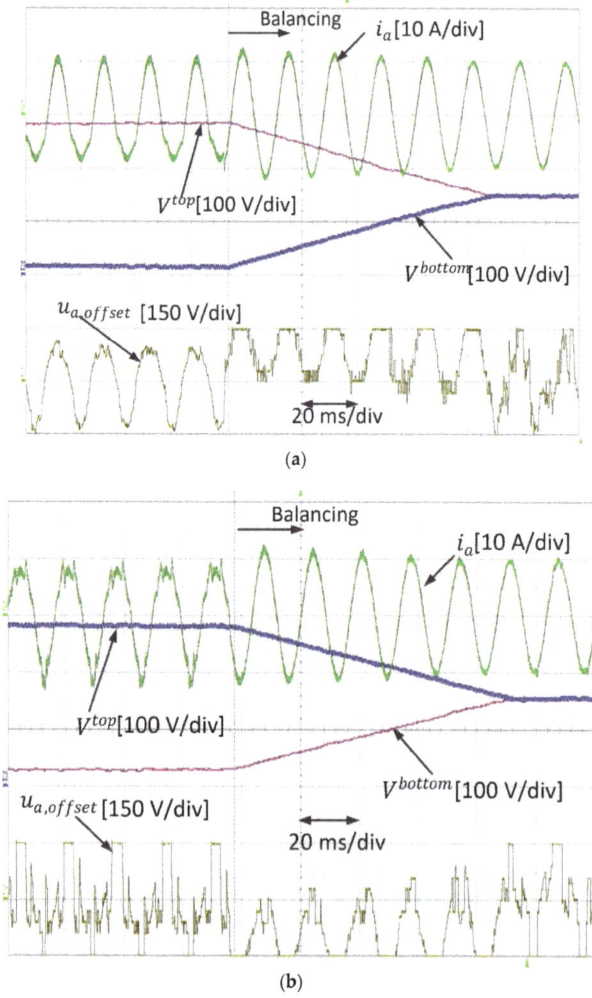

Figure 13. Experimental results of capacitance voltage balancing for deadbeat DSVM-MPC. (**a**) Conventional balancing method [20]. (**b**) Proposed balancing method.

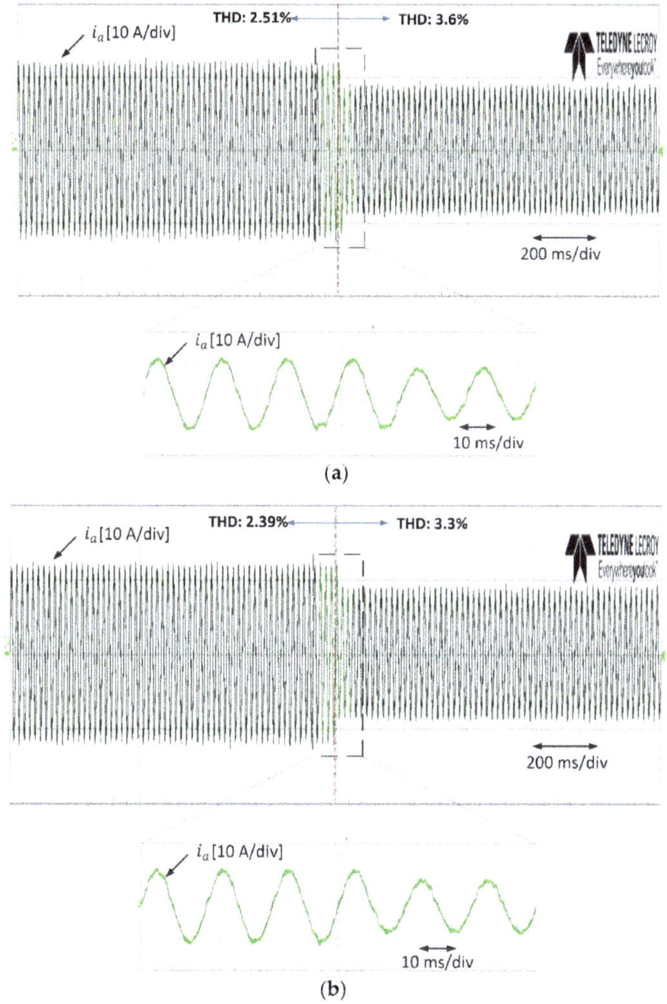

Figure 14. Experimental results of current dynamic response for deadbeat DSVM-MPC for the (**a**) conventional balancing method [20] and (**b**) proposed balancing method.

Figure 15 shows the comparison of the hardware computation time of the deadbeat DSVM-MPC system, without weighting factors, for the control method in Reference [20] and the proposed method. As can be seen, the computation time required by the proposed deadbeat DSVM-MPC method was reduced by 12.30% when compared to the method from Reference [20]. This indicates the simplicity of the proposed algorithm.

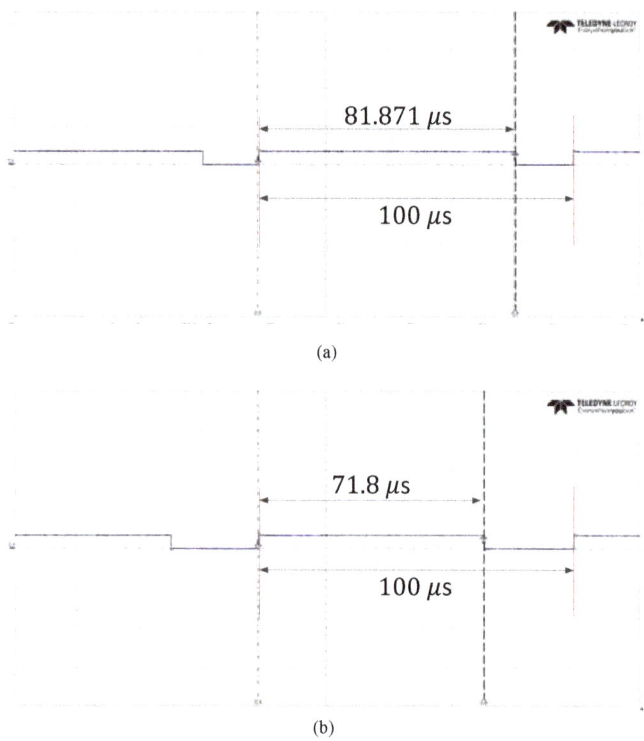

Figure 15. The hardware execution time. (**a**) Deadbeat FS-MPC with extra cost function [20]. (**b**) Proposed method.

7. Conclusions

This paper proposed a deadbeat DSVM-MPC with a simplified neutral-point balancing method for enhancement of the output current quality in three-level grid-connected voltage source inverters. The DSVM-MPC method produces various virtual voltage vectors by subdividing the space vector diagram, and selects the optimal voltage vector that minimizes the error with respect to the reference current. To alleviate the computational burden, a deadbeat control was applied to restrict the optimal region of candidate voltage vectors. In addition, this work introduced simplified neutral-point capacitance voltage balancing using a modified DPWM method without a need for an extra cost function, thus reducing the total computation time. The reduction in the computation time may be advantageous for incorporating additional estimation or protection algorithms. The modified DPWM method had a higher efficiency than the conventional CPWM method, owing to the reduced number of switching operations.

Author Contributions: Supervision, K.-B.L.; writing—original draft, I.M.A.

Funding: This work was supported by "Human Resources Program in Energy Technology" of the Korea Institute of Energy Technology Evaluation and Planning (KETEP), granted financial resource from the Ministry of Trade, Industry & Energy, Republic of Korea and National Research Foundation of Korea (NRF) funded by the Ministry of Science and ICT for First-Mover Program for Accelerating Disruptive Technology Development. (No. 20194030202370, NRF-2018M3C1B9088457).

Conflicts of Interest: The authors declare no conflict of interest.

References

1. Ma, W.; Sun, P.; Zhou, G.; Sailijiang, G.; Zhang, Z.; Liu, Y. A Low-Computation Indirect Model Predictive Control for Modular Multilevel Converters. *J. Power Electron.* **2019**, *19*, 529–539.
2. Rodriguez, J.; Kazmierkowski, M.P.; Espinoza, J.R.; Zanchetta, P.; Abu-Rub, H.; Young, H.A.; Rojas, C.A. State of the art of finite control set model predictive control in power electronics. *IEEE Trans. Ind. Inform.* **2013**, *9*, 1003–1016. [CrossRef]
3. Alsofyani, I.M.; Kim, S.; Lee, K. Finite Set Predictive Torque Control Based on Sub-divided Voltage Vectors of PMSM with Deadbeat Control and Discrete Space Vector Modulation. In Proceedings of the 2019 IEEE Applied Power Electronics Conference and Exposition (APEC), Anaheim, CA, USA, 17–21 March 2019; pp. 1853–1857.
4. Zhao, B.; Li, H.; Mao, J. Double-Objective Finite Control Set Model-Free Predictive Control with DSVM for PMSM Drives. *J. Power Electron.* **2019**, *19*, 168–178.
5. El-naggar, M.F.; Elgammal, A.A.A. Multi-Objective Optimal Predictive Energy Management Control of Grid-Connected Residential Wind-PV-FC-Battery Powered Charging Station for Plug-in Electric Vehicle. *J. Electr. Eng. Technol.* **2018**, *13*, 742–751.
6. Kwak, S.; Moon, U.-C.; Park, J.-C. Predictive-control-based direct power control with an adaptive parameter identification technique for improved AFE performance. *IEEE Trans. Power Electron.* **2014**, *29*, 6178–6187. [CrossRef]
7. Cortes, P.; Rodriguez, J.; Vazquez, S.; Franquelo, L.G. Predictive control of a three-phase UPS inverter using two steps prediction horizon. In Proceedings of the 2010 IEEE International Conference on Industrial Technology, Vina del Mar, Chile, 14–17 March 2010; pp. 1283–1288.
8. Vazquez, S.; Montero, C.; Bordons, C.; Franquelo, L.G. Design and experimental validation of a model predictive control strategy for a VSI with long prediction horizon. In Proceedings of the IECON 2013-39th Annual Conference of the IEEE Industrial Electronics Society, Vienna, Austria, 10–13 November 2013; pp. 5788–5793.
9. Karamanakos, P.; Geyer, T.; Oikonomou, N.; Kieferndorf, F.D.; Manias, S. Direct model predictive control: A review of strategies that achieve long prediction intervals for power electronics. *IEEE Ind. Electron. Mag.* **2014**, *8*, 32–43. [CrossRef]
10. Xie, W.; Wang, X.; Wang, F.; Xu, W.; Kennel, R.M.; Gerling, D.; Lorenz, R.D. Finite-control-set model predictive torque control with a deadbeat solution for PMSM drives. *IEEE Trans. Ind. Electron.* **2015**, *62*, 5402–5410. [CrossRef]
11. Park, J.Y.; Sin, J.; Bak, Y.; Park, S.-M.; Lee, K.-B. Common-mode Voltage Reduction for Inverters Connected in Parallel Using an MPC Method with Subdivided Voltage Vectors. *J. Electr. Eng. Technol.* **2018**, *13*, 1212–1222.
12. Vazquez, S.; Leon, J.I.; Franquelo, L.G.; Carrasco, J.M.; Martinez, O.; Rodriguez, J.; Cortes, P.; Kouro, S. Model predictive control with constant switching frequency using a discrete space vector modulation with virtual state vectors. In Proceedings of the 2009 IEEE International Conference on Industrial Technology, Gippsland, Australia, 10–13 February 2009.
13. Wang, Y.; Wang, X.; Xie, W.; Wang, F.; Dou, M.; Kennel, R.M.; Lorenz, R.D.; Gerling, D. Deadbeat model-predictive torque control with discrete space-vector modulation for PMSM drives. *IEEE Trans. Ind. Electron.* **2017**, *64*, 3537–3547. [CrossRef]
14. Kouro, S.; Cortes, P.; Vargas, R.; Ammann, U.; Rodriguez, J. Model predictive control—A simple and powerful method to control power converters. *IEEE Trans. Ind. Electron.* **2009**, *56*, 1826–1838. [CrossRef]
15. Li, J.; Chang, X.; Yang, D.; Liu, Y.; Jiang, J. Deadbeat and Hierarchical Predictive Control with Space-Vector Modulation for Three-Phase Five-Level Nested Neutral Point Piloted Converters. *J. Power Electron.* **2018**, *18*, 1791–1804.
16. Wang, C.; Li, Y. Analysis and calculation of zero-sequence voltage considering neutral-point potential balancing in three-level NPC converters. *IEEE Trans. Ind. Electron.* **2010**, *57*, 2262–2271. [CrossRef]
17. Lee, J.-S.; Lee, K.-B. Time-offset injection method for neutral-point AC ripple voltage reduction in a three-level inverter. *IEEE Trans. Power Electron.* **2016**, *31*, 1931–1941. [CrossRef]
18. Moon, H.-C.; Lee, J.-S.; Lee, J.-H.; Lee, K.-B. MPC-SVM method with subdivision strategy for current ripples reduction and neutral-point voltage balance in three-level inverter. In Proceedings of the 2017 IEEE Energy Conversion Congress and Exposition (ECCE), Cincinnati, OH, USA, 1–5 October 2017; pp. 191–196.

19. Wang, C.; Li, Z.; Xin, H. Neutral-point Voltage Balancing Strategy for Three-level Converter based on Disassembly of Zero Level. *J. Power Electron.* **2019**, *19*, 79–88.
20. Lee, J.; Lee, J.; Moon, H.; Lee, K. An Improved Finite-Set Model Predictive Control Based on Discrete Space Vector Modulation Methods for Grid-Connected Three-Level Voltage Source Inverter. *IEEE J. Emerg. Sel. Top. Power Electron.* **2018**, *6*, 1744–1760. [CrossRef]
21. Shueai Alnamer, S.; Mekhilef, S.; Bin Mokhlis, H. A Four-Level T-Type Neutral Point Piloted Inverter for Solar Energy Applications. *Energies* **2018**, *11*, 1546. [CrossRef]
22. Choi, U.M.; Lee, H.H.; Lee, K.B. Simple neutral-point voltage control for three-level inverters using a discontinuous pulse width modulation. *IEEE Trans. Energy Convers.* **2013**, *28*, 434–443. [CrossRef]
23. Tcai, A.; Alsofyani, I.M.; Seo, I.-Y.; Lee, K.-B. DC-link Ripple Reduction in a DPWM-Based Two-Level VSI. *Energies* **2018**, *11*, 3008. [CrossRef]
24. Tcai, A.; Alsofyani, I.M.; Lee, K. DC-link Ripple Reduction in a DPWM-based Two-Level VSC. In Proceedings of the IEEE Energy Conversion Congress and Exposition (ECCE), Portland, OR, USA; 2018; pp. 1483–1487.
25. Lee, K.-B.; Lee, J.-S. *Reliability Improvement Technology for Power Converters*; Springer: Singapore, 2017.
26. Wang, X.; Zhao, J.; Wang, Q.; Li, G.; Zhang, M. Fast FCS-MPC-Based SVPWM Method to Reduce Switching States of Multilevel Cascaded H-Bridge STATCOMs. *J. Power Electron.* **2019**, *19*, 244–253.

© 2019 by the authors. Licensee MDPI, Basel, Switzerland. This article is an open access article distributed under the terms and conditions of the Creative Commons Attribution (CC BY) license (http://creativecommons.org/licenses/by/4.0/).

Article

A Modified DSC-Based Grid Synchronization Method for a High Renewable Penetrated Power System Under Distorted Voltage Conditions

Tie Li [1,2], Yunlu Li [1], Junyou Yang [1,*], Weichun Ge [2] and Bo Hu [2]

1 School of Electrical Engineering, Shenyang University of Technology, Shenyang 110870, China; tony_fe@126.com (T.L.); liyunlu@sut.edu.cn (Y.L.)
2 Liaoning Province Electric Power Company, Shenyang 110006, China; gwc@ln.sgcc.com.cn (W.G.); hubo@ln.sgcc.com.cn (B.H.)
* Correspondence: junyouyang@sut.edu.cn

Received: 30 September 2019; Accepted: 18 October 2019; Published: 23 October 2019

Abstract: With the increasing penetration of renewable energy, a weak grid with declining inertia and distorted voltage conditions becomes a significant problem for wind and solar energy integration. Grid frequency is prone to deviate from its nominal value. Grid voltages become more easily polluted by unbalanced and harmonic components. Grid synchronization technique, as a significant method used in wind and solar energy grid-connected converters, can easily become ineffective. As probably the most widespread grid synchronization technique, phase-locked loop (PLL) is required to detect the grid frequency and phase rapidly and precisely even under such undesired conditions. While the amount of filtering techniques can remove disturbances, they also deteriorate the dynamic performance of PLL, which may not meet the standard requirements of grid codes. The objective of this paper is to propose an effective PLL to tackle this challenge. The proposed PLL is based on quasi-type-1 PLL (QT1-PLL), which provides a good filtering capability by using a moving average filter (MAF). To accelerate the transient behavior when disturbance occurs, a modified delay signal cancellation (DSC) operator is proposed and incorporated into the filtering stage of QT1-PLL. By using modified DSCs and MAFs in a cascaded way, the settling time of the proposed method is reduced to around one cycle of grid fundamental frequency without degrading any disturbance rejection capability. To verify the performance, several test cases, which usually happen in high renewable penetrated power systems, are carried out to demonstrate the effectiveness of the proposed PLL.

Keywords: phase-locked loop (PLL); synchronization; hybrid filter

1. Introduction

In modern power systems, more and more renewable energies such as solar and wind energy are integrated with the power grid through grid-connected power converters. To use this energy in more efficient ways, several advanced techniques, such as high voltage direct current transmission (HVDC) [1], flexible AC transmission systems (FACTS) [2] and energy storage systems (ESSs) [3], are developed. Some power electronic converter-based applications on the customer side, such as constant power load (CPL) and micro-grids with ESSs, are also widely used for some purposes, like peak load shaving. All the above yields a more complicated power system with a large amount of power converters. A typical power system integrating high penetrated renewable energy sources is depicted in Figure 1. The irregular phenomenon and uncertainty of wind and solar energy may cause undesired grid conditions [4]. Sub- and super-synchronous oscillations may emerge in wind farms if the controller is not well-designed [5]. In recent years, these issues happened several times in Oklahoma, USA [6] and Hebei and Xinjiang, China [7,8], and caused many wind turbines to be tripped off. This highly impacted the operation of power distribution and renewable energy utilization. The grid frequency is prone to deviate from its nominal value since the inertia of the grid decreases. Grid voltages contain an amount of undesired components, such as fundamental frequency negative sequences (FFNSs) and harmonic components, which results in unbalanced and distorted grid conditions. It is an essential requirement for a renewable energy power converter to maintain stable operation with high performance under such conditions. To achieve this goal, a proper grid synchronization method is needed for all grid-connected applications. It is a big challenge for a grid synchronization method to extract grid frequency and phase information under such adverse conditions.

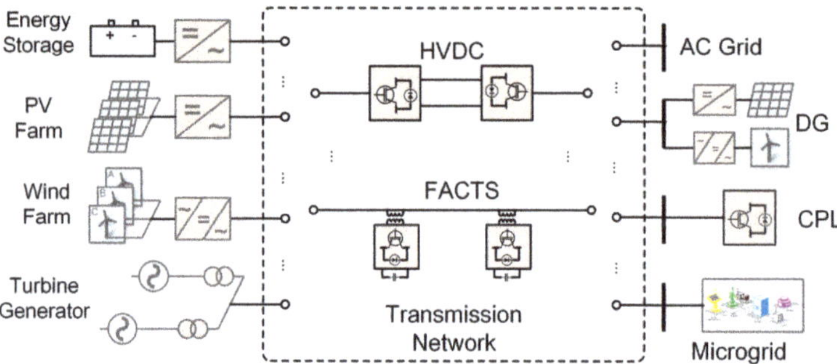

Figure 1. Block schematic of a typical power system with high penetration of renewable energy.

Among various grid synchronization techniques, phase-locked loop (PLL) is the most widely used method in grid phase and frequency detection area for its robust performance and simple implementation [9–12]. Figure 2 shows the most common employment of PLLs in practical applications. After extracting grid voltage information at the common coupling point (PCC), PLL sends estimated grid phases to the controller of the grid power converter (GPC). For a three phase power system, synchronous reference frame PLL (SRF-PLL) is an effective and typical method under ideal conditions [13,14]. However, with more utilization of renewable energy sources, a power system turns "weak" [15–18]. The grid voltages usually contain fundamental frequency negative sequence components (FFNSs) and other harmonic components. Furthermore, grid frequency does not always stay at its nominal value. To maintain the phase tracking accuracy under distorted grid voltages conditions, various filtering technique were used to remove disturbances at the cost of slowing down the transient response, which may violate the requirement on settling time in common grid codes [19–23].

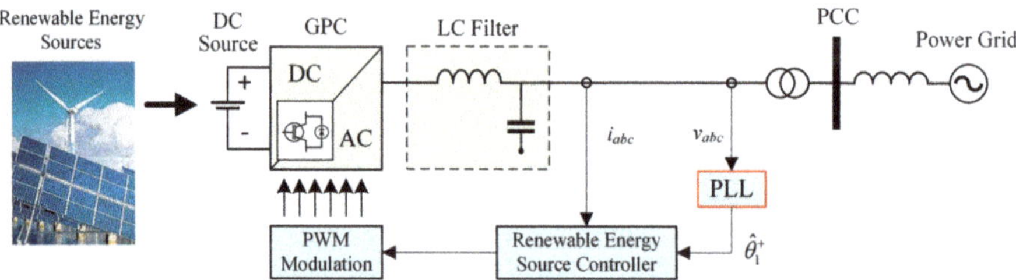

Figure 2. Block schematic of a typical grid-connected power converter system.

The block diagram of SRF-PLL is depicted in Figure 3. Park transformation is utilized as a phase detector (PD). An integrator is employed as a voltage controller oscillator (VCO). A proportional-integral controller (PI) is used to mitigate the phase-tracking error. To enhance the disturbance rejection capability and keep the phase tracking accuracy of SRF-PLL, a low pass filter (LPF) or moving average filter (MAF) is employed in the inner loop. However, LPF-based PLL can only attenuate but not eliminate the disturbance component. To achieve good disturbance rejection, the bandwidth of LPF-based PLL must be significantly reduced, which results in a large settling time [23]. Compared with LPF, MAF can totally remove a specific set of disturbances, which depends on its window length (T_ω). However, T_ω deteriorates the dynamic response since T_ω has to be 0.01 s (half grid period) to eliminate all disturbance [24].

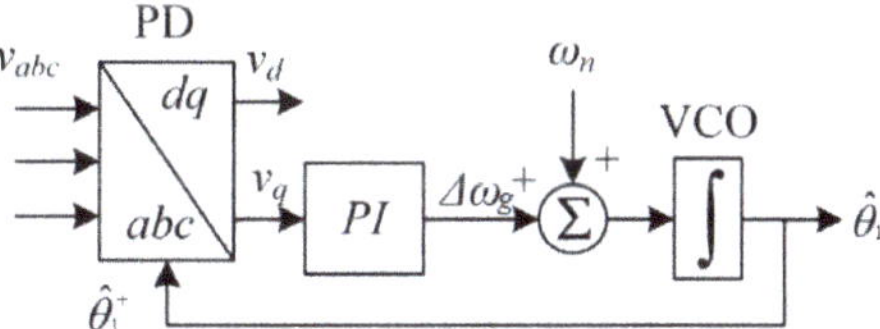

Figure 3. Block diagram of SRF-phase-locked loop (PLL).

To tackle the problem mentioned above, many advanced methods were proposed in recent years. According to their various filtering methods, these advanced PLLs can be classified into two categories: LPF-based PLLs and MAF-based PLLs. To improve the performance of LPF-based PLLs, dual second-order generalized integrators (DSOGIs) [25], multiple complex-coefficient filters (MCCFs) [26], multiple reference frame-based filter structures (MRFs) [27] and decoupled double synchronous reference frame-based filter structures (DDSRFs) [28] were arranged before applying Park transformation. The basic idea of these PLLs is to make the filtering stage act as a hybrid filter that consists of notch filters and LPFs. Notch filters are responsible for eliminating FFNS components and other important harmonics. LPFs are used to attenuate other disturbances. These methods are not suitable when grid voltages contain several significant components. Only the disturbance components coincident with the specific frequency in a notch filter can be totally removed. Compared with LPF-based PLLs, MAF-based PLLs can provide ideal filtering capabilities at the cost of increasing time delay. With a T_ω time delay, MAF can eliminate all n/T_ω ($n = 1, 2, 3, \ldots$) frequency components. The delay signal cancellation (DSC) operator is another effective filter, the behavior of which is similar to MAFs. However, single DSCs cannot eliminate all desired disturbances. To overcome this weakness, several DSCs with different T_ω are usually employed to build an entire filtering stage in cascaded [29–31]. Consequently, the time delay of the filtering stage is the sum of delay introduced by all DSCs. Too many DSCs also increases the computational burden and implementation complexity.

Besides using an advanced filter, another way to improve PLLs' performance is to change the control structure. In [32,33], a secondary control path is built to accelerate the transient behavior. However, an inappropriate design of a secondary control path may give rise to the stability problem. It also increases the order of open loop transfer functions and the implementation complexity of a system [34]. Recently, a PLL with a new structure named quasi-type 1 PLL (QT1 PLL) was proposed in [35]. Compared with the traditional type-2 SRF-PLL, QT1-PLL provides a feed-forward control path to the output. This makes QT1-PLL able to track phase precisely without using integral operations in the controller. One more open-loop pole is provided at the origin point, which accelerate the dynamic response. Since the filtering stage of QT1-PLL is built by MAFs, QT1-PLL can also offer a satisfied filtering capability. It is a good idea to do some further performance improvement of PLL based on the QT1-PLL structure.

To improve the dynamic performance without degrading PLLs' filtering capability, this paper propose a new PLL based on the QT1-PLL structure. In order to provide a fast transient response, a hybrid filtering stage is designed and arranged at the inner loop of the proposed PLL. The proposed hybrid filtering stage consists of a modified DSC (MDSC) and MAFs with narrowed T_ω. Our basic idea is to eliminate two sets of disturbance components by using MDSCs and MAFs, separately. Different from the conventional DSC-based PLL, there is only one MDSC unit in our method, which is easy for digital implementation. To demonstrate the effectiveness, an experimental case study is carried out when grid voltage conditions are under phase jump, frequency jump, frequency ramp change and harmonic polluted voltage conditions, which usually happens to high renewable energy-penetrated power system.

This paper is organized as follows. In Section 2, the modified DSC is presented based on the analysis of DSCs. The hybrid filtering stage and new PLLs are proposed in Section 3. In Section 4, the mathematics model is established. Based on this model, the parameters are designed based on analysis of the system. In Section 5, the performance of the proposed method is validated by a comprehensive case study.

2. Modified Delay Signal Cancellation Operator

The DSC operator has been widely studied in much literature [36]. In the Laplace domain, most of the existing DSC operators can be written as:

$$\mathrm{DSC}_n(s) = \frac{1 + e^{\frac{j2\pi}{n}} e^{-\frac{T}{n}s}}{2} \tag{1}$$

To achieve a desired performance, several DSCs have to be connected in cascaded. For instance, five DSCs with different value of n (n = 2, 4, 8, 16, 32) were arranged at the pre-filtering stage in [37]. The typical block diagram of the control strategy is shown in Figure 4. Too many DSCs used in PLLs results in complicated implementation. It is a normal idea to simplify the system by reducing the number of DSCs.

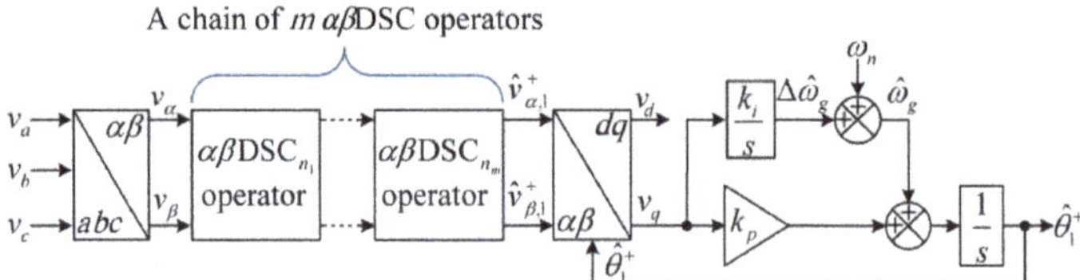

Figure 4. The common block diagram of using delay signal cancellation (DSC)s in PLL.

Observing Equation (1), it can be found that there is only one parameter n which can decide the characteristics of DSC_n. To make its property more flexible, DSC_n is modified to the following form with two parameters.

$$\mathrm{MDSC}_{m,n}(s) = \frac{1 + e^{\frac{j2\pi}{m}} e^{-\frac{T}{n}s}}{2} \tag{2}$$

In Equation (2), T is the grid period (0.02 s for 50 Hz power system). m is used to shift the original DSC_n frequency characteristics along the frequency axis. n is a parameter that can decide the time delay inside the DSC_n. After using Euler transformation, the implementation of $\mathrm{MDSC}_{m,n}$ is shown in Figure 5. It should be noted that the input of $\mathrm{MDSC}_{m,n}$ is a vector with two dimensions. The output of $\mathrm{MDSC}_{m,n}$ is also a 2D vector. The effect of m can be considered as a rotating operation to the input vector.

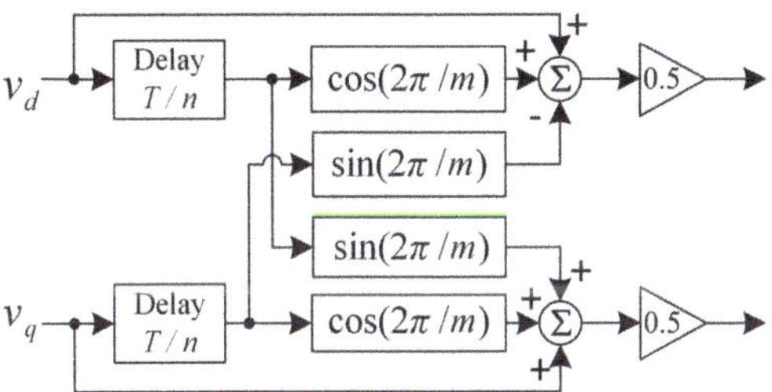

Figure 5. Time-domain implementation of a modified delay signal cancellation (MDSC) operator.

To examine the effect of two parameters (m, n), two bode diagrams are depicted in Figures 6 and 7. In Figure 6, n is set to 4. The solid lines, dashed lines and dotted lines correspond to different $\text{MDSC}_{m,n=4}$ with $m = 4, 2, 4/3$, respectively. The frequency characteristic of $\text{MDSC}_{m=4,n=4}$ can be considered as that of $\text{MDSC}_{m=1,n=4}$ shifted $\frac{n}{Tm} = 50$ Hz along the positive frequency direction. $\text{MDSC}_{m=2,n=4}$, $\text{MDSC}_{m=4/3,n=4}$ corresponds to $\frac{n}{Tm} = 100$ Hz and $\frac{n}{Tm} = 150$ Hz, respectively. This nice property can be used to arrange the notch frequency of MDSCs by setting an appropriate m.

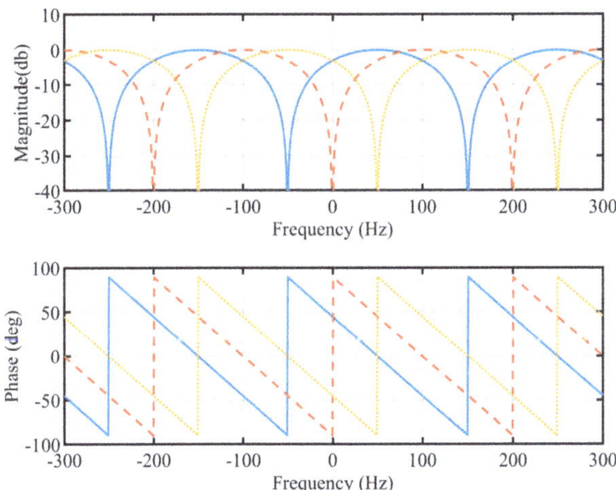

Figure 6. The frequency characteristic of MDSCs with different m ($n = 4$).

In Figure 7, m is 2. The solid lines, dashed lines and dotted lines are the bode diagrams of $\text{MDSC}_{m=2,n}$ with $n = 2, 4, 8$, respectively. With different value of n, the interval of the notch frequency of an MDSC can be changed. This property can be used to eliminate a set of specific harmonic frequency components. In the next section, the design procedure of the proposed hybrid filtering stage is presented based on the analysis of MDSCs above.

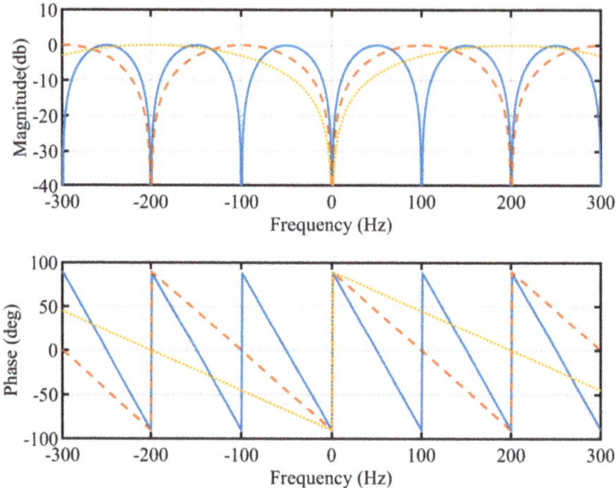

Figure 7. The frequency characteristic of MDSCs with different n (m = 2).

3. The Proposed PLL Structure

In this section, the voltage sequence component of non-ideal grid voltages is analyzed at first to provide the basis for the design procedure of the proposed PLL. To achieve our objective, a hybrid filtering stage based on MAFs and MDSCs is suggested and analyzed in this section. Then, it is incorporated into a QT1-PLL structure.

3.1. The Component Analysis of Distorted Grid Voltages

Under distorted grid voltage conditions, three-phase grid voltages contain fundamental frequency positive sequence (FFPS), fundamental frequency negative sequences (FFNS) and other harmonic sequence components. FFPS components can be written as follows:

$$\begin{aligned} v_{a,1}(t) &= V_1^+ \sin(\omega t) \\ v_{b,1}(t) &= V_1^+ \sin(\omega t - \tfrac{2\pi}{3}) \\ v_{c,1}(t) &= V_1^+ \sin(\omega t + \tfrac{2\pi}{3}) \end{aligned} \tag{3}$$

where V_1^+ represents the amplitude of FFPS and ω is the grid frequency. Then, n order harmonic sequence components can be written as:

$$\begin{aligned} v_{a,n}(t) &= V_n \sin(n\omega t + \varphi_n) \\ v_{b,n}(t) &= V_n \sin(n\omega t + \varphi_n - \tfrac{2\pi}{3}) \\ v_{c,n}(t) &= V_n \sin(n\omega t + \varphi_n + \tfrac{2\pi}{3}) \end{aligned} \tag{4}$$

By using the symmetrical component method, all voltage components can be considered as the sum of positive sequences, negative sequences and zero sequences. Applying Clark transformation to three phase voltages yields:

$$\begin{bmatrix} v_\alpha(t) \\ v_\beta(t) \\ v_0(t) \end{bmatrix} = \begin{bmatrix} T_{\alpha\beta} \end{bmatrix} \begin{bmatrix} V_{1,4,7,\ldots}^+ \sin(n\omega t + \phi_n) & +V_{2,5,8,\ldots}^- \sin(n\omega t + \phi_n) & +V_{3,6,9,\ldots}^0 \sin(n\omega t + \phi_n) \\ V_{1,4,7,\ldots}^+ \sin(n\omega t + \phi_n - \frac{2\pi}{3}) & +V_{2,5,8,\ldots}^- \sin(n\omega t + \phi_n + \frac{2\pi}{3}) & +V_{3,6,9,\ldots}^0 \sin(n\omega t + \phi_n) \\ V_{1,4,7,\ldots}^+ \sin(n\omega t + \phi_n + \frac{2\pi}{3}) & +V_{2,5,8,\ldots}^- \sin(n\omega t + \phi_n - \frac{2\pi}{3}) & +V_{3,6,9,\ldots}^0 \sin(n\omega t + \phi_n) \end{bmatrix} \tag{5}$$

In Equation (5), $T_{\alpha\beta}$ is the transfer matrix of Clark transformation. After applying Clark transformation, Equation (5) is as follows:

$$\begin{bmatrix} v_\alpha(t) \\ v_\beta(t) \\ v_0(t) \end{bmatrix} = \begin{bmatrix} v^+_{\alpha,1,7,13,\ldots}(t) \\ v^+_{\beta,1,7,13,\ldots}(t) \\ 0 \end{bmatrix} + \begin{bmatrix} v^-_{\alpha,5,11,\ldots}(t) \\ v^-_{\beta,5,11,\ldots}(t) \\ 0 \end{bmatrix} + \begin{bmatrix} 0 \\ 0 \\ v^0_{3,6,9,\ldots}(t) \end{bmatrix} \quad (6)$$

Observing Equation (6), it can be found that there is no triple odd harmonic in v_α and v_β. In the $\alpha\beta$-frame, only n= +1, −5, +7, −11, +13, … order sequence components exist. By using Park transformation, the components in the $\alpha\beta$-frame turn out to be n = −2, ±6, ±12, … order and DC components. The voltage sequence components can be summarized in Table 1. It should be noticed that the sign of frequency represents the rotating direction of the voltage sequence vector. A negative frequency means the voltage vector rotates in a counterclockwise direction.

Table 1. Voltage sequence components in grid voltages.

Harmonic Order	…	+1	−1	−5	+7	−11	+13	…
$\alpha\beta$-frame (Hz)	…	50	−50	−250	350	−550	650	…
Harmonic order	…	0	−2	−6	+6	−12	+12	…
dq-frame (Hz)	…	0	−100	−300	300	−600	600	…

3.2. The Hybrid Filtering Stage

According to the analysis of components in distorted grid voltages, a hybrid filter is well designed to eliminate the undesired components listed in Table 1. The hybrid filtering stage consists of an MDSC operator and two MAFs, which are arranged after Park transformation at the inner loop of the proposed PLL. MDSC is responsible for rejecting the FFNS component. Other dominant components are eliminated by two MAFs.

To achieve this goal, m and n, which are the parameters in MDSC, are chosen to be m = 4 and n = 8, respectively. The window length of MAF is set to be 1/300 s. The gain and phase of $\text{MDSC}_{m=4,n=8}$ can be calculated by following equations.

$$\left| \text{MDSC}(j\omega) \begin{matrix} \\ m=4 \\ n=8 \end{matrix} \right| = \left| \frac{1}{2} + j\frac{1}{2}\left(\cos\frac{0.02}{8}\omega - j\sin\frac{0.02}{8}\omega\right) \right| \quad (7)$$

$$\angle \text{MDSC}(j\omega) \begin{matrix} \\ m=4 \\ n=8 \end{matrix} = \tan^{-1}\left(\frac{\cos 0.0025\omega}{1+\sin 0.0025\omega}\right) \quad (8)$$

The bode diagram of MAF and $\text{MDSC}_{m=4,n=8}$ are depicted in Figure 8. The frequency characteristic reveals that $\text{MDSC}_{m=4,n=8}$ (solid lines) can eliminate the FFNS component. Furthermore, MAF in Figure 8 can remove ±300 Hz, ±600 Hz, … sequence components. One thing should be noticed. Since the MDSC is implemented in a dq-frame, the FFPS component turns to be a DC component (0 Hz). According to Equations (7) and (8), the gain of $\text{MDSC}_{m=4,n=8}$ at 0 Hz is 0.707. the phase of FFPS at 0 hz is +45°. The impact on amplitude and phase of input vector will be discussed and compensated below. The bode plot of the entire hybrid filtering stage is depicted in Figure 9. The components list in Table 1 is totally removed. A +45° phase deviation exists at 0 Hz owing to $\text{MDSC}_{m=4,n=8}$.

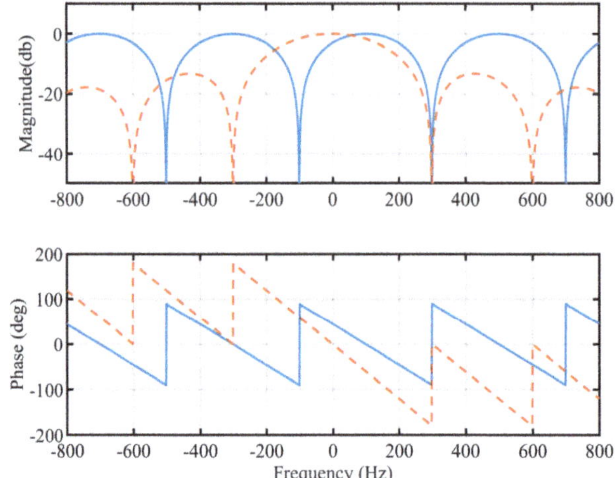

Figure 8. Bode diagram of $\mathrm{MDSC}_{m=4,n=8}$ and moving average filter (MAF).

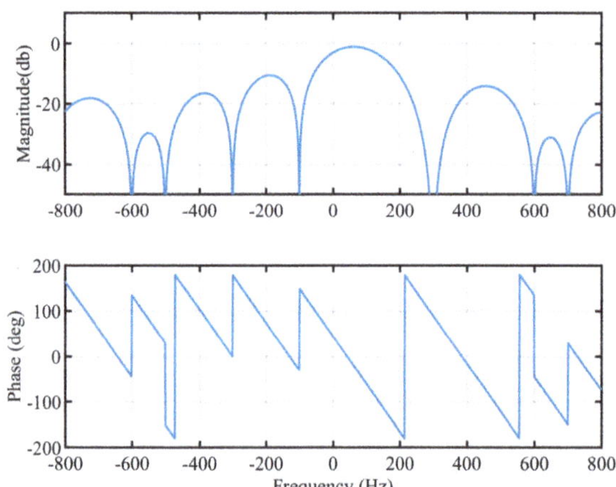

Figure 9. Bode diagram of $\mathrm{MDSC}_{m=4,n=8}$ and MAF.

3.3. The Proposed PLL Structure

The proposed PLL structure is based on the QT1-PLL structure which is depicted in Figure 10. ω_n is the nominal frequency of the grid. $\hat{\omega}_g$ is the estimated grid frequency. $\hat{\theta}_1^+$ is the estimated phase of FFPS. The arc tangent function can remove the impact of amplitude variation of the input voltage vector. k is the only control parameter of QT1-PLL.

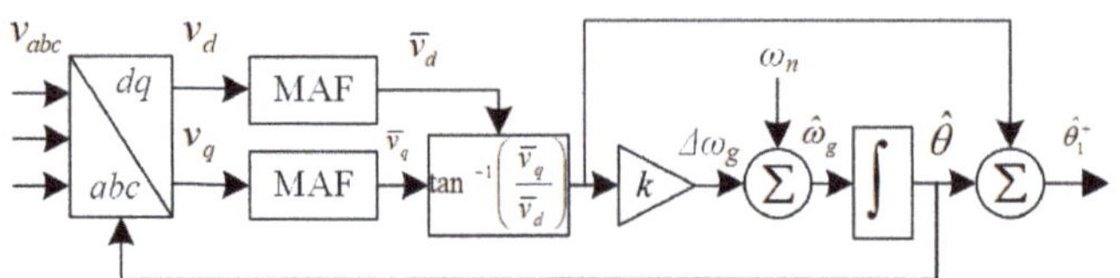

Figure 10. Block diagram of quasi-type-1 (QT1)-PLL.

Replacing the MAFs in QT1-PLL by the proposed hybrid filtering stage yields the proposed PLL structure as shown in Figure 11. While arc tangent function can remove the impact of amplitude variation, phase deviation still exists. Hence, to compensate a $\pi/4$ phase deviation introduced by $MDSC_{m=4,\,n=8}$, which is mentioned above, $-\pi/4$ is added at the output of proposed PLL. k is the only one control parameter to be designed.

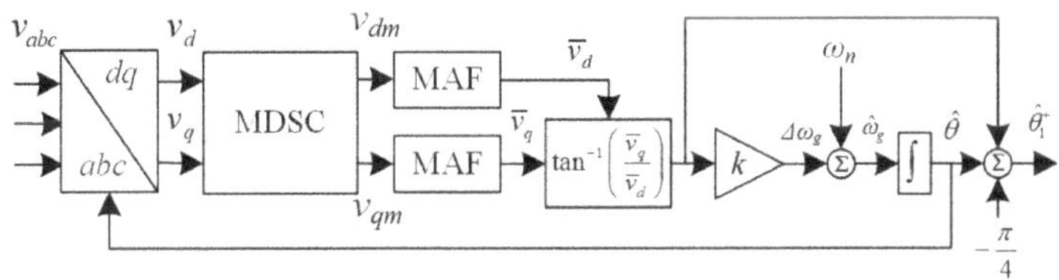

Figure 11. Block diagram of the proposed PLL.

4. Mathematics Model and Parameters Design Procedure

According to Figure 11, the mathematics model is derived in this section. To achieve a desired dynamic response, the parameter design procedure is also given. After providing the mathematics model, the stability of the system is also examined.

4.1. Mathematics Model

After Clark transformation, the FFPS component in $\alpha\beta$-frame can be written as:

$$\begin{aligned} v_\alpha(t) &= V_1^+ \cos(\theta_1^+) \\ v_\beta(t) &= V_1^+ \sin(\theta_1^+) \end{aligned} \tag{9}$$

By using Park transformation, the FFPS component in the dq-frame becomes:

$$\begin{bmatrix} v_d(t) \\ v_q(t) \end{bmatrix} = \begin{bmatrix} \cos(\hat{\theta}) & \sin(\hat{\theta}) \\ -\sin(\hat{\theta}) & \cos(\hat{\theta}) \end{bmatrix} \begin{bmatrix} v_\alpha(t) \\ v_\beta(t) \end{bmatrix} = V_1^+ \begin{bmatrix} \cos(\theta_1^+ - \hat{\theta}) \\ \sin(\theta_1^+ - \hat{\theta}) \end{bmatrix} \tag{10}$$

According to Figure 5, after $MDSC_{m=4,n=8}$, the FFPS component turns out to be:

$$\begin{bmatrix} v_{dm}(t) \\ v_{qm}(t) \end{bmatrix} = 0.5 \begin{bmatrix} 1 & -e^{-\frac{T}{8}s} \\ e^{-\frac{T}{8}s} & 1 \end{bmatrix} \begin{bmatrix} v_d(t) \\ v_q(t) \end{bmatrix} = 0.5 \begin{bmatrix} v_d(t) - v_q(t-\frac{T}{8}) \\ v_q(t) + v_d(t-\frac{T}{8}) \end{bmatrix} \tag{11}$$

Since two MAFs are arrange at each control path, the arc tangent function can be considered as arranged after $MDSC_{m=4,n=8}$. Then, the arc tangent operation can be expressed as:

$$\arctan\left(\frac{v_{qm}(t)}{v_{dm}(t)}\right) = \arctan\left(\frac{\cos(\theta_1^+ - \hat{\theta} - \frac{\pi}{2}) + \cos(\theta_1^+(t-\frac{T}{8}) - \hat{\theta}(t-\frac{T}{8}))}{-\sin(\theta_1^+ - \hat{\theta} - \frac{\pi}{2}) - \sin(\theta_1^+(t-\frac{T}{8}) - \hat{\theta}(t-\frac{T}{8}))}\right) \tag{12}$$

By applying trigonometric operation, Equation (12) turns out to be:

$$\arctan\left(\frac{v_{qm}(t)}{v_{dm}(t)}\right) = \arctan\left(\tan\left(\frac{\theta_1^+ - \hat{\theta} - \frac{\pi}{2} + \theta_1^+(t-\frac{T}{8}) - \hat{\theta}(t-\frac{T}{8})}{2} - \frac{\pi}{2}\right)\right) \tag{13}$$

Therefore, Equation (13) can be written as:

$$\arctan\left(\frac{v_{qm}(t)}{v_{dm}(t)}\right) = \frac{\theta_1^+ + \theta_1^+(t-\frac{T}{8}) - \hat{\theta} - \hat{\theta}(t-\frac{T}{8})}{2} + \frac{\pi}{4} \tag{14}$$

According to the derivation of Equations (9)–(14), the mathematics model of the proposed PLL is depicted in Figure 12. $D'(s)$ is the disturbance components injected into the input voltages. $R(s)$ is defined as follows:

$$R(s) = \frac{1}{2} + \frac{1}{2}e^{-\frac{T}{8}s} \tag{15}$$

Compared with the model of other existing PLLs, our mathematics model is not a small-signal model since the arc tangent function extracts phase information directly without any linearization procedure.

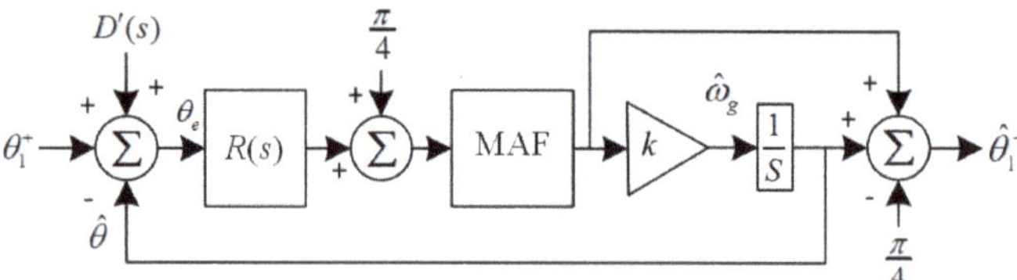

Figure 12. Mathematics model of the proposed PLL.

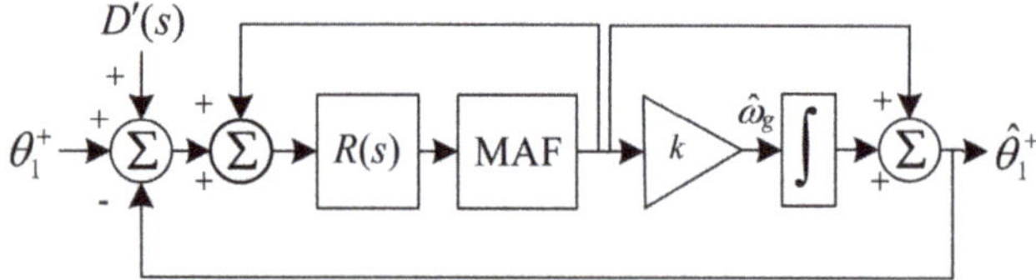

Figure 13. Simplified model of the proposed PLL.

4.2. Parameter Design Guidelines and Stability Analysis

To transform the proposed PLL into a traditional form of a closed-loop feedback system, block diagram algebra is utilized. The block diagram in Figure 12 is transformed to a simplified schematic shown in Figure 13. Hence, the open-loop transfer function can be written as:

$$G_{ols}(s) = \frac{\hat{\theta}_1^+(s)}{\theta_1^+(s) - \hat{\theta}_1^+(s)} = \left(\frac{R(s)\text{MAF}(s)}{1 - R(s)\text{MAF}(s)}\right)\left(\frac{s+k}{s}\right) \tag{16}$$

Then, the transfer function of phase-error can be expressed as:

$$G_e(s) = \frac{\theta_e}{\theta_1^+} = \frac{1}{1 + G_{ols}(s)} \tag{17}$$

In this paper, a parameter is chosen by its impact on the settling time of the phase-error transfer function. With different values of k, the settling time of $G_e(s)$ is examined. The phase-error transfer function under these two conditions is expressed by:

$$\Theta_e^{\Delta\theta}(s) = \frac{\Delta\theta}{s}G_e(s) \tag{18}$$

$$\Theta_e^{\Delta\omega}(s) = \frac{\Delta\omega}{s^2}G_e(s) \tag{19}$$

To calculate the settling time, inverse Laplace transformation is applied to Equations (18) and (19). Two curves of settling time as a function of k are depicted in Figure 14. When phase error is less than 2% of step change, transient response is considered to be over. To make a trade-off under both conditions, k is selected to be 148 to achieve an optimal dynamic performance for both conditions. The settling time is around one grid period.

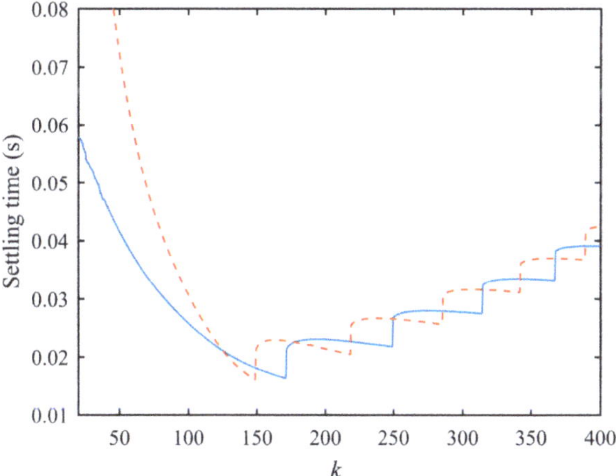

Figure 14. The settling time of the proposed PLL with different values of k under. phase jump (solid line) and frequency step-change (dashed line) conditions.

Since the model of the proposed PLL contains a time delay unit, the system turns out to be a non-minimum phase system. To examine the stability, nyquist stabilization criterion is employed in this paper instead of using a bode diagram. The nyquist diagram of $G_{ols}(s)$ is depicted in Figure 15. The nyquist curve does not surround the $(-1, j0)$ point, which means the closed-loop feedback system of $G_{ols}(s)$ is stable. The gain stability margin (GM) is 16.5 dB at 162 Hz. The phase stability margin (PM) is 45° at 56.6 Hz.

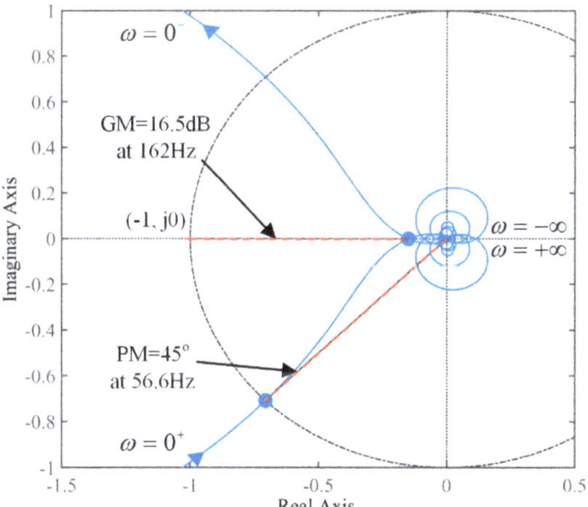

Figure 15. The Nyquist diagram of $G_{ols}(s)$.

The bode diagram of the proposed PLL and QT1-PLL is depicted in Figure 16. It can be seen that the crossover frequency of the proposed PLL is larger than that in the QT1-PLL. This yields faster transient behavior for the proposed method. It is noted that the 100 Hz component is only attenuated in the bode diagram. This does not reveal the filtering capability at 100 Hz, since the diagram is based on the model whose input is grid phase, not grid voltages. The filtering performance is already analyzed in Section 3. Experiments are also carried out to verify the filtering capability in the next section.

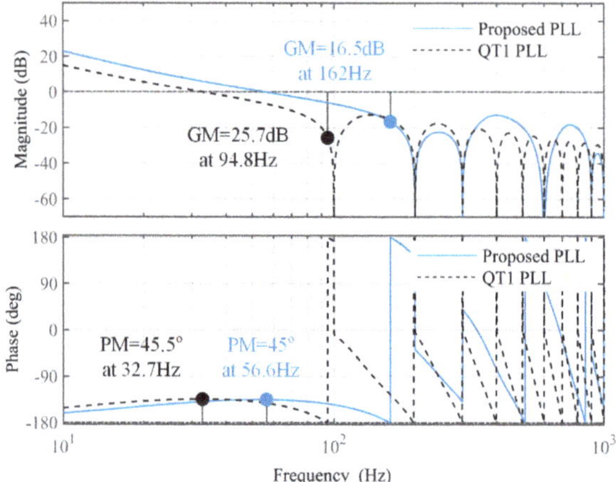

Figure 16. Bode diagram of open-loop system: Proposed PLL (solid lines), QT1-PLL (dashed lines).

4.3. Assessment of Model Accuracy

To validate the analysis above, a simulation is implemented under two conditions. The transient behaviors of phase-errors are depicted in Figure 17. This figure shows that the phase-errors of proposed PLL and its mathematics model are equal to each other during two step change in phase and frequency. The design procedure based on the mathematics model is reasonable.

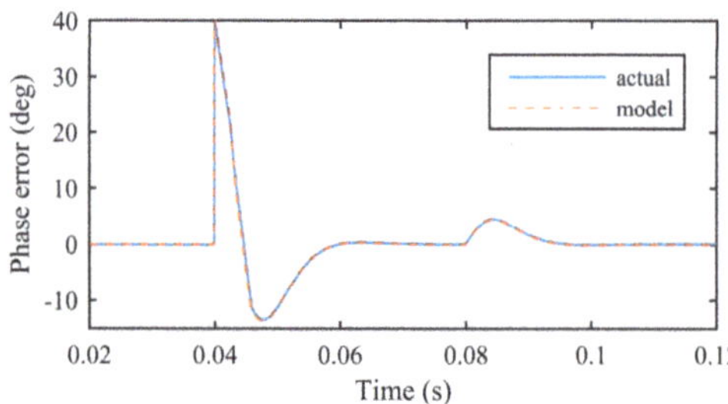

Figure 17. The transient response of the proposed PLL (solid line) and its mathematics model (dashed line) under +40° phase jump (at 0.04 s) and +5 Hz frequency jump (at 0.08 s).

5. Experimental Results

In this section, experiments are carried out under several distorted grid conditions. Different from a traditional grid, when a grid is dominated by renewable energy sources, called a "weak grid", phase jump is inevitable during the process of grid faults, sudden large load tripping and other transient behavior. Wind power fluctuation is an important issue with the increasing penetration of wind power plants, which usually cause grid frequency deviation. Thus, the proposed method is also evaluated under frequency step change and frequency ramp change conditions. Owing to the existence of a large amount of power electronics elements, harmonic disturbances are injected by power converter-based equipment such as HVDC, MMC, etc. Sub/super-synchronous oscillations arising from inappropriate system configuration can also pollute grid voltages. Hence, the proposed PLL is also examined under unbalanced and harmonic distorted voltage conditions.

In this section, experiments are implemented on a real-time experimental platform to examine the performance of the proposed PLL. Three phase voltage signals are generated by a personal computer with a data acquisition board. Through the Digital-Analogy output ports, the voltages signals are exported. PLL is implemented on a digital signal processor (DSP TMS320F28335) board. After receiving voltage signals, the estimated phase and frequency are exported through DA ports on a DSP board. Oscilloscope is used to capture all waveforms.

The grid nominal frequency is 50 Hz. The sampling frequency of a digital system is 10 kHz. The zero-order hold method is used for discretization. Besides this, the proposed PLL, EGDSC-PLL [37] is also implemented as a traditional DSC-based PLL for comparative study. QT1-PLL [36] and MAF-PLL [38] are also compared to assess the performance. Figure 18 shows the setup of the experimental platform.

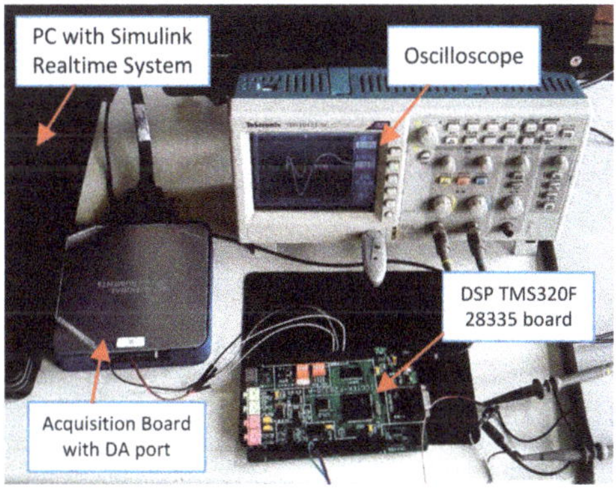

Figure 18. The experimental setup.

5.1. Test Case 1: Phase Jump

The performance is evaluated when phase jump occurs in grid voltages. Figure 19 shows the waveform of three phase voltages. A +40° phase jump came up during the experiments. Figure 20 shows the transient behavior after a phase jump. It is observed that the proposed PLL provides the fastest transient response. The settling times of phase-error and estimated frequency are around one grid period. The dynamic behavior of QT1-PLL takes over 30 ms to converge. The dynamic performance of EGDSC-PLL and MAF-PLL are even more unacceptable, owing to their more than two grid period settling time. According to the requirement of a transient response in some grid codes [21,22], an accurate estimation of grid voltage information should be extracted after undesired injection of disturbance within 25 ms. EGDSC-PLL and MAF-PLL can definitely not fulfill this requirement.

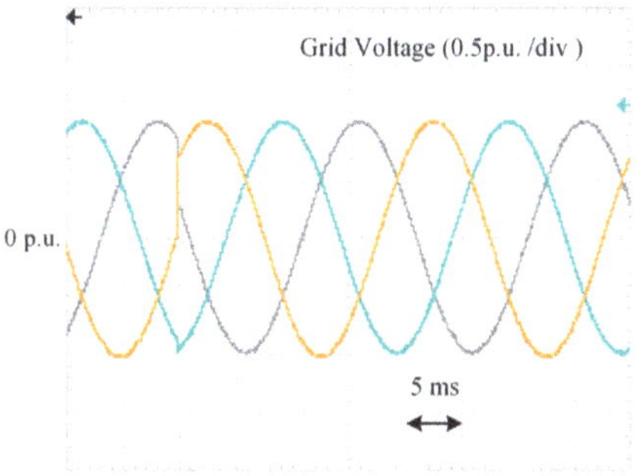

Figure 19. Grid voltages under +40° phase jump condition.

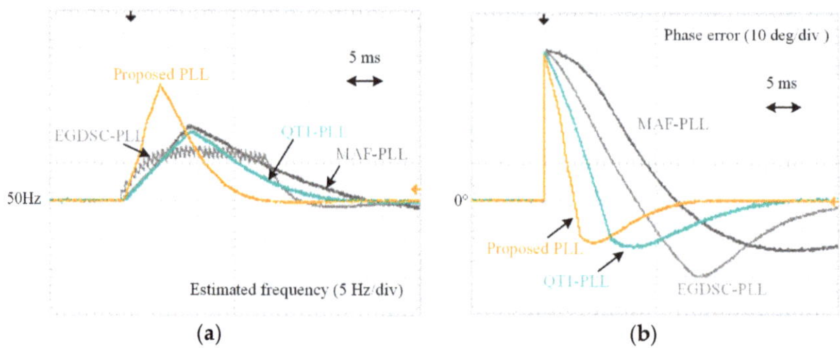

Figure 20. The experimental results of test case 1. (**a**) Estimated frequency and (**b**) phase error.

5.2. Test Case 2: Frequency Step Change

Test case 2 is carried out when grid voltages are under +5 Hz frequency step change. Figure 21 depicts the waveform of three phase voltages. The waveform of estimated frequency and phase error are shown in Figure 22. It is observed that the proposed PLL can accurately estimate grid frequency in less than one period. The phase error of the proposed PLL can converge to zero within only 15 ms. The settling time of QT1-PLL is around one period in both estimated frequency and phase error, which is acceptable for most practical application. While EGDSC-PLL can track grid frequency as fast as QT1-PLL, the phase error takes over 30 ms to converge.

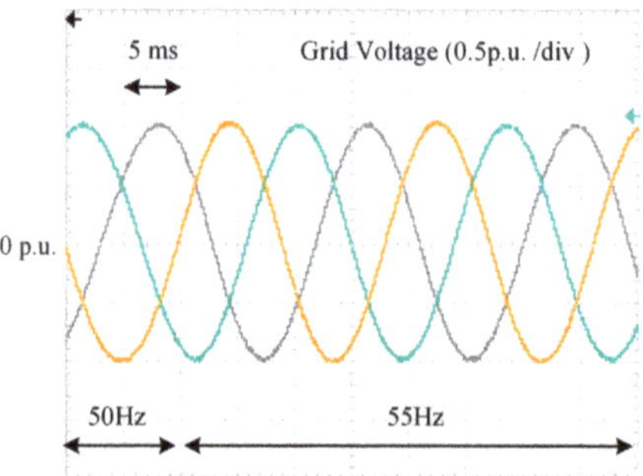

Figure 21. Grid voltages under +5 Hz frequency step change condition.

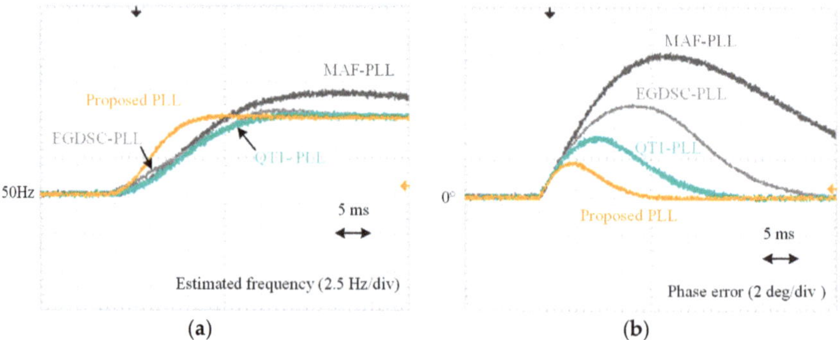

Figure 22. The experimental results of test case 2. (**a**) Estimated frequency and (**b**) phase error.

5.3. Test Case 3: Frequency Ramp Change

In practical conditions, grid frequency can hardly have a step change. Most of time, grid frequency varies bit by bit continuously. Consequently, a ramp change is implemented to the grid frequency to examine PLLs' performance. The grid frequency rises from 50 Hz to 55 Hz with a +100 Hz/s ramp rising rate. The whole behavior last for 50 ms. Figure 23 shows the waveform of grid voltages during this rising procedure. Figure 24 shows the transient behavior of four advanced PLLs. It can be seen that the proposed PLL provides 0.5° phase error during the frequency rising procedure. The phase error of QT1-PLL is less than 2°, which is also acceptable. By contrast, the phase errors of EGDSC-PLL and MAF-PLL are too big, which means PLL cannot provide acceptable phase information during the frequency changing procedure.

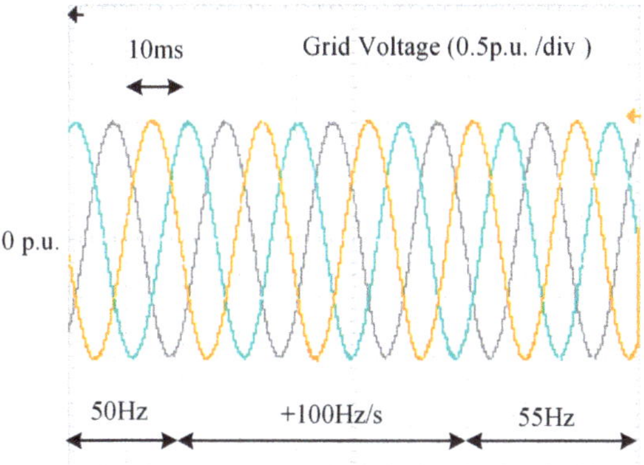

Figure 23. Grid voltages under +5 Hz frequency step change conditions.

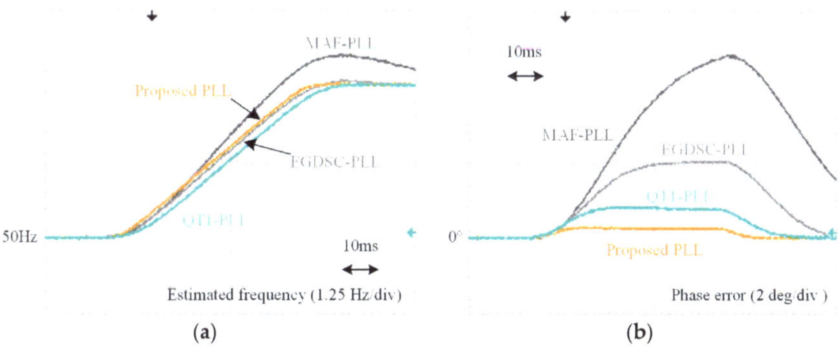

Figure 24. The experimental results of test case 3. (**a**) Estimated frequency and (**b**) phase error.

5.4. Test Case 4: Unbalanced and Distorted Grid Voltages

To examine the filtering capability, the proposed PLL is evaluated under unbalanced and distorted grid voltages condition. A frequency step change occurs during the experimental procedure. The parameters of voltage components in the polluted grid voltages are listed in Table 2. To achieve a satisfactory performance, the delay and MAF units in the proposed PLL, QT1-PLL and MAF-PLL are frequency adaptive. The corresponding implementation of a frequency adaptive structure can be found in [36] and [38]. While EGDSC-PLL can also improve its performance by making DSCs adaptive, too many DSCs used in its filtering stage increases its computational burden dramatically. Hence, EGDSC-PLL with non-adaptive DSCs is more practical for the comparison.

Table 2. Parameters of grid voltages.

Voltage Component (in $\alpha\beta$-frame)	Amplitude (p.u.)
Fundamental positive sequence (+1st order)	1
Fundamental negative sequence (−1st order)	0.1
5th harmonic negative sequence (−5th order)	0.1
7th harmonic positive sequence (+7th order)	0.05
11th harmonic negative sequence (−11th order)	0.05
13th harmonic positive sequence (+13th order)	0.05

The polluted grid voltages are depicted in Figure 25. The transient responses of four PLLs are depicted in Figure 26. It can be observed that all PLLs can eliminate disturbances completely when grid frequency is at its nominal value. When grid frequency jumps from 50 Hz to 55 Hz, transient behavior occurs for every PLL. Thanks to frequency adaptive implementation, the proposed PLL, QT1-PLL and MAF-PLL can still provide a satisfactory filtering capability after frequency change. However, oscillation of phase error occurs in the transient behavior of EGDSC-PLLs for its non-adaptive DSCs. While this problem can be solved by applying adaptive DSCs, a huge computational burden is still a big problem for designers to solve.

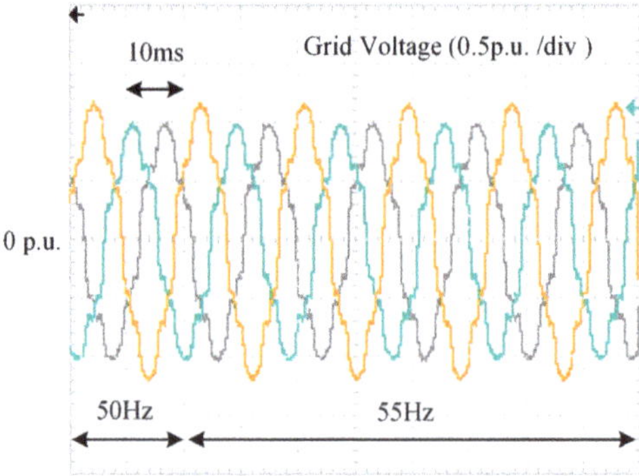

Figure 25. Grid voltages under +5 Hz frequency step change conditions.

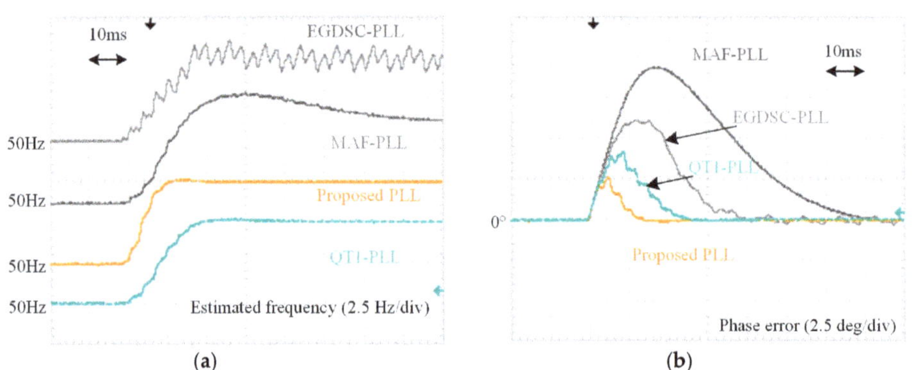

Figure 26. The experimental results of test case 3. (**a**) Estimated frequency and (**b**) phase error.

6. Conclusions

With more and more wind and solar energy plants connected to power systems, grid voltages are prone to be distorted. To keep grid frequency and phase tracking accuracy during distorted and even faulty conditions without degrading dynamic performance, a hybrid filter-based PLL is proposed. By using well designed MDSCs and MAFs with narrowed window length, time delay introduced by filtering stage is reduced. The hybrid filtering stage is incorporated with a QT1-PLL structure. A comprehensive experimental study is implemented to examine the effectiveness. Test cases concerning several conditions, which are easily triggered in high renewable penetration power systems, are carried out. Experimental results illustrate that the proposed PLL can provide a more satisfactory dynamic response than other traditional methods. The filtering capability and implementation complexity are also better than the conventional DSC-based PLLs.

Author Contributions: T.L. and J.Y. proposed the original idea and analyzed the experimental results. Y.L. built the experimental setup and performed the experiments. T.L. wrote the full manuscript. W.G. and B.H. supervised the experiments.

Funding: This work was supported by scientific research fund from Liaoning Electric Power Co., Ltd., State Grid Co. of China.

Acknowledgments: This work was with technical assistance from Liaoning Electric Power Co., Ltd., State Grid Co. of China.

Conflicts of Interest: The authors declare no conflict of interest.

Nomenclature

The following nomenclatures are used in this manuscript:

v_{abc}	Three-phase grid voltage
v_d, v_q	The d-q-axis voltage components after Park transformation
v_α, v_β	The α-β-axis components of grid voltage after Clarke transformation
$\hat{v}^+_{\alpha,1}, \hat{v}^+_{\beta,1}$	The α-β-axis components after prefiltering stage
$\overline{v}_d, \overline{v}_q$	The d-q-axis voltage components after filtering processing
v_{dm}, v_{qm}	The d-q-axis voltage components after MDSC
ω_n	The fundamental nominal angular frequency of grid voltage
$\Delta\omega_g$	The error of estimated angular frequency of grid voltage
$\hat{\omega}_g$	The estimated angular frequency of grid voltage
θ^+_1	The angular phase of input grid voltage
$\hat{\theta}^+_1$	The estimated phase of grid voltage
θ_e	The phase-tracking error
T_ω	The window length of MAF
$\Delta\theta^+_1$	The value of phase jump
k	The only control parameter in the proposed PLL

References

1. Kunjumuhammed, L.; Pal, B.C.; Gupta, R.; Dyke, K.J. Stability analysis of a PMSG-based large offshore wind farm connected to a VSC-HVDC. *IEEE Trans. Energy Convers.* **2017**, *32*, 1166–1176. [CrossRef]
2. Zhang, X.; Shi, D.; Wang, Z.; Zeng, B.; Wang, X.; Tomsovic, K.; Jin, Y. Optimal allocation of series FACTS devices under high penetration of wind power within a market environment. *IEEE Trans. Power Syst.* **2018**, *33*, 6206–6217. [CrossRef]
3. Zheng, Y.; Zhao, J.; Song, Y.; Luo, F.; Meng, K.; Qiu, J.; Hill, D.J. Optimal operation of battery energy storage system considering distribution system uncertainty. *IEEE Trans. Energy Convers.* **2018**, *9*, 1051–1060. [CrossRef]
4. Mahmud, M.A.; Hossain, M.J.; Pota, H.R.; Roy, N.K. Robust nonlinear controller design for three-phase grid-connected photovoltaic systems under structured uncertainties. *IEEE Trans. Power Del.* **2014**, *29*, 1221–1230. [CrossRef]
5. Chowdhury, M.A.; Shafiullah, G.M. SSR mitigation of series-compensated DFIG wind farms by a nonlinear damping controller using partial feedback linearization. *IEEE Trans. Power Syst.* **2018**, *33*, 2528–2538. [CrossRef]

6. Vanfretti, L.; Baudette, M.; Domínguez-García, J.L.; Almas, M.S.; White, A.; Gjerde, J.O. A phasor measurement unit based fast real-time oscillation detection application for monitoring wind-farm-to-grid sub-synchronous dynamics. *Electr. Mach. Power Syst.* **2015**, *44*, 123–134. [CrossRef]
7. Liu, H.; Xie, X.; Zhang, C.; Li, Y.; Liu, H.; Hu, Y. Quantitative SSR analysis of series-compensated DFIG-based wind farms using aggregated RLC circuit model. *IEEE Trans. Power Syst.* **2015**, *30*, 2772–2779. [CrossRef]
8. Liu, H.; Xie, X.; Liu, W. An oscillatory stability criterion based on the unified *dq*-frame impedance network model for power systems with high-penetration renewables. *IEEE Trans. Power Syst.* **2018**, *33*, 3472–3485. [CrossRef]
9. Golestan, S.; Guerrero, J.M.; Vidal, A.; Yepes, A.G.; Doval-Gandoy, J.; Freijedo, F.D. Small-signal modeling, stability analysis and design optimization of single-phase delay-based PLLs. *IEEE Trans. Power Electron.* **2016**, *31*, 3517–3527. [CrossRef]
10. Hui, N.; Wang, D.; Li, Y. An efficient hybrid filter-based phase-locked loop under adverse grid conditions. *Energies* **2018**, *11*, 703. [CrossRef]
11. Li, Y.; Yang, J.; Wang, H.; Ge, W.; Ma, Y. A hybrid filtering technique-based PLL targeting fast and robust tracking performance under distorted grid conditions. *Energies* **2018**, *11*, 973. [CrossRef]
12. Li, Y.; Yang, J.; Wang, H.; Ge, W.; Ma, Y. Leveraging hybrid filter for improving quasi-type-1 phase locked loop targeting fast transient response. *Energies* **2018**, *11*, 2472. [CrossRef]
13. Kulkarni, A.; John, V. Analysis of bandwidth-unit vector distortion trade off in PLL during abnormal grid conditions. *IEEE Trans. Ind. Electron.* **2013**, *60*, 5820–5829. [CrossRef]
14. Batista, Y.N.; De Souza, H.E.P.; Neves, F.A.S.; Filho, R.F.D.; Bradaschia, F. Variable-structure generalized delayed signal cancellation PLL to improve convergence time. *IEEE Trans. Ind. Electron.* **2015**, *62*, 7146–7150. [CrossRef]
15. Rodriguez, P.; Luna, A.; Munoz-Aguilar, R.S.; Etxeberria-Otadui, I.; Teodorescu, R.; Blaabjerg, F. A stationary reference frame grid synchronization system for three-phase grid-connected power converters under adverse grid conditions. *IEEE Trans. Power Electron.* **2012**, *27*, 99–112. [CrossRef]
16. Burgos-Mellado, C.; Costabeber, A.; Sumner, M.; Cárdenas-Dobson, R.; Sáez, D. Small-signal modeling and stability assessment of phase-locked loops in weak grids. *Energies* **2019**, *12*, 1227. [CrossRef]
17. Yang, L.; Chen, Y.; Wang, H.; Luo, A.; Huai, K. Oscillation suppression method by two notch filter for parallel inverters under weak grid conditions. *Energies* **2018**, *11*, 3441. [CrossRef]
18. Li, M.; Zhang, X.; Zhao, W. A novel stability improvement strategy for multi-inverter system in a weak grid utilizing dual-model control. *Energies* **2018**, *12*, 2144. [CrossRef]
19. E.On Netz. *Grid Code-High and Extra High Voltage*; Netz Gmbh: Bayreuth, Germany, 2006; Available online: http://www.pvupscale.org/IMG/pdf/D4_2_DE_annex_A-3_EON_HV_grid__connection_requirements_ENENARHS2006de.pdf (accessed on 16 April 2018).
20. Ministerio de Industria, Turismo y Comercio. *Requisitos de Respuesta Frente a Huecos de Tension de las Instalaciones Eolicas*; Comisión Nacional de Energía: Madrid, Spain, 2006.
21. IEEE Std. *IEEE Standard for Interconnecting Distributed Resources with Electric Power Systems*; IEEE: New York, NY, USA, 2003.
22. National Grid Electricity Transmission. *The Grid Code: Revision 31*; National Grid Electricity Transmission: Warwick, UK, 2008.
23. Li, Y.; Wang, D.; Han, W.; Tan, S.; Guo, X. Performance improvement of quasi-type-1 PLL by using a complex notch filter. *IEEE Access* **2016**, *4*, 6272–6282. [CrossRef]
24. Golestan, S.; Guerrero, J.M.; Vidal, A.; Yepes, A.G.; Doval-Gandoy, J. PLL with MAF-based prefiltering stage: Small-signal modeling and performance enhancement. *IEEE Trans. Power Electron.* **2016**, *31*, 4013–4019. [CrossRef]
25. Rodriguez, P.; Teodorescu, R.; Candela, I.; Timbus, A.V.; Liserre, M.; Blaabjerg, F. New positive-sequence voltage detector for grid synchronization of power converters under faulty grid conditions. In Proceedings of the IEEE Power Electronics Specialists Conference, Jeju, Korea, 18–22 June 2006; pp. 1–7.
26. Guo, X.; Wu, W.; Chen, Z. Multiple-complex coefficient-filter-based phase-locked loop and synchronization technique for three-phase grid-interfaced converters in distributed utility networks. *IEEE Trans. Ind. Electron.* **2011**, *58*, 1194–1204. [CrossRef]

27. Xiao, P.; Corzine, K.A.; Venayagamoorthy, G.K. Multiple reference frame-based control of three-phase PWM boost rectifiers under unbalanced and distorted input conditions. *IEEE Trans. Power Electron.* **2008**, *23*, 2006–2017. [CrossRef]
28. Rodríguez, P.; Pou, J.; Bergas, J.; Candela, J.I.; Burgos, R.P.; Boroyevich, D. Decoupled double synchronous reference frame PLL for power converters control. *IEEE Trans. Power Electron.* **2007**, *22*, 584–592. [CrossRef]
29. Golestan, S.; Ramezani, M.; Guerrero, J.M.; Monfared, M. Dq-frame cascaded delay signal cancellation-based pll: Analysis, design, and comparison with moving average filter-based PLL. *IEEE Trans. Power Electron.* **2015**, *3*, 1618–1632. [CrossRef]
30. Wang, Y.F.; Li, Y.W. Grid synchronization pll based on cascaded delay signal cancellation. *IEEE Trans. Power Electron.* **2011**, *26*, 1987–1997. [CrossRef]
31. Wang, Y.F.; Li, Y.W. Three-phase cascaded delay signal cancellation pll for fast selective harmonic detection. *IEEE Trans. Ind. Electron.* **2013**, *60*, 1452–1463. [CrossRef]
32. Rani, B.I.; Aravind, C.K.; Ilango, G.S.; Nagamani, C. A three phase PLL with a dynamic feed forward frequency estimator for synchronization of grid connected converters under wide frequency variations. *Int. J. Electr. Power Energy Syst.* **2012**, *41*, 63–70. [CrossRef]
33. Geng, H.; Sun, J.; Xiao, S.; Yang, G. Modeling and implementation of an all digital phase-locked-loop for grid-voltage phase detection. *IEEE Trans. Ind. Inf.* **2013**, *9*, 772–780. [CrossRef]
34. Golestan, S.; Ramezani, M.; Guerrero, J. An analysis of the PLLs with secondary control path. *IEEE Trans. Ind. Electron.* **2014**, *61*, 4824–4828. [CrossRef]
35. Golestan, S.; Freijedo, F.D.; Vidal, A.; Guerrero, J.M.; Doval-Gandoy, J. A quasi-type-1 phase-locked loop structure. *IEEE Trans. Power Electron.* **2014**, *29*, 6264–6270. [CrossRef]
36. Golestan, S.; Guerrero, J.M.; Vasquez, J.C. Hybrid adaptive/nonadaptive delay signal cancellation-based phase-locked loop. *IEEE Trans. Ind. Electron.* **2017**, *64*, 470–479. [CrossRef]
37. Golestan, S.; Freijedo, F.D.; Vidal, A.; Yepes, A.G.; Guerrero, J.M.; Doval-Gandoy, J. An efficient implementation of generalized delayed signal cancellation PLL. *IEEE Trans. Power Electron.* **2016**, *31*, 1085–1094. [CrossRef]
38. Golestan, S.; Ramezani, M.; Guerrero, J.M.; Freijedo, F.D.; Monfared, M. Moving average filter based phase-locked loop: Performance analysis and design guidelines. *IEEE Trans. Power Electron.* **2014**, *29*, 2750–2763. [CrossRef]

© 2019 by the authors. Licensee MDPI, Basel, Switzerland. This article is an open access article distributed under the terms and conditions of the Creative Commons Attribution (CC BY) license (http://creativecommons.org/licenses/by/4.0/).

Article

A Novel Hybrid Converter Proposed for Multi-MW Wind Generator for Offshore Applications

Muhammad Luqman [1,2], Gang Yao [1,2,*], Lidan Zhou [1,2], Tao Zhang [1,2] and Anil Lamichhane [1,2]

1 Key Laboratory of Control of Power Transmission and Conversion (Ministry of Education), Shanghai Jiao Tong University, Shanghai 200240, China; Luqman@sjtu.edu.cn (M.L.); zhoulidan@sjtu.edu.cn (L.Z.); roxaszt@sjtu.edu.cn (T.Z.); anillamichhane@sjtu.edu.cn (A.L.)

2 Department of Electrical Engineering, School of Electronic Information and Electrical Engineering Shanghai Jiao Tong University, Shanghai 200240, China

* Correspondence: yaogangth@sjtu.edu.cn; Tel.: +86-139-1669-1886

Received: 6 October 2019; Accepted: 24 October 2019; Published: 1 November 2019

Abstract: Modern multi-MW wind generators have used multi-level converter structures as well as parallel configuration of a back to back three-level neutral point clamped (3L-NPC) converters to reduce the voltage and current stress on the semiconductor devices. These configurations of converters for offshore wind energy conversion applications results in high cost, low power density, and complex control circuitry. Moreover, a large number of power devices being used by former topologies results in an expensive and inefficient system. In this paper, a novel bi-directional three-phase hybrid converter that is based on a parallel combination of 3L-NPC and '*n*' number of Vienna rectifiers have been proposed for multi-MW offshore wind generator applications. In this novel configuration, total power equally distributes by sharing of total reference current in each parallel-connected generator side power converter, which ensures the lower current stress on the semiconductor devices. Newly proposed topology has less number of power devices compared to the conventional configuration of parallel 3L-NPC converters, which results in cost-effective, compact in size, simple control circuitry, and good performance of the system. Three-phase electric grid is considered as a generator source for implementation of a proposed converter. The control scheme for a directly connected three-phase source with a novel configuration of a hybrid converter has been applied to ratify the equal power distribution in each parallel-connected module with good power factor and low current distortion. A parallel combination of a 3L-NPC and 3L-Vienna rectifier with a three-phase electric grid source has been simulated while using MATLAB and then implemented it on hardware. The simulation and experimental results ratify the performance and effectiveness of the proposed system.

Keywords: low-cost hybrid converter; bi-directional converter; parallel configuration of converters; converter for multi-MW wind generator; offshore wind energy converter applications

---

1. Introduction

From the last few decades, power sectors are considering accommodating clean and sustainable energy sources to minimize the adverse effect of non-sustainable energy. Among the sustainable energy sources, wind energy has become a globally more attractive solution for high power electricity production [1]. According to a report, the State of Alaska Legislature has committed to supplying 50% of its energy from renewable sources by 2025 [2]. Meanwhile, China has also set her target to rise supply from renewable energy up to 20% in the national energy consumption by 2030 [3]. Therefore, with the passage of time, the demand for high power generator in manufacturing and installing industries is also growing for high power production.

Wind energy conversion systems (WECSs) have been installed in 80 countries and their total expected capacity will reached 800 GW by 2021 [4,5]. Nowadays, the most commonly deployed high capacity wind energy conversion systems occurs in the form of onshore and offshore wind farms. Such a developing tendency in wind energy area attracts the manufacturing companies to build individual high power wind turbine generator. Therefore, from the last few years', generator designing industries are manufacturing high power generators up to 10 MW, which will become double after few years [6,7].

For the wind energy conversion system, a power electronics converter plays an important role as a machine side frond-end converter [8,9]. Generated alternating energy from the wind needs to be converted into stable direct current for appropriate use in numerous applications, such as standalone and grid-connected systems [10,11]. Therefore, generator side converters are considered the most accountable part for efficiency, power density, control circuitry, and cost of the entire wind energy conversion system.

Among power electronics converters, there are various sorts of rectifiers that have used for generator side energy conversion, such as, passive, active, and hybrid rectifier, as reviewed in [12,13]. Among the family of passive rectifiers, diode full-bridge rectifier remained more common in the past as a front end rectifier with harmonics and power factor problems, but these issues lower the efficiency of the whole system. Therefore, to overcome the aforementioned problems, another two-level six switch active rectifier, as shown in Figure 1a, became more popular as a front end rectifier whereas both discussed types of rectifiers were best designed for low power applications up to KVs [13]. Thus, for medium or high power applications, multilevel converter as well as parallel configuration of neutral point clamped (NPC) converter, as shown in Figure 1b, became more attractive in wind energy conversion systems (WECSs) to reduce the voltage and current stress on the semiconductor devices [14–16].

Nowadays, a three-level NPC converter as a front end rectifier is commonly used with a few hundreds of KW wind generator. While, in multi-MW wind generator, parallel arrangement of 3L-NPC converters, as shown in Figure 2, became a more attractive solution for reducing the current stress on power devices by equal sharing of total current in all parallel-connected converters. Otherwise, high rated semiconductor devices would be required that can bear thousands of amperes and KVolt, which are expensive and not common in markets. Besides, the parallel configuration of NPC converters also increases the complexity in aspects of a large number of active switches and control circuitry, which affects cost and efficiency. Therefore, another type of active rectifier, known as the Vienna rectifier analogous to a t-type inverter, as shown in Figure 1c, has been considered as a generator side unidirectional rectifier that is based on numerous advantages, as mentioned in [17].

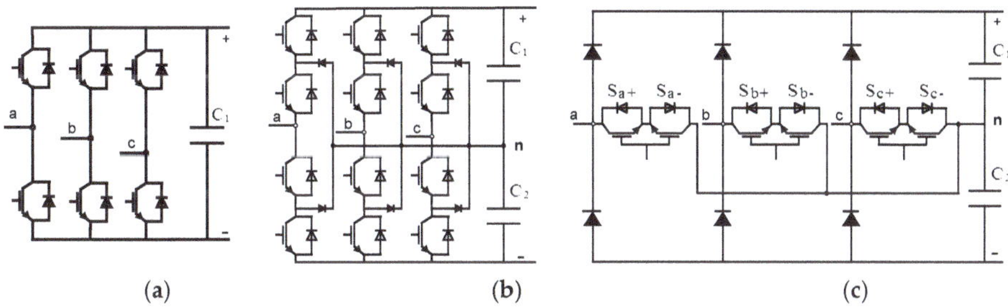

Figure 1. (**a**) Two-level converter; (**b**) Three-Level NPC converter; (**c**) Three-level Vienna rectifier.

In the aspects of power flow direction, NPC has bi-directional, whereas the Vienna rectifier has unidirectional power flow options. In the case of wind energy integration with the electric grid, occasionally voltage sag occurs because of rapid change in rotor speed, which demands stabilizing power operation. In addition, it also sometimes requires a little power for field excitation of the wind generator. Accordingly, a full-scale bi-directional power flow converter helps to deal with such problems [18–21].

A new bi-directional configuration of converter has been proposed for high power offshore WEC applications by keeping in view the above-mentioned issues for bidirectional power flow in multi-MW WECS. In this fresh topology, a single module of three-phase/three-level NPC converter connected in parallel with 'n' number of three-phase Vienna rectifiers has been designed. In this configuration of converters, NPC works as a bi-directional power flow, while Vienna rectifiers work as a unidirectional active-rectifier having a lower number of active switches and lower cost as compared to 3L-NPC [22]. The functioning of a deliberated parallel-connected converters ratifies for a higher range of wind generator applications by dropping current stress on power devices in each parallel converters [23,24]. Hence, the overall proposed hybrid configuration uses less number of switches with high power density as compared to existing parallel-connected three-level neutral point clamped converters. The major advantages of the considered system and its control scheme are:

- reduction of switching loss;
- reduction in number of switches;
- works with unity PF and less THD;
- cost-effective bi-directional converter;
- reduction of current rating power devices; and,
- improved the converter handling power capacity up to multi-MW.

Different configurations of machine side converters for wind energy transformation are discussed in Section 2; the proposed bi-directional hybrid converter scheme is presented in Section 3; the control scheme and working of the proposed system are discussed in Section 4; and, the MATLAB based simulation results of suggested hybrid converter are shown in Section 5. In Section 6, the DSP based experimental results are presented. Finally, in Section 7, the conclusion is discussed.

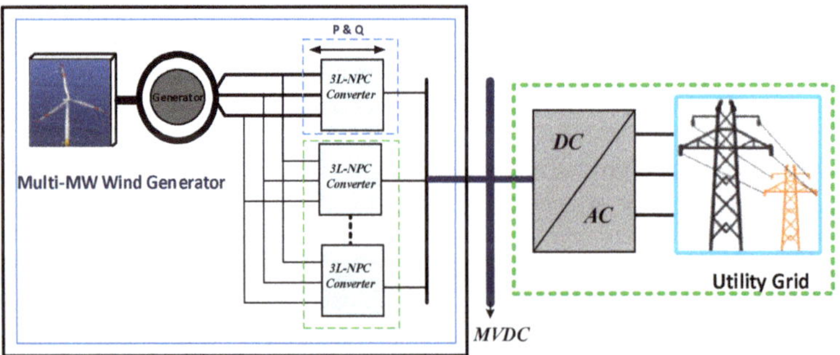

Figure 2. Conventional scheme of 3L-NPC converter with multi-MW wind generator.

2. Front End Active Rectifiers for WECS

In this section, the most commonly used active rectifiers and number of devices used by them have been studied. Due to the inefficient and low power applications of passive rectifiers, they are not considered as a good option in WEC applications.

Therefore, active rectifiers that are based on control switches became more captivating for efficient energy conversion in WECSs [13,25]. Among the family of active rectifiers, the two-level converter was most commonly used as a front end rectifier in energy conversion progression. On the other hand, such a low power application rectifier has higher harmonics and EMI. Although, in the case of medium or high power wind generators up to 10 MW or more, it cannot endure the voltage and current stress, because of the low power handling capability of the power devices [15,24]. Hence, predefined multilevel or parallel configuration of converters was considered as a vigorous solution, as discussed in Section 1, to overcome these issues. From the last few years, a three-level back-to-back neutral point diode clamped (NPC) converters are extensively using as a front end rectifier as a machine side converter. This type of dual-mode converter can work as an inverter as well as a rectifier mode, which mainly depends on its modulation signal. Additionally, a three-level NPC converter or its parallel configuration also handle more power with low power handling devices [26–29].

The rest of NPC circuit, there is another three-level unidirectional rectifier having less number of active switches and easy to implement as a machine side rectifier that is called a Vienna rectifier. It also has a boosting and continuous input current ability. According to the efficiency point of view, a comparative analysis of all three topologies of Vienna rectifier has been discussed in [30], which shows that three-phase Vienna with a t-type inverter shape with six control switches, as shown in Figure 1c, has more efficiency than all other topologies. This simplest three-phase power circuit shows less number of power switches, easy to control, and cost-effective circuit. It has several advantages, such as high efficiency, operation at high power factor, and being reliable to implement with high power density [22,30].

Table 1 mentions the number of devices used by all discussed circuits.

Table 1. Comparison of the number of devices used.

Converter Type	Number of Diodes	Number of IGBTs	Number of IGBTs Drivers	Total Number of Devices
Two-level	6	6	6	18
Three-level NPC	18	12	12	42
3L-Vienna rectifier	12	6	6	24

Based on a comparative study of devices used by discussed converters, as mentioned in Table 1 and aforementioned problems of converters, two parallel combinations of 3L-NPC converters use almost twice the number of power devices as compared to two parallel-connected Vienna rectifiers. Therefore, a parallel configuration of two 3L-NPC will make the circuit complex, expensive, and with low power density. Thus, a new hybrid bi-directional converter for high power applications with high power density and cost-effective is recommended, as shown in the next section.

3. Proposed Converter Configuration

Figure 3 shows a new hybrid converter directly connected with multi-MW wind generator. In this arrangement, 3L-NPC works as a bi-directional converter for motoring as well as generating action, intended for electrically field excitation to produce starting torque. In the case of electrical field excitation in the generator, 3L-NPC works as an inverter mode, on the other hand, it works as a rectifier mode, just like all interleaved unidirectional Vienna rectifiers operation, as mentioned in the proposed scheme. Out of $(n + 1)$ parallel-connected power converters, the projected system only uses a single unit of three-phase 3L-NPC converter and 'n' number of Vienna rectifiers, as shown in Figure 3. A deliberated system, in fact, is a parallel operation of circuits, as mentioned in Figure 2b,c, which results in low THD and good power factor. The total power equally distributes in each power electronics building block of the deliberated system by $1/(n + 1)$, which helps to select the lower power rating semiconductor devices. Hence, the intended power converter interface best fit for multi-MW generators. A recommended bidirectional power flow system is specially designed for high power wind generators, which are often used as directly connected offshore wind turbines in wind energy conversion solicitations [22,31].

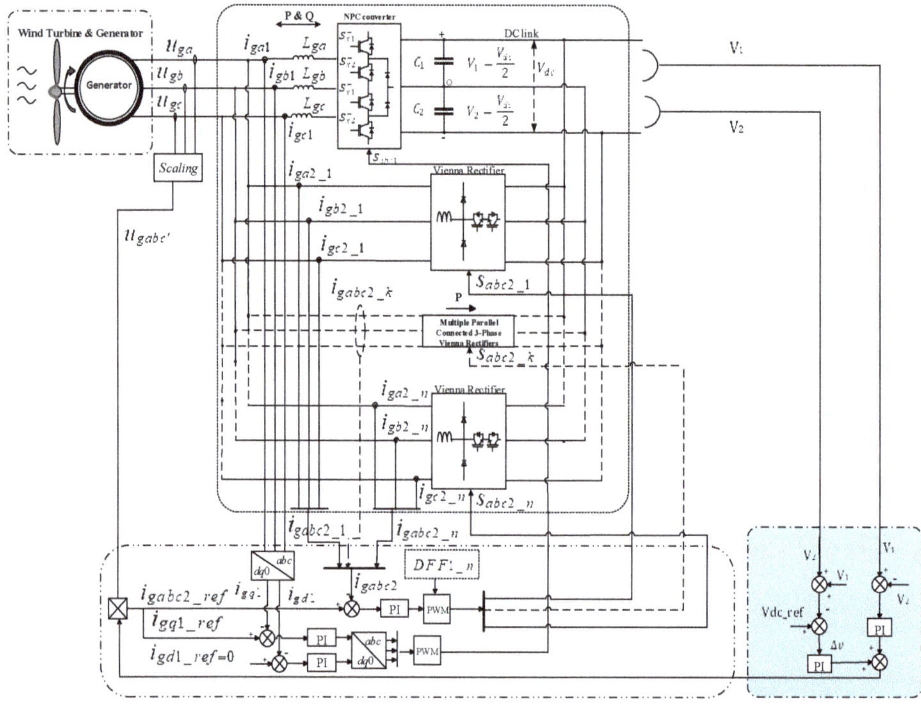

Figure 3. Proposed hybrid converter for multi-MW wind generator.

Parallel Operation of Converters

A balanced three-phase electric grid has been considered as a generated supply for the validation of the proposed converter operation. Mathematical form of voltages in a three-phase system can be written as:

$$U_{ga} = ESin(wt)$$
$$U_{gb} = ESin(wt - 120) \tag{1}$$
$$U_{gc} = ESin(wt + 120)$$

where U_{ga}, U_{gb}, and U_{gc} are the phase voltages of the deliberated electric grid, while 'E' and 'w' represent the peak voltage and angular frequency of the input source.

In the proposed configuration, NPC works as a full-scale converter for the considered three-phase power feedback system, while others interleaved Vienna rectifiers work as a unidirectional power flow from generator to the DC link. In the planned system, the total output power of the generator divided by a factor of $1/(n + 1)$ for each converter. Where '1' in the denominator represent to the NPC, while 'n' represent the number of other parallel connected Vienna rectifiers.

The uppermost NPC converter, as depicted in Figure 3, have dual nature of operation (rectifier & inverter) while all other parallel-connected circuits only work as a unidirectional rectifier mode of operation. Phase voltages of NPC converter as a rectifier mode are mentioned in Table 2 while assuming ideal power switches with their following switching states:

$$d_x = \begin{cases} 1 \; S_{x1}^{+}, S_{x2}^{+} : ON \quad S_{x1}^{-}, S_{x2}^{-} : OFF \\ 0 \; S_{x1}^{-}, S_{x2}^{+} : ON \quad S_{x1}^{+}, S_{x2}^{-} : OFF \\ -1 \; S_{x1}^{-}, S_{x2}^{-} : ON \quad S_{x1}^{+}, S_{x2}^{+} : OFF \end{cases} \tag{2}$$

Whereas, $x = a, b, c$ for three-phase system. There are three kinds of switching states for a single leg 3L-NPC. Therefore, overall states for three-phase converter are '3^3 =27', which can be expressed as:

$$U_{gx0} = \begin{cases} V_1 & if \ \ d_x = 1 \\ 0 & if \ \ d_x = 0 \\ -V_2 & if \ \ d_x = -1 \end{cases} \tag{3}$$

Table 2. Configurations and state of 3L-NPC converter.

Conf.	S_{x1}^{+}	S_{x2}^{+}	S_{x1}^{-}	S_{x2}^{-}	V_{out}
1st 1100	1	1	0	0	$+V_{dc}/2$
2nd 0110	0	1	1	0	0
3rd 0011	0	0	1	1	$-V_{dc}/2$

Moreover, for 'n' number of parallel-connected unidirectional Vienna rectifiers, the most efficient topology, like t-type inverter with six active switches, has been used. Assuming the balanced grid supply voltage at the input of three-phase Vienna rectifier the terminal voltages with switching states and polarity of the phase current can be expressed as:

$$\begin{aligned} U_{gA0} &= \frac{V_{dc}}{2} sgn(i_{sa2_n})(1 - S_a) \\ U_{gB0} &= \frac{V_{dc}}{2} sgn(i_{sb2_n})(1 - S_b) \\ U_{gC0} &= \frac{V_{dc}}{2} sgn(i_{sc2_n})(1 - S_c) \end{aligned} \tag{4}$$

where S_a, S_b, and S_c are the switching states that switched between 1 and 0. Table 3 mentions eight different switching conditions [32–34].

Table 3. Switching conditions for three-phase Vienna Rectifier.

S_a	S_b	S_c	V_{an}	V_{bn}	V_{cn}
0	0	0	+Vo/2	−Vo/2	−Vo/2
0	0	1	+Vo/2	−Vo/2	0
0	1	0	+Vo/2	0	−Vo/2
0	1	1	+Vo/2	0	0
1	0	0	0	−Vo/2	−Vo/2
1	0	1	0	−Vo/2	0
1	1	0	0	0	−Vo/2
1	1	1	0	0	0

Vienna rectifiers have boosting ability with the inductive filter at the input side of the circuit, which can be calculated while using Equation (5) [35].

$$L_i = \frac{V_{bus}}{8 * F_{sw} * \Delta I_{pp\ \max}} \tag{5}$$

where L_i is the input inductor, V_{bus} is a dc bus voltage, F_{SW} is the value of switching frequency, and ΔI_{ppmax} is a maximum ripple current value.

Finally, the output of the rectifier terminals has connected with the common dc-link capacitors to obtain the regulated suppressed ripple content and for power flow control problems. The dc-link consists of two equally sized capacitors that were also used to provide a low inductive path for the turned-off current. The dc-link capacitance can be calculated by Equation (6).

$$C_{DC} = \frac{2\tau S_N}{V_{DC^2}} \quad (6)$$

where S_N is a nominal apparent power of the converter, τ is the time constant that usually considered as less than 5 ms, and V_{DC} is the total dc-link voltage [36]. Equation (7) determines the individual capacitor sizes.

$$\frac{1}{C_{DC}} = \frac{1}{C_{DC1}} + \frac{1}{C_{DC2}} \quad (7)$$

having the same magnitude of current. Moreover, the dc-link voltage is to be kept constant for the input variations due to wind or load side dynamics. The overall control strategy of the proposed structure is demonstrated in the next section.

4. Control Strategy for the Proposed Converter

Wind energy produces dynamic alternating energy that is not acceptable by appliances and for long-distance transmission. Therefore, some control schemes on directly connected wind generator side converter require application to achieve a regulated energy. Some aspects of the modified hybrid control strategy for the new system were taken into account, such as good power factor, low THD of input current, and equal sharing of power in each parallel-connected module. The detail of the converter control is discussed in the following subsections:

4.1. NPC Converter Control

Figure 4 shows a simplified machine-side 3L-NPC converter control strategy. Voltage equation considering the input inductive filter in stationary 'abc' frame can be written as:

$$L_{gx}\frac{di_x}{dt} = U_x - U_{gx} \quad (8)$$

Whereas, $x = a, b, c$

u_x = converter input voltage

u_{gx} = grid supply voltage

L_{gx} = *Input inductor*

A commonly used voltage oriented control (VOC) has been applied for the 3L-NPC converter control. The classical type proportional-integral controllers (PI) method has been adopted to obtain good performance on dc values with small steady-state error [37]. Therefore, the stationary frame requires conversion into a synchronous d–q reference frame. Moreover, active power control is achived by setting the d-axis along with the grid voltage amplitude, whereas the q-axis has been set to zero. Active and reactive power can be controlled by the d and q-axis of the current, as illustrated in Equations (9) and (10).

$$P_g = \frac{3}{2} v_{gd} i_{gd} \quad (9)$$

$$Q_g = -\frac{3}{2} v_{gd} i_{gd} \quad (10)$$

Park and Clark's transformation is required to transform a stationary frame to a synchronous frame, and vice versa, to implement the VOC method. For the balanced amplitude of the sinusoidal waveform, the transformation from 'dq0' frame to 'abc' frame have done while using Equation (11).

$$\begin{pmatrix} u_{ga} \\ u_{gb} \\ u_{gc} \end{pmatrix} = \begin{pmatrix} Cos(\theta_e) & -Sin(\theta_e) & \frac{1}{\sqrt{2}} \\ Cos\left(\theta_e - \frac{2\pi}{3}\right) & -Sin\left(\theta_e - \frac{2\pi}{3}\right) & \frac{1}{\sqrt{2}} \\ Cos\left(\theta_e + \frac{2\pi}{3}\right) & -Sin\left(\theta_e + \frac{2\pi}{3}\right) & \frac{1}{\sqrt{2}} \end{pmatrix} \begin{pmatrix} u_{gd} \\ u_{gq} \\ 0 \end{pmatrix} \tag{11}$$

Similarly, voltage vectors in the 'abc' frame can also be directly transformed into the 'dq0' frame, according to the multiplication of the matric, as given by Equation (12).

$$\begin{pmatrix} u_{gd} \\ u_{gq} \\ 0 \end{pmatrix} = \frac{2}{3} \begin{pmatrix} Cos(\theta_e) & Cos\left(\theta_e - \frac{2\pi}{3}\right) & Cos\left(\theta_e + \frac{2\pi}{3}\right) \\ -Sin(\theta_e) & -Sin\left(\theta_e - \frac{2\pi}{3}\right) & -Sin\left(\theta_e + \frac{2\pi}{3}\right) \\ \frac{1}{\sqrt{2}} & \frac{1}{\sqrt{2}} & \frac{1}{\sqrt{2}} \end{pmatrix} \begin{pmatrix} u_{ga} \\ u_{gb} \\ u_{gc} \end{pmatrix} \tag{12}$$

The VOC approach consists of two closed-loop controllers, one for dc-link (outer loop) and the second one for current control (inner loop). In the outer loop control, two dc voltage controller have been taken to ensure the voltage balancing across each capacitors, among two voltage controllers one controller take the input error signal after comparing the reference dc voltage and the sum of measured individual capacitor voltages, while the other controller takes the error signal that was generated by the voltage difference across each capacitor to suppress the zero leakage current. The error that was generated by both controller was sent to two separate PI controllers for tuning. Finally, the sum of two PI controller results in a reference current i_{gd1_ref}. Meanwhile, i_{gq1} is set zero to deal with active power only. In Figure 4, both i_{sd1} and i_{sq1} are the feedback currents that were obtained by transforming the grid current into a rotating reference frame for inner loop control. The voltage controller sets the reference current i_{sd1_ref}. Subsequently, the differences between the reference and actual currents are sent to the inner loop of the PI controller. The output of these current controller was provided to the PWM generator to obtain the appropriate signal. The outputs of the PWM generator were applied to the NPC switches for proper operation.

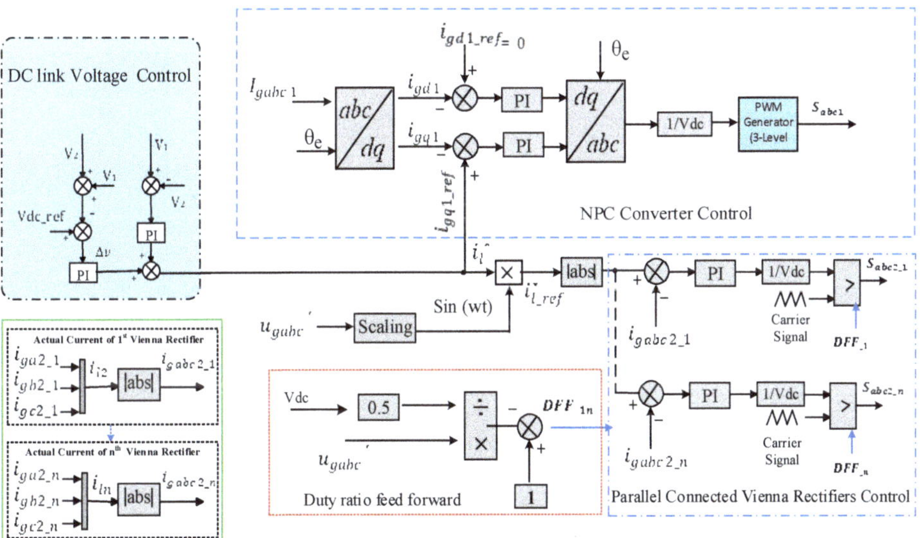

Figure 4. Control strategy of a proposed hybrid converter.

4.2. Vienna Rectifier Control

The voltage equations for input side of "n" number of Vienna rectifier can be written as

$$U_{ga} = L\frac{di_{sa2\_n}}{dt} + U_{gA0}$$
$$U_{gb} = L\frac{di_{sb2\_n}}{dt} + U_{gB0} \quad (13)$$
$$U_{gc} = L\frac{di_{sc2\_n}}{dt} + U_{sC0}$$

where U_{gA0}, U_{gB0}, and U_{gC0} are the input voltages of the Vienna rectifier and $n = 1,2,3 \ldots$.

Here, a simple current average control scheme has been implemented, where the current obtained from the voltage controller was set as a reference current i_l^*, which further divides in NPC and interleaved Vienna rectifiers. In Figure 4, the final reference current $i_{l\_ref}^*$ for all interleaved rectifiers was attained by multiplying i_l^* with $\sin(wt)$. Whereas, $\sin(wt)$ represents the unit amplitude waveform of the generator voltage. The error obtained by taking the difference between the reference current i_l^* and measured current i_{sabc2_1} is taken as an input to the current control. The PI controller is also selected for the inner current control of Vienna rectifier, which gives the output, like PWM duty cycle. Additionally, the duty ratio feedforward (DFF1_n) method has also been applied to improve the THD and waveform of the current at zero crossing. While designing the parameters of the PI controller bandwidth of the current control was kept wider than the outer voltage loop [37].

5. Simulation Results

A 2 KW three-phase MATLAB based simulation studies of two parallel-connected 3L-NPC and 3L-Vienna rectifier, as shown in Figure 5, has been performed in order to verify the performance of the proposed converter.

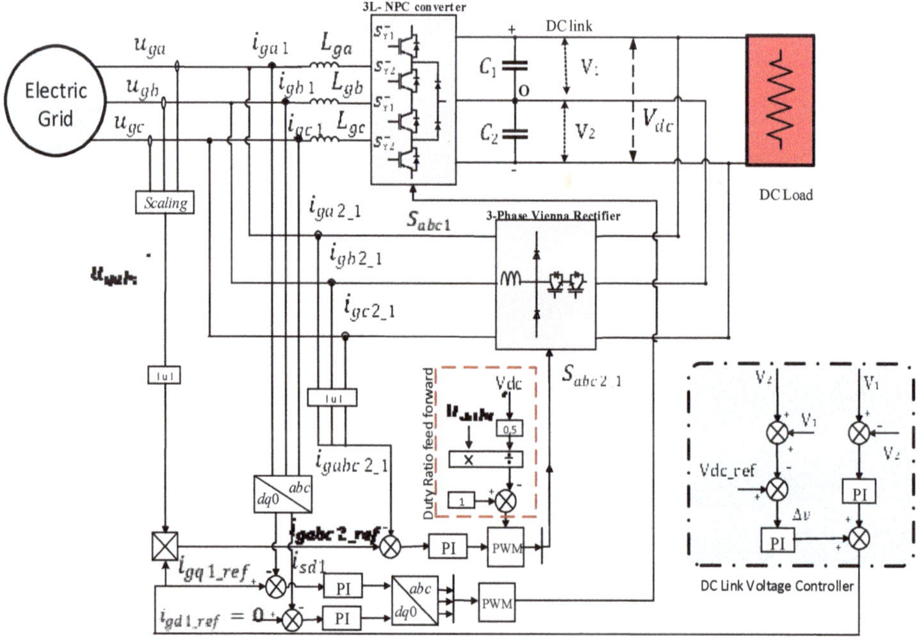

Figure 5. Diagram of simulated and Implemented system.

The recommended converter is best designed for multi-MW wind generator that is most prevalent in offshore wind energy applications. For the purpose of validation three-phase, electric grid as a generated source was considered to implement the proposed converter. All of the parameters that were used by the simulation are mentioned in Table 4.

Figure 6a represents an applied three-phase voltage of 50 Vrms from the grid and Figure 6e,i denote waveforms of voltage across two parallel-connected 3L-NPC and Vienna rectifier, respectively. Figure 6b illustrates the total current drawn from the source, which is exactly the sum of both the NPC and Vienna rectifier. Similarly, half of the total current in Figure 6f,j justify the division of current equally in each parallel connected converter that also satisfies the control main purpose. Figure 6c,g,k show that the phase among applied voltage and current are the same, which means their power factor nearly unity. Figure 6d,h,l are a representation of THD of total current, bidirectional converter current and the Vienna rectifier current, respectively, which also results that the Vienna rectifier has the lowest THD when compared to NPC converter. Therefore, the Vienna rectifier was also selected for the collection of efficient DC energy from wind energy.

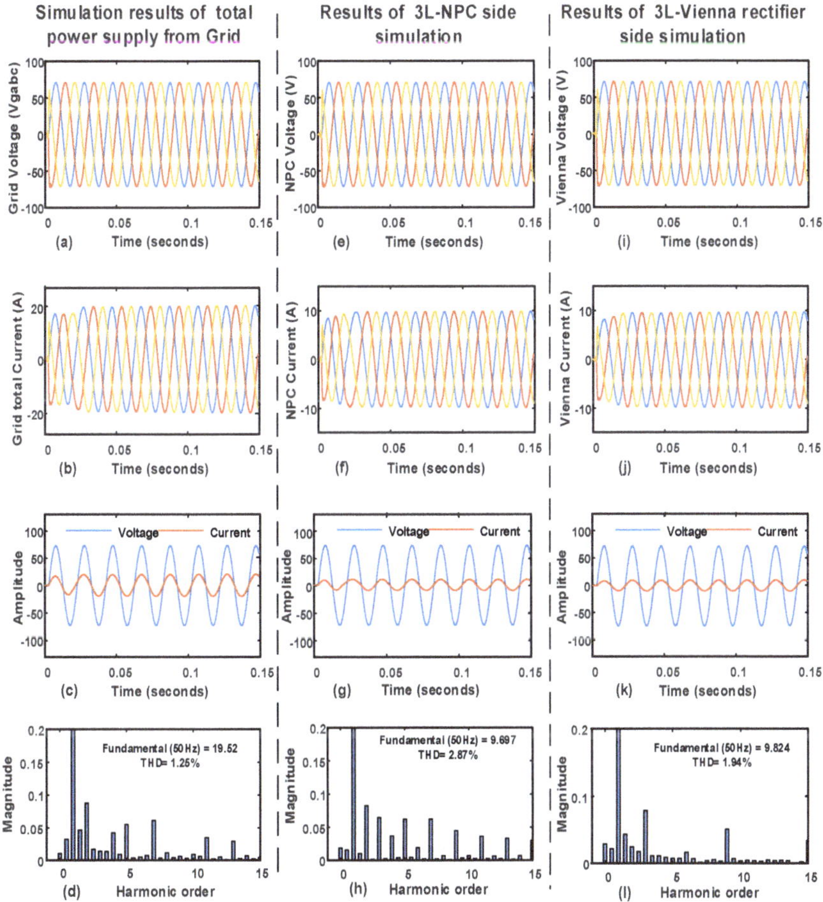

Figure 6. Simulation results of a 2-KW proposed hybrid converter.

Table 4. Data used for Simulation and Experiment.

Parameter	Value
Total power	2 KW
Grid side voltage	50 Vrms
DC- link voltage	200 V
Grid frequency	50 Hz
Switching frequency	15 KHz
Inductor (3L-NPC/Vienna)	300 uH
C1/C2 capacitor	9020 uF

The blue and red legends in Figure 7a represent voltage across C1 and C2, while the yellow label represents a total DC-link voltage. The total dc current drawn by the resistive load is shown in Figure 7b.

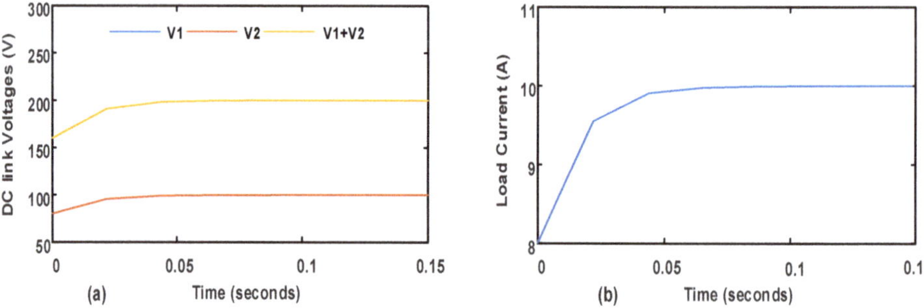

Figure 7. Simulation results of DC link voltage control and load current.

6. Experimental Results

The proposed hybrid converter consists of a three-phase Vienna rectifier that was interleaved with 3L-NPC converter with total power of 2-KW was designed and implemented as per the specifications mentioned in Table 4. Figure 8 shows the requisite experimental results of the system, as depicted in Figure 5. As a Page: 9total grid supply voltage to the parallel connected converters will be same as showm in is Figure 8a–c. Figure 8a–d epitomize the results of total power that is supplied by the grid to both parallel-connected converters. 3.2% THD of total current also meet the IEEE standard. Figure 8b,f,j show the waveforms of total current, 3L-NPC and Vienna rectifier respectively. Half of the total current in 3L-NPC and Vienna rectifier show that the total system power divided among parallel-connected converters.

The detailed data in Figure 8c,g,k also verify that the total power has equally distributed in both parallel-connected converters, which verifies that the control algorithm works well in power-sharing. The currents drawn by NPC and Vienna rectifier, as shown in Figure 8h,l with THD 4.7% and 2.5%, respectively, also follow the IEEE standard.Moreover, the Vienna rectifier has less THD as compared to 3L-NPC, which also leads towards increased efficiency.

In Figure 9. the blue line shows the experimental results of the total dc-link voltage, while the yellow line represents a load current, as mentioned in table-IV. Finally, the whole setup of the implemented converter is shown in Figure 10 by labeling the main parts.

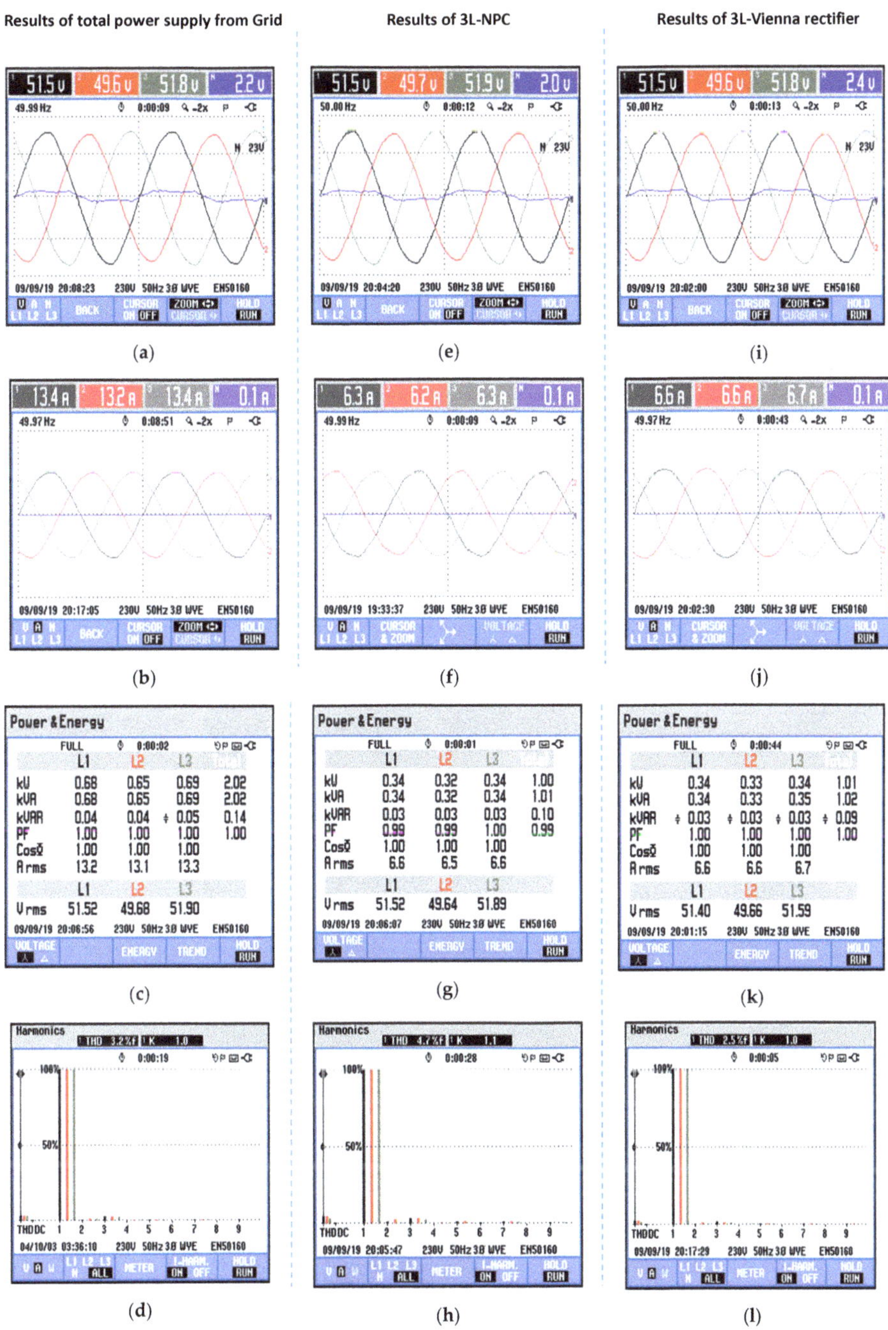

Figure 8. Experimental results of a 2-KW proposed hybrid converter.

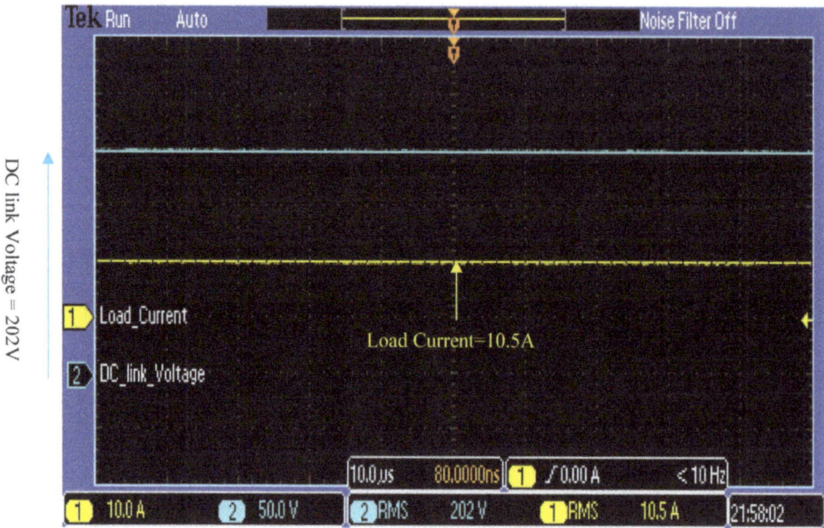

Figure 9. Experimental results of DC link voltage and load current.

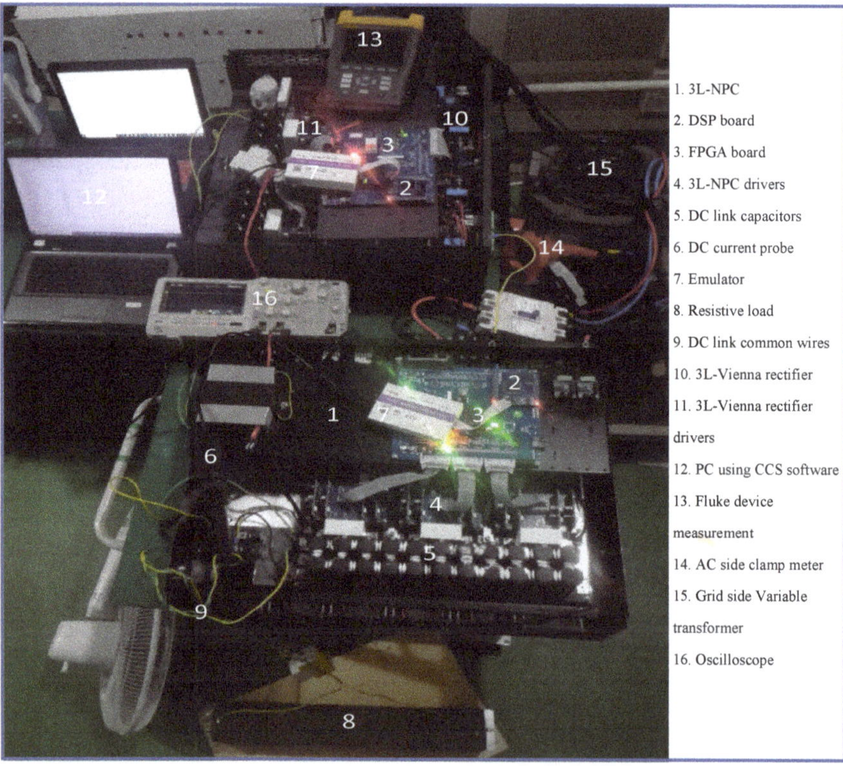

Figure 10. Experimental setup of two parallel-connected 3L-NPC and Vienna rectifier.

7. Conclusions

A novel bi-directional converter, especially for multi-MW wind generator for offshore wind energy conversion applications, was designed. This converter has the ability to handle a wide range of power in MW with thousands of ampere current by adding parallel circuitry of 'n' number of Vienna rectifiers with a single module of three-phase 3L-NPC converter. Moreover, the designed converter has a lower number of power devices, which leads towards cost-effectivness, a reduction in switching losses, as well as high power density system as compared to the conventional parallel-connected 3L-NPC converters. A simple hybrid control scheme consists of a VOC for 3L-NPC and current average control technique with the addition of duty ratio feed-forward for Vienna rectifier control to improve the current distortion was also investigated. Equally distributed power in Vienna rectifier and 3L-NPC converter verified the control performance. The proposed converter with a control scheme also verified that the Vienna rectifier has unity power factor (PF) and low THD factor as compared to 3L-NPC. The simulation and experimental results of a deliberated 2-KW system verified the fast dynamic response, good power factor (PF), and current THD less than 5%.

Author Contributions: M.L. and G.Y. conceived and designed the study; M.L. preformed experiments and edited the draft with guidance from G.Y and L.Z. L.Z. checked the language and style of the manuscript. T.Z. and A.L. made a contribution to the revisions of manuscript.

Funding: This research was supported by the Shanghai Natural Science Foundation (SNSF) under grant 18ZR1418400.

Acknowledgments: We thank all the journal editors and the reviewers for their valuable feedback and constructive comments that have contributed to improving this manuscript.

Conflicts of Interest: The authors declare no conflict of interest.

References

1. Burton, D.S.T.; Jenkins, N.; Bossanyi, E. *Wind Energy Handbook*; John Wiley & Sons Ltd.: Chichester, UK, 2001.
2. Doubrawa Moreira, P.; Scott, G.N.; Musial, W.D.; Kilcher, L.F.; Draxl, C.; Lantz, E.J. *Offshore Wind Energy Resource Assessment for Alaska*; No. NREL/TP-5000-70553; National Renewable Energy Lab.(NREL): Golden, CO, USA, 2018.
3. National Development and Reform Commission. Available online: http://finance.sina.com.cn/roll/2017-04-25/doc-ifyepnea4974842.shtml (accessed on 25 April 2017).
4. Wang, S.; Wang, S. Impacts of wind energy on environment: A review. *Renew. Sustain. Energy Rev.* **2015**, *49*, 437–443. [CrossRef]
5. Global Wind Energy Council (GWEC). Global Wind Report: Annual Market Update. April 2017. Available online: http://www.gwec.net (accessed on 14 January 2018).
6. Blaabjerg, F.; Ma, K. Future on power electronics for wind turbine systems. *IEEE J. Emerg. Sel. Top. Power Electron.* **2013**, *1*, 139–152. [CrossRef]
7. Blaabjerg, F.; Liserre, M.; Ma, K. Power electronics converters for wind turbine systems. *IEEE Trans. Ind. Appl.* **2012**, *48*, 708–719. [CrossRef]
8. Dos Santos, E.C.; Rocha, N.; Jacobina, C.B. Suitable single-phase to three-phase AC-DC-AC power conversion system. *IEEE Trans. Power Electron.* **2015**, *30*, 860–870. [CrossRef]
9. Wang, G.; Konstantinou, G.; Townsend, C.D.; Pou, J.; Vazquez, S.; Demetriades, G.D.; Agelidis, V.G. A review of power electronics for grid connection of utility-scale battery energy storage systems. *IEEE Trans. Sustain. Energy* **2016**, *7*, 1778–1790. [CrossRef]
10. Schnitzer, D.; Lounsbury, D.; Carvallo, J.; Deshmukh, R.; Apt, J.; Kammen, D.M. *Microgrids for Rural Electrification: A Critical Review of Best Practices Based on Seven Case Studies*; UN Found: New York, NY, USA, 2014.
11. Shaahid, S.; El-Amin, I. Techno-economic evaluation of off-grid hybrid photovoltaic–diesel–battery power systems for rural electrification in Saudi Arabia—A way forward for sustainable development. *Renew. Sustain. Energy Rev.* **2009**, *13*, 625–633. [CrossRef]
12. Sokolovs, A.; Grigans, L. Front-end converter choice considerations for PMSG-based micro-wind turbines. In Proceedings of the 56th International Scientific Conference on Power and Electrical Engineering of Riga Technical University (RTUCON), Riga. Latvia, 14 October 2015; pp. 1–6.

13. Vilathgamuwa, D.M.; Jayasinghe, S.D.G. Rectifier systems for variable speed wind generation-a review. In Proceedings of the 2012 IEEE International Symposium on Industrial Electronics, Hangzhou, China, 28–31 May 2012; pp. 1058–1065.
14. Salgado-Herrera, N.; Campos-Gaona, D.; Anaya-Lara, O.; Medina-Rios, A.; Tapia-Sánchez, R.; Rodríguez-Rodríguez, J. THD reduction in wind energy system using type-4 wind turbine/PMSG applying the active front-end converter parallel operation. *Energies* **2018**, *11*, 2458. [CrossRef]
15. Ma, K.; Blaabjerg, F. The impact of power switching devices on the thermal performance of a 10 MW wind power NPC converter. *Energies* **2012**, *5*, 2559–2577. [CrossRef]
16. Ventosa-Cutillas, A.; Montero-Robina, P.; Umbría, F.; Cuesta, F.; Gordillo, F. Integrated Control and Modulation for Three-Level NPC Rectifiers. *Energies* **2019**, *12*, 1641. [CrossRef]
17. Luqman, M.; Yao, G.; Zhou, L.; Lamichhane, A. Analysis of Variable Speed Wind Energy Conversion System With PMSG and Vienna Rectifier. In Proceedings of the 2019 14th IEEE Conference on Industrial Electronics and Applications, Xi'an, China, 19–21 June 2019; pp. 1296–1301.
18. Nasiri, M.; Mohammadi, R. Peak Current Limitation for Grid Side Inverter by Limited Active Power in PMSG-Based Wind Turbines During Different Grid Faults. *IEEE Trans. Sustain. Energy* **2017**, *26*, 3–12. [CrossRef]
19. Hou, C.C.; Cheng, P.T. Experimental Verification of the Active Front-End Converters Dynamic Model and Control Designs. *IEEE Trans. Power Electron.* **2011**, *26*, 1112–1118. [CrossRef]
20. Shen, L.; Bozhko, S.; Asher, G.; Patel, C.; Wheeler, P. Active DC-Link Capacitor Harmonic Current Reduction in Two-Level Back-to-Back Converter. *IEEE Trans. Power Electron.* **2016**, *31*, 6947–6954. [CrossRef]
21. Blaabjerg, F.; Ma, K. High power electronics: Key technology for wind turbines. *Power Electron. Renew. Energy Syst. Transp. Ind. Appl.* **2014**, *18*, 136–159.
22. Luqman, M.; Yao, G.; Zhou, L.; Yang, D.; Lamichhane, A. Study and Implementation of a Cost-Effective 3L-Active Rectifier for DC Collection in WECS. EDP Sciences. In Proceedings of the 2019 4th International Conference on Advances in Energy and Environment Research (ICAEER 2019), Shanghai, China, 16–18 August 2019.
23. Xu, Z.; Li, R.; Xu, D. Control of parallel multi-rectifiers for a direct drive permanent-magnet wind power generator. *IEEE Trans. Ind. Appl.* **2013**, *49*, 1687–1696. [CrossRef]
24. Khan, S.U.; Maswood, A.I.; Tariq, M.; Tafti, H.D.; Tripathi, A. Parallel Operation of Unity Power Factor Rectifier for PMSG Wind Turbine System. *IEEE Trans. Ind. Appl.* **2019**, *55*, 721–731. [CrossRef]
25. Baroudi, J.A.; Dinavahi, V.; Knight, A.M. A review of power converter topologies for wind generators. *Renew. Energy* **2007**, *32*, 2369–2385. [CrossRef]
26. Liserre, M.; Cárdenas, R.; Molinas, M.; Rodriguez, J. Overview of Multi-MW Wind Turbines and Wind Parks. *IEEE Trans. Ind. Electron.* **2011**, *58*, 1081–1095. [CrossRef]
27. Carrasco, J.M.; Franquelo, L.G.; Bialasiewicz, J.T.; Galvan, E.; Guisado, R.C.P.; Martín Prats, M.D.; Leon, J.I.; Moreno-Alfonso, N. PowerElectronic Systems for the Grid Integration of Renewable Energy Sources: A Survey. *IEEE Trans. Ind. Electron.* **2006**, *53*, 1002–1016. [CrossRef]
28. Chen, Z.; Guerrero, J.M.; Blaabjerg, F. A Review of the State of the Art of Power Electronics for Wind Turbines. *IEEE Trans. Power Electron.* **2009**, *24*, 1859–1875. [CrossRef]
29. Wu, B.; Lang, Y.; Zargari, N.; Kouro, S. *Power Conversion and Control of Wind Energy Systems*; John Wiley & Sons: Hoboken, NJ, USA, 2011 26 September; Volume 76.
30. Thandapani, T.; Karpagam, R.; Paramasivam, S. Comparative study of VIENNA rectifier topologies. *Int. J. Power Electron.* **2015**, *7*, 147–165. [CrossRef]
31. Lamichhane, A.; Zhou, L.; Yao, G.; Luqman, M. LCL Filter Based Grid-Connected Photovoltaic System with Battery Energy Storage. In Proceedings of the 2019 14th IEEE Conference on Industrial Electronics and Applications, Xi'an, China, 19–21 June 2019.
32. Hang, L.; Zhang, M. Constant power control-based strategy for Vienna-type rectifiers to expand operating area under severe unbalanced grid. *IET Power Electron.* **2014**, *7*, 41–49. [CrossRef]
33. Teshnizi, H.M.; Moallem, A.; Zolghadri, M.; Ferdowsi, M. A dual-frame hybrid vector control of vector modulated VIENNA I rectifier for unity power factor operation under unbalanced mains condition. In Proceedings of the 2008 Twenty-Third Annual IEEE Applied Power Electronics Conference and Exposition, Austin, TX, USA, 24 February 2008.

34. Youssef, N.B.H.; Al-Haddad, K.; Kanaan, H.Y. Implementation of a new linear control technique based on experimentally validated small-signal model of three-phase three-level boost-type Vienna rectifier. *IEEE Trans. Ind. Electron.* **2008**, *55*, 1666–1676. [CrossRef]
35. Chen, H.; Aliprantis, D.C. Analysis of squirrel-cage induction generator with Vienna rectifier for wind energy conversion system. *IEEE Trans. Energy Convers.* **2011**, *26*, 967–975. [CrossRef]
36. Win, M.M.; Tin, T. Engineering, Voltage Source Converter Based HVDC Overhead Transmission System. *IJSETR Int. J. Sci. Eng. Technol. Res.* **2014**, *3*, 3909–3914.
37. Kwon, Y.D.; Park, J.H.; Lee, K.B. Improving Line Current Distortion in Single-Phase Vienna Rectifiers Using Model-Based Predictive Control. *Energies* **2018**, *11*, 1237. [CrossRef]

© 2019 by the authors. Licensee MDPI, Basel, Switzerland. This article is an open access article distributed under the terms and conditions of the Creative Commons Attribution (CC BY) license (http://creativecommons.org/licenses/by/4.0/).

Article

Reduction of DC Current Ripples by Virtual Space Vector Modulation for Three-Phase AC–DC Matrix Converters

Hoang-Long Dang, Eun-Su Jun and Sangshin Kwak *

School of Electrical and Electronics Engineering, Chung-Ang University, Heukseok-dong, Dongjak-gu, Seoul 06974, Korea; danghoanglong2692@gmail.com (H.-L.D.); elizame3@naver.com (E.-S.J.)

* Correspondence: sskwak@cau.ac.kr; Tel.: +82-2-820-5346

Received: 25 October 2019; Accepted: 11 November 2019; Published: 13 November 2019

Abstract: In this paper, the virtual space vector modulation for the AC–DC (alternating current–direct current) matrix converters is proposed to reduce the DC current ripples in the whole modulation index range. In the proposed method, each virtual vector is synthesized by the two nearest original active vectors. To synthesize the current reference vector, two virtual vectors and one zero vector are used in every switching period. The main principle of the proposed method is to reduce the dwelling period of the largest active current vector in each sector. In addition, the optimized switching patterns are proposed to further reduce the DC current ripples at both high- and low-power operation. Finally, simulation and experimental results are illustrated to validate the effectiveness of the proposed strategy.

Keywords: AC-DC matrix converter; virtual space vector; DC ripple reduction

---

1. Introduction

In recent years, the AC–DC (alternating current–direct current) matrix converters (MC) have received significant attention in various fields. The AC–DC MC is derived from indirect MCs and inherited several advantages of the MCs, such as bidirectional power flow, sinusoidal input waveforms, controllable input power factor, high power density, and compact design [1–5]. There are various applications of the AC–DC MCs in various fields such as electric vehicles, photovoltaic generation systems, grid-connected converters, microgrids, fuel cell power systems, and battery chargers. [6–9]. Basically, it is a single-stage bidirectional current source AC–DC converter, which rectifies the sinusoidal AC signals to the pure DC signals. Different modulation strategies have been applied for the AC–DC MCs, such as the Alesina–Venturini [5], the pulse-width modulation (PWM) [10,11], the model predictive control (MPC) [12–14], and the space vector modulation (SVM) [15–21]. Predictive control strategies for the AC–DC MCs under unbalanced grid voltage were proposed in [12] and [13]. Literature [14] presented a unity power factor predictive control method for the AC–DC MC. The most wide modulation control strategy for the AC–DC MCs has been considered the SVM method. A unity power factor fuzzy battery charger using the ultra-sparse matrix rectifier was designed and implemented in [15] with only three switches; however, it is a unidirectional converter. The direct power factor control strategy for the three-phase AC–DC MCs was illustrated in [16] based on applying the reduced general direct SVM approach of the AC–AC MC theory. Modulation and control strategies of the AC–DC MC for the battery energy storage system applications were investigated in [17] and [18]. Literature [19] studied the optimal zero-vector configuration to reduce the output inductor current ripple for the space-vector-modulated AC–DC MCs. The optimized modulation strategy to deduce the charging current ripple for vehicle to grid (V2G) applications using the AC–DC MC was presented in [20]. An input power factor control method was studied in [21] based on the concept of the virtual capacitor. A controlled rectifier was implemented in [22] using the AC–AC MC theory. Literature [23] presented a digitally controlled switch mode power supply based on the MC without any modulation block. Dynamic characteristics of the matrix rectifier researches were studied in [24]. One of the most important issues for the AC–DC MC operation is DC ripples. Generally, increasing the switching frequency or increasing the inductor size of output filter can reduce the DC ripples. However, higher switching frequency leads to higher switching losses and larger size of output inductor results in the increase of size and cost of the converter. There are several approaches based on the SVM algorithm to reduce DC ripples [19,20,25]. In [19], an optimal zero-vector configuration was proposed to reduce DC ripples, however, the effectiveness of this approach is mainly maintained at low modulation because a period of zero-vector at high-modulation operation is very short compared with periods of two active vectors. Thus, this optimal configuration does not effectively reduce the DC ripples in wide operation ranges. The approach [20] proposed a sectional optimized modulation strategy, which can reduce DC ripples within the whole operation range by dividing many different sectors, and different groups of vectors are selected to synthesize the current vector. However, the zero vector configuration is not optimized at low-modulation operations. A recent work in [25] reduced DC ripples by dividing 12 different sectors and using only active vectors to synthesize the current vector. Since only active vectors are used, the operation of this method is not guaranteed at low-modulation operation.

In this paper, the virtual space vector modulation (VSVM) for the AC–DC MC is proposed to reduce the DC current ripples within the whole modulation range. Previously, the VSVM concept was proposed to suppress the common-mode voltage of a two-level voltage source inverter (VSI) [26], and balance the neutral-point potential of a three-level neutral-point-clamped (NPC) inverter [27,28]. However, none of these approaches have been tried to reduce the DC current ripples using the VSVM, at the best knowledge of the authors. In this proposed VSVM method, each virtual vector is synthesized by two nearest active vectors, and each virtual sector is defined with the area between two virtual vectors. The current reference vector is synthesized by two virtual vectors and one zero vector in every switching period. The main principle of the proposed VSVM is reducing the dwelling period of the largest active current vector in each sector. In addition, the optimized switching patterns are proposed to further reduce the DC current ripples at both high- and low-power operations. Simulation and experimental results are demonstrated to verify validity and effectiveness of the proposed VSVM for reducing the DC current ripples of the AC–DC MCs.

2. Topology and Modulations of AC–DC Matrix Converter

2.1. The Topology of AC–DC Matrix Converter

The topology of the AC–DC MC is shown in Figure 1. It is made up of an array of six bidirectional power semiconductor switches, with the ability to conduct current in both directions. Each bidirectional switch is generally constructed by two insulated-gate bipolar transistors (IGBTs) connected in series with a common emitter. An input filter is used to suppress the high-frequency harmonic generated by the operation of converter and the grid. At the output terminal, the LC filter is used to smooth the output current. The AC–DC MC is powered by the AC voltage sources; thus the AC voltages are not allowed to be shorted. In addition, because of the inductive nature of the load, the load terminal must never be opened. Therefore, the AC–DC MC operates by connecting only one bidirectional switch in the upper-arm and only one bidirectional switch in the lower-arm at any instant. This leads to there being nine switching current vectors for the operation of the AC–DC MC.

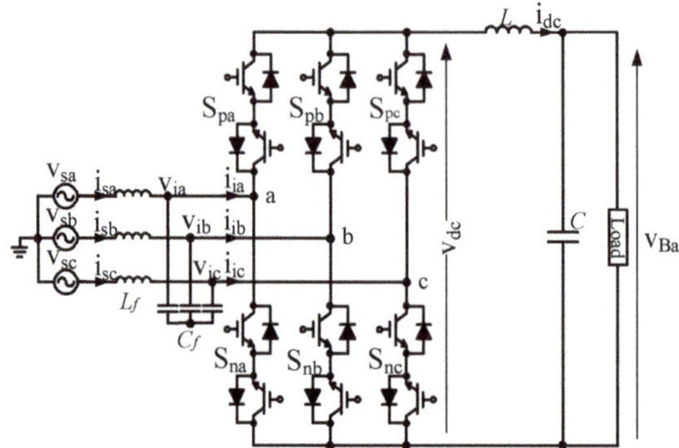

Figure 1. Topology of AC–DC (alternating current–direct current) matrix converter.

2.2. Modulations of AC–DC Matrix Converter

2.2.1. Conventional Space Vector Modulation

The current space vector diagram of conventional space vector modulation (C-SVM) for the AC–DC MC is illustrated in Figure 2. In order to synthesize the desired input current reference vector $\vec{I}_{ref}$, C-SVM uses the two nearest active vectors and one zero vector among six active vectors $\vec{I}_1 \sim \vec{I}_6$ and three zero vectors $\vec{I}_7 \sim \vec{I}_9$, according to the sector location of the input current reference vector. Each sector is denoted as the area between two active vectors, for example, the area between two active vectors $\vec{I}_1$ and $\vec{I}_2$ is sector I, the area between two active vectors $\vec{I}_2$ and $\vec{I}_3$ is sector II, and so on. As it can be seen from Figure 2, the input current reference vector locates in sector I, thus two active vectors $\vec{I}_1$, $\vec{I}_2$ and one zero vector are used to synthesize the desired current reference vector. The selection of zero vector is based on the constraint to minimize the switching frequency of bidirectional switches so that the commutation between two switching states involves only two switches in two phase-legs of the converter, one switch is turned on, and one switch is turned off at the same time. Among three zero vectors, only zero vector $\vec{I}_7$ satisfies the requirements for switching devices. Therefore, zero vector $\vec{I}_7$ and two active vectors $\vec{I}_1$, $\vec{I}_2$ are selected.

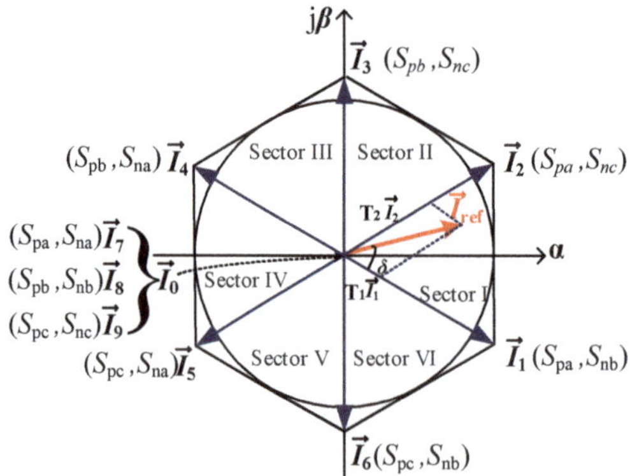

Figure 2. Space vector diagram of conventional space vector modulation (C-SVM) for AC–DC matrix converter.

The duty cycles d_1, d_2, and d_0 of the active and zero vectors, when the input current reference vector is in the sector I, are given by:

$$d_1 = m_i \sin\left(\frac{\pi}{3} - \delta\right) \quad (1)$$

$$d_2 = m_i \sin(\delta) \quad (2)$$

$$d_0 = 1 - d_1 - d_2 \quad (3)$$

where m_i: modulation index; $m_i = i_{i1}/i_{dc}$ $m_i \in [0, 1]$; i_{i1}: the peak value of the fundamental-frequency component in i_i; δ : input current reference vector angle, $\delta \in \left[0, \frac{\pi}{3}\right]$.

The durations T_1, T_2, and T_0 of the active vectors $\vec{I}_1$, $\vec{I}_2$, and zero vector $\vec{I}_0$ are respectively expressed as:

$$T_1 = d_1 T_s \quad (4)$$

$$T_2 = d_2 T_s \quad (5)$$

$$T_0 = T_s - T_1 - T_2 \quad (6)$$

where T_s: switching period: $T_s = \frac{1}{f_s}$; f_s: switching frequency.

The calculation of the duty cycles and durations, when the input current reference vector passes through other sectors one by one, is obtained in a similar algorithm as sector I.

2.2.2. Proposed Virtual Space Vector Modulation

(a) Sector division and duty cycles.

In the proposed VSVM, each virtual vector is synthesized by the two nearest active vectors, and each virtual sector is denoted as the area between two virtual vectors. There are totally six virtual vectors $\vec{I}_a \sim \vec{I}_f$ and six virtual sectors. The concept of virtual vector in this paper is similar to the virtual vector concept in [26]. Figure 3 presents the virtual sector divisions according to the virtual vectors and the synthesis of input current reference vector in the virtual sector I of the proposed method. As it can be seen from Figure 3b, when the input current reference vector $\vec{I}_{ref}$ is in virtual sector I, two virtual vectors $\vec{I}_a$, $\vec{I}_b$ and one zero vector are used to synthesize the reference vector depending on the magnitude of modulation index m_i. The selection of zero vector in the proposed VSVM is similar to zero vector selection of the C-SVM method.

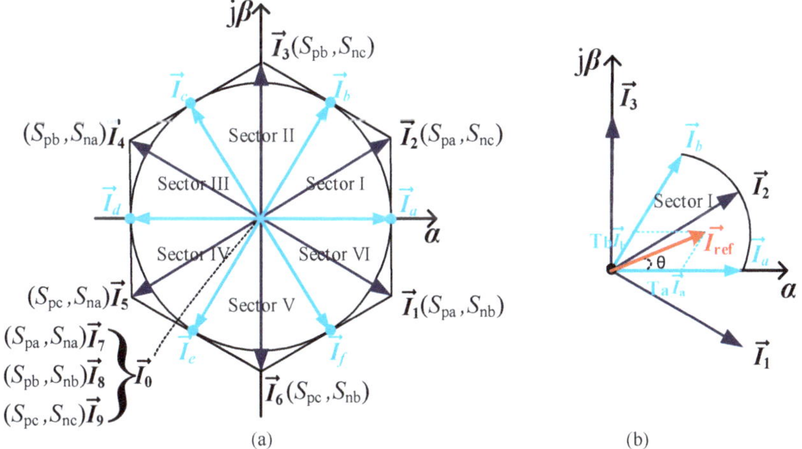

Figure 3. Space vector diagram of proposed virtual space vector modulation (VSVM) for AC–DC matrix converter. (**a**) Sector division; (**b**) Input current reference vector synthesis.

The duty cycles d_a, d_b, d_0 of virtual vectors $\vec{I}_a$, $\vec{I}_b$ and zero vector $\vec{I}_0$, when the input current reference vector is in the virtual sector I, are expressed as:

$$d_a = m_i \sin\left(\frac{\pi}{3} - \theta\right) \tag{7}$$

$$d_b = m_i \sin(\theta) \tag{8}$$

$$d_0 = 1 - d_a - d_b. \tag{9}$$

where m_i: modulation index; $m_i = i_{i1}/i_{dc}$; $m_i \in [0, 1]$; i_{i1}: the peak value of the fundamental-frequency component in i_i; θ: input current reference vector angle $\theta \in \left[0, \frac{\pi}{3}\right]$.

The durations T_a, T_b, T_0 of virtual vectors $\vec{I}_a$, $\vec{I}_b$ and zero vector are determined as:

$$T_a = d_a T_s \tag{10}$$

$$T_b = d_b T_s \tag{11}$$

$$T_0 = T_s - T_a - T_b. \tag{12}$$

The virtual vectors $\vec{I}_a$, $\vec{I}_b$ are synthesized by three original active vectors $\vec{I}_1$, $\vec{I}_2$, and $\vec{I}_3$. The dwell times T_1, T_2, T_3 of three original active vectors can be derived by:

$$T_1 = \frac{T_a}{2} \tag{13}$$

$$T_2 = \frac{T_a}{2} + \frac{T_b}{2} \tag{14}$$

$$T_3 = \frac{T_b}{2}. \tag{15}$$

The dwell times of other virtual and original active vectors in remaining sectors are determined in the similar manner of the sector I.

(b) Switching patterns.

The effectiveness of the proposed VSVM not only depends on the modulation of virtual vectors, but also on the switching patterns.

Figure 4 presents the switching patterns of conventional VSVM (C-VSVM) in [26]. This switching pattern, which is a seven-segment pattern, is not optimized for reducing the DC current ripple. In this paper, the optimized switching patterns are proposed to further reduce the DC current ripple at both high- and low-modulation operation. The proposed switching patterns of the VSVM strategy for AC–DC MC under different modulation index ranges in sector I are illustrated in Figure 5. At low modulation index operation, the switching period of the zero vector is the longest switching period compared with other switching periods of the remaining active vectors. The longer the switching period of the zero vector, the higher the current ripple due to the longer period of decreasing output DC current. Most of the approaches for DC ripple reduction are not optimized for the zero vector at low-modulation operation. Thus, the optimized switching patterns for the zero vector are proposed in this paper to further reduce the DC current ripple of AC–DC MC.

(c) Controller system.

The control block diagram of the proposed control strategy is illustrated in Figure 6. The battery voltage is sensed and compared with the reference voltage, then passed through the proportional-integral (PI) controller and compared with the DC current or using direct DC current reference depending on the selection of constant voltage (CV) control mode or constant current (CC) control mode. After that, the signal is passed through the PI controller to obtain the modulation index. At the input terminal, three-phase source voltages, three phase source currents are sensed and passed through the $\alpha\beta$-transformation, then combined with the PLL block for the calculation of input current. Finally, the VSVM algorithm is applied to compute the duty cycles for AC–DC MC.

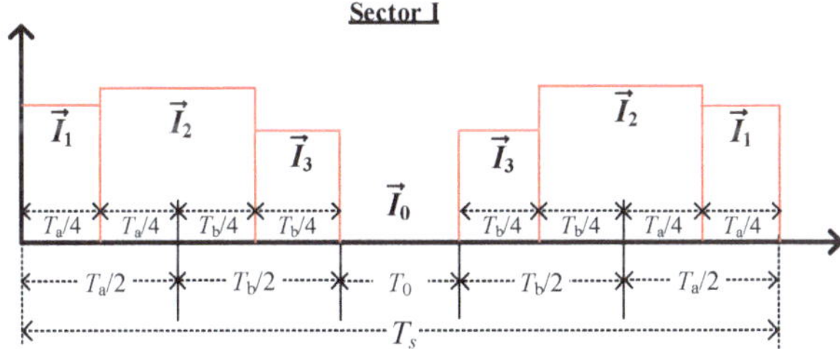

Figure 4. Switching patterns of C-VSVM (conventional VSVM) for AC–DC matrix converter.

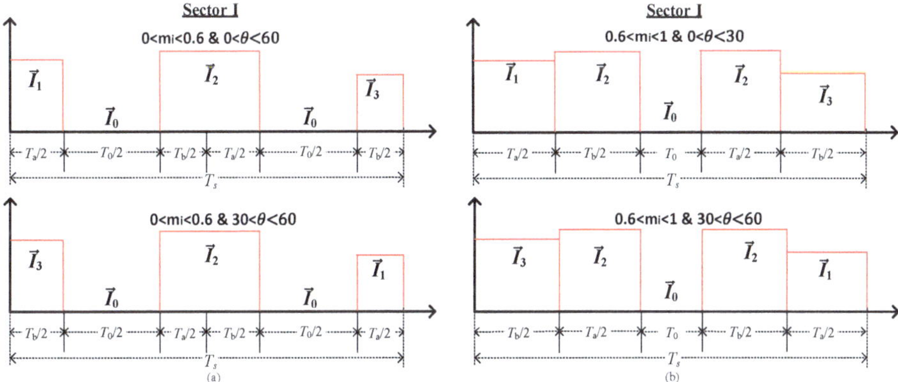

Figure 5. Switching patterns of proposed VSVM for AC–DC matrix converter under different modulation index and input current reference angle. (**a**) Low modulation index; (**b**) High modulation index.

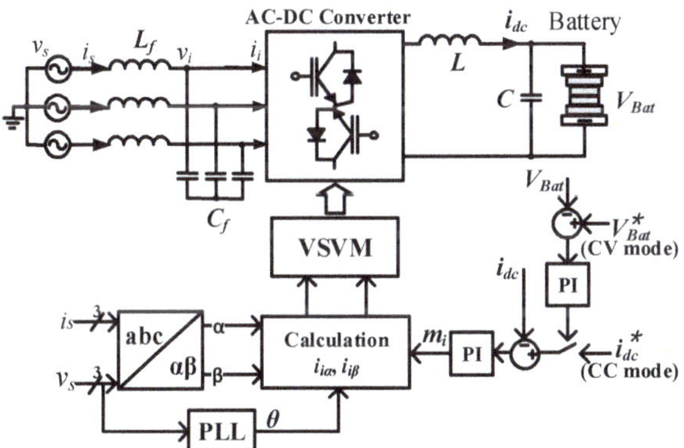

Figure 6. Control block diagram of proposed VSVM for AC–DC matrix converter.

3. DC Current Ripple Analysis

3.1. DC Current Ripple Analysis of C-SVM

In this analysis, the voltage drops by the power devices are neglected and the battery voltage is assumed constant. The load side model of the converter is given by:

$$v_{dc} = V_{Bat} + L\frac{di_{dc}}{dt}. \tag{16}$$

Hence, the DC current ripple in one switching period can be obtained from (16) as follows:

$$\Delta i_{dc} = \frac{v_{dc} - V_{Bat}}{L}T_s. \tag{17}$$

From (17), it is clear that the DC current ripple depends on the instantaneous output voltage v_{dc} of converter, switching period T_s, and output inductor L. Increasing the size of output inductor or switching frequency ($T_s = 1/f_s$) can reduce DC current ripple, however, the size, cost, and switching losses of converter are increased. These solutions are not preferred in this paper.

In one switching period of C-SVM, two active vectors and one zero vector are used to synthesize the input current reference vector. DC current ripples of each vector in the sector I are derived by:

$$\Delta i_{dc1} = \frac{v_{ab} - V_{Bat}}{L}T_1 \tag{18}$$

$$\Delta i_{dc2} = \frac{v_{ac} - V_{Bat}}{L}T_2 \tag{19}$$

$$\Delta i_{dc0} = \frac{v_0 - V_{Bat}}{L}T_0. \tag{20}$$

According to the location of input current reference vector in Figure 2 and the bilateral symmetric switching pattern, the peak-to-peak DC current ripple in one switching period is Δi_{dc0}, as shown in Figures 7a and 8a.

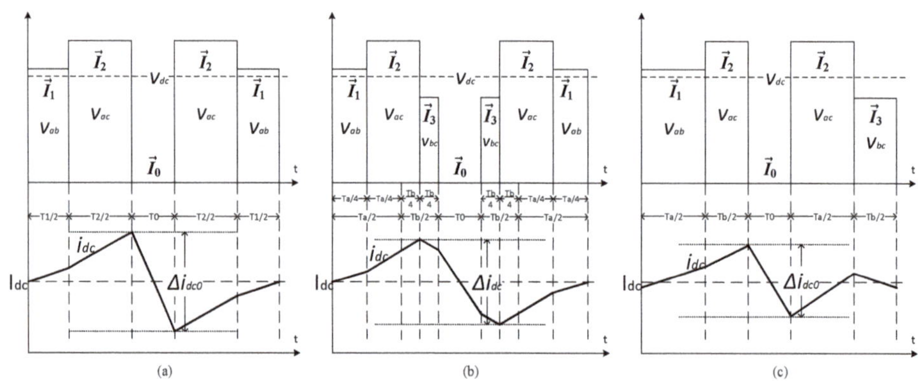

Figure 7. DC current ripple waveforms under different modulation strategies for AC–DC matrix converter at high modulation index. (**a**) C-SVM; (**b**) C-VSVM; (**c**) Proposed VSVM.

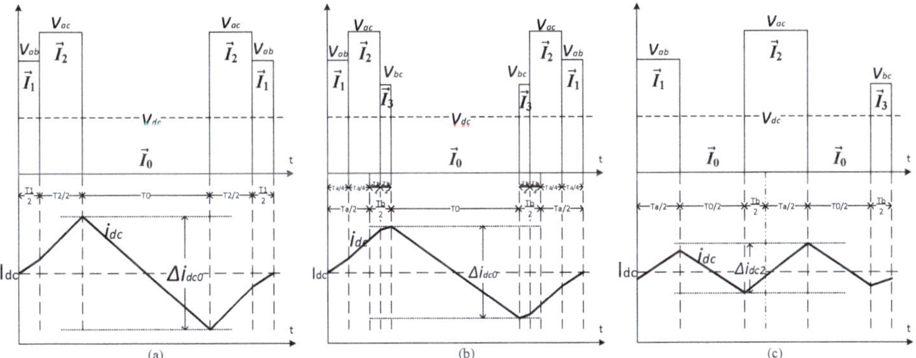

Figure 8. DC current ripple waveforms under different modulation strategies for AC–DC matrix converter at low modulation index. (**a**) C-SVM; (**b**) C-VSVM; (**c**) Proposed VSVM.

3.2. DC Current Ripple Analysis of Proposed VSVM

In one switching period of the proposed VSVM, two active vectors and one or two zero vectors are used to synthesize the input current reference vector.

DC current ripples of each vector in the virtual sector I are derived by:

$$\Delta i_{dc1} = \frac{v_{ab} - V_{Bat}}{L}\left(\frac{T_a}{2}\right) \tag{21}$$

$$\Delta i_{dc2} = \frac{v_{ac} - V_{Bat}}{L}\left(\frac{T_a}{2} + \frac{T_b}{2}\right) \tag{22}$$

$$\Delta i_{dc3} = \frac{v_{bc} - V_{Bat}}{L}\left(\frac{T_b}{2}\right) \tag{23}$$

$$\Delta i_{dc0} = \frac{v_0 - V_{Bat}}{L} T_0. \tag{24}$$

According to the location of the input current reference vector in Figure 3 and the switching pattern, the peak-to-peak DC current ripples in one switching period are Δi_{dc0} at high-modulation operation, as shown in Figure 7c, and Δi_{dc2} at low-modulation operation, as shown in Figure 8c.

The goal of VSVM is reducing the switching period of the switching vector, which has the longest switching period, according the amplitude of input current reference vector in one switching period. In C-SVM, it can be seen from Figure 2 that the switching period of active vector $\vec{I}_2$ is longest and greater than the switching period of active vector $\vec{I}_1$, expressed as:

$$T_2 > T_1 \tag{25}$$

hence

$$T_2 > \frac{T_1 + T_2}{2} \tag{26}$$

then

$$T_2 > \frac{T_s - T_0}{2}. \tag{27}$$

In the proposed VSVM, applying a similar manner as in C-SVM and using (13–15), the switching period of active vector $\vec{I}_2$ is expressed as:

$$T_2 = T_1 + T_3 \tag{28}$$

hence

$$T_2 = \frac{T_1 + T_2 + T_3}{2} \quad (29)$$

then

$$T_2 = \frac{T_s - T_0}{2}. \quad (30)$$

As it can be seen from Figure 7, the switching period of zero vector of C-SVM, C-VSVM, and proposed VSVM is almost the same, however, the longest switching period is active vector $\vec{I}_2$ and reduced by VSVM. Thus, the increasing of DC current ripple is lower than in the conventional strategy, resulting in a reduction in the DC current ripple of VSVM at high-modulation operation. The proposed switching patterns further reduce the DC current ripple compared with the conventional switching pattern of VSVM. In addition, the switching period of zero vector is the longest period at low-modulation operation, thus the optimized switching patterns for zero vector are proposed in this paper to further reduce the DC current ripple of AC–DC MC. Figure 8 presents the DC current ripples under different modulation control strategies at low-modulation operation. As it can be seen from Figure 8, the switching period of zero vector is divided into two intervals by the proposed switching pattern. Hence, the continuous reduction of DC current is avoided compared with the switching patterns of C-SVM and C-VSVM, resulting in the reduction in DC current ripple.

4. Simulation and Experimental Results

4.1. Simulation

In order to verify the validity and effectiveness of the proposed control strategy, simulations were carried out with the parameters in Table 1 using PSIM software. Figure 9 shows the comparison of three-phase currents, A-phase voltage, DC current, and DC current reference at 6 A. It can be seen that the proposed VSVM control strategy effectively reduces the DC current ripple of AC–DC MC compared with C-SVM and C-VSVM strategies in the range of high-modulation operation. Figure 10 presents the zoom-in on DC current and DC output voltage in one switching period of the sector I under different modulation control strategies. The switching period of the largest line-to-line voltage vector was significantly reduced by the proposed VSVM compared with C-SVM. Therefore, the increasing of DC current is reduced before decreasing when zero vector is applied to the converter. Both switching patterns of the VSVM methods effectively reduce the DC current ripple. The proposed optimized switching patterns were applied, and the DC current ripple of AC–DC MC was successfully further reduced while accurately tracking its reference, which is in agreement with the current ripple analysis in Figure 7. The peak-to-peak values of the DC current ripples of the C-SVM, the C-VSVM, and the proposed VSVM operating at a high modulation index are 2.9 A, 2.4 A, and 1.65A, as shown in Figure 10, respectively. As a result, comparing with the C-SVM and the C-VSVM, the proposed VSVM can reduce the DC current ripples by 43.1% and 31.25%, respectively.

Table 1. Parameters for AC–DC matrix converter.

Parameters	Value
Source phase voltage (v_s)	100 V
Source frequency (f_i)	60 Hz
Input filter inductance (L_f)	2.5 mH
Input filter capacitance (C_f)	60 µF
Output filter inductance (L)	1 mH
Output filter capacitance (C)	40 µF
Load resistance (R)	20 Ω
Sampling frequency (f_s)	10 kHz

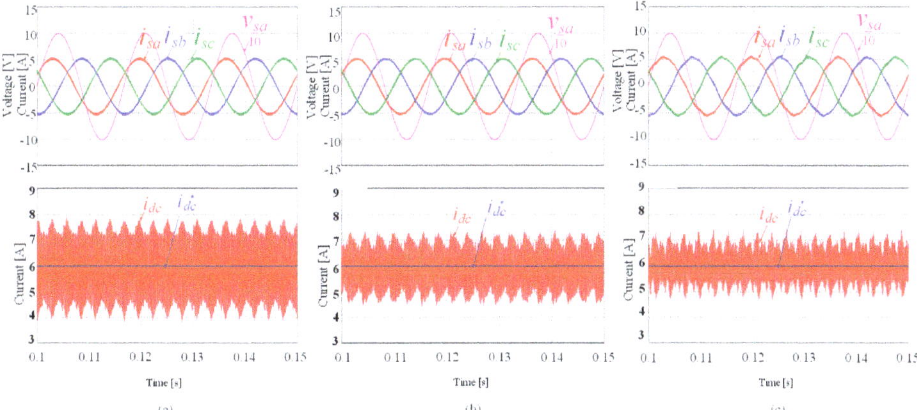

Figure 9. Three-phase currents, A-phase voltage, DC current, and DC current reference under different modulation strategies for AC–DC matrix converter at high-modulation operation. (**a**) C-SVM; (**b**) C-VSVM; (**c**) Proposed VSVM.

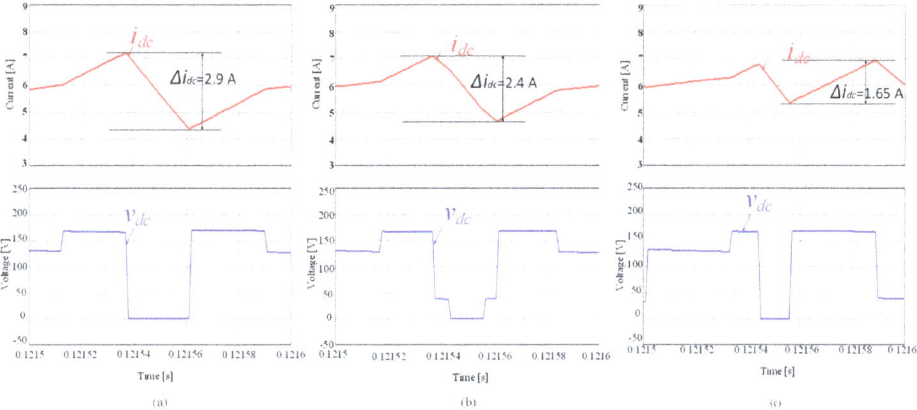

Figure 10. The zoom-in of DC current and DC output voltage under different modulation strategies for AC–DC matrix converter at high-modulation operation. (**a**) C-SVM; (**b**) C-VSVM; (**c**) Proposed VSVM.

The total distortion harmonics (THD) of A-phase source current of C-SVM and VSVM methods are shown in Figure 11. The THD value of the proposed VSVM method is slightly higher than that of the C-SVM method because the four vectors are used to synthesize the input current reference vector in the proposed VSVM to decrease the dc current ripples, including three active vectors and one zero vector, compared with two optimal active vectors and one zero vector of the C-SVM. The distance between the reference current vector and one additional stationary vector used for generating a virtual vector in the proposed VSVM is larger than that of the C-SVM. Besides, the symmetrical switching pattern is applied to guarantee low input current distortion in the C-SVM, whereas the proposed VSVM utilizes unsymmetrical switching patterns to further reduce the dc current ripples. Based on the above factors, the THD of three-phase currents of the proposed VSVM becomes slightly higher than that of the conventional methods, at costs of the reduction of the dc current ripples, although the envelope waveforms of the three-phase input currents are still sinusoidal. The simulation waveforms of three-phase source currents, A-phase source voltage, DC current, and DC current reference at 2 A of AC–DC MC under the C-SVM strategy, the C-VSVM strategy, and the proposed VSVM strategy are illustrated in Figure 12. It can be seen that the effectiveness of the proposed VSVM method in the reduction of DC current ripple correctly operates within the whole range of modulation while maintaining the high performance of AC–DC MC compared with the C-SVM and the C-VSVM methods. Without the proposed optimized switching pattern at low-power mode, the C-VSVM slightly reduces ripple compared with the C-SVM. The zoom-in of DC currents, DC output voltages in one switching period under the proposed VSVM and the C-SVM are presented in Figure 13. The optimized switching patterns for zero vector are applied when the converter operates at low modulation range. The switching period of zero vector is rearranged to avoid the continuous decreasing of DC current. Therefore, the DC current ripple is further reduced by the proposed VSVM strategy, which agrees with the theoretical analysis in Figure 8. In addition, the peak-to-peak values of the DC current ripples of the C-SVM, the C-VSVM, and the proposed VSVM at a low modulation index are 3.15 A, 2.88 A, and 2.04 A, respectively. Thus, it can be known that comparing with the C-SVM and the C-VSVM, the reduction of the DC current ripples by the proposed VSVM can be obtained by 35.23% and 29.17%, respectively. The transient state performances of AC–DC MC under the C-SVM, the C-VSVM, and the proposed VSVM methods are illustrated in Figure 14. The proposed method successfully reduces the DC current ripple in both high- and low-power operation compared with the C-SVM and the C-VSVM methods. The simulation results show the effectiveness of the proposed method in the reduction of DC current ripple compared with the conventional methods.

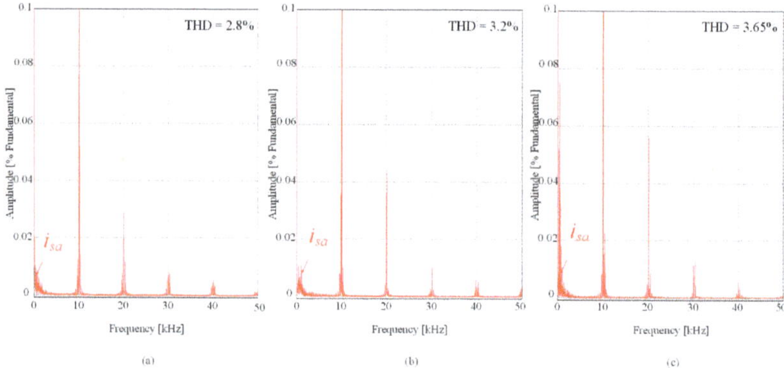

Figure 11. THD (total distortion harmonics) of A-phase source current under different modulation strategies for AC–DC matrix converter at high-modulation operation. (**a**) C-SVM; (**b**) C-VSVM; (**c**) Proposed VSVM.

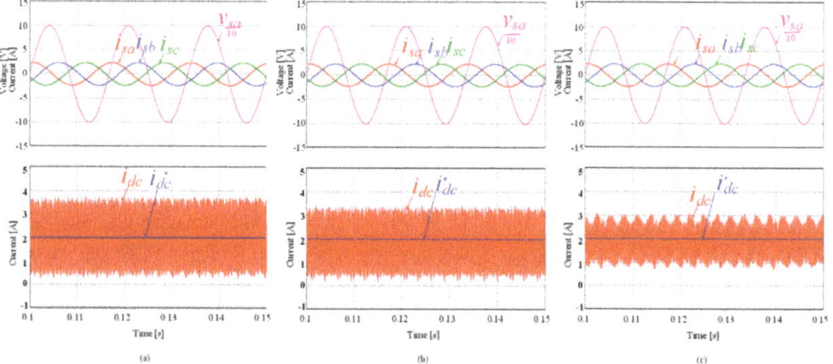

Figure 12. Three-phase currents, A-phase voltage, DC current, and DC current reference under different modulation strategies for AC–DC matrix converter at low-modulation operation. (**a**) C-SVM; (**b**) C-VSVM; (**c**) Proposed VSVM.

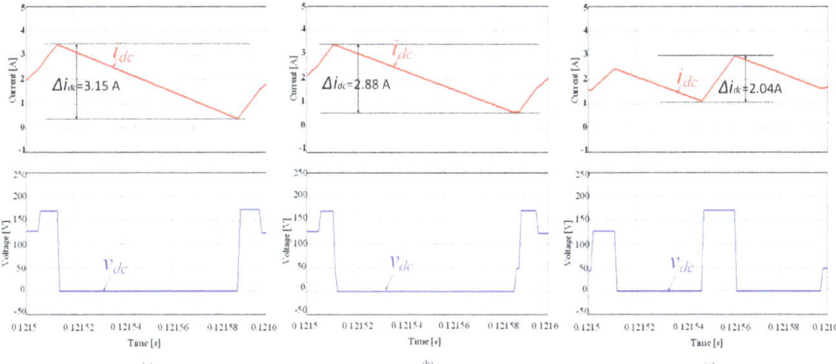

Figure 13. The zoom-in of DC current and DC output voltage under different modulation strategies for AC–DC matrix converter at low-modulation operation. (**a**) C-SVM; (**b**) C-VSVM; (**c**) Proposed VSVM.

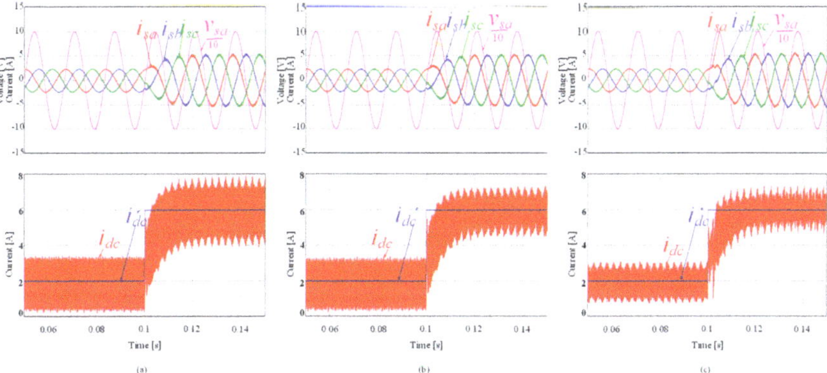

Figure 14. Three-phase currents, A-phase voltage, DC current, and DC current reference under different modulation strategies for AC–DC matrix converter at transient state. (**a**) C-SVM; (**b**) C-VSVM; (**c**) Proposed VSVM.

4.2. Experiment

In order to validate the effectiveness of the proposed control strategy in a real system, an AC–DC MC prototype was constructed with six bidirectional switches, which are built by two insulated-gate bipolar transistors (IGBTs) modules (IXA37IF1200HJ), connected in series with a common emitter to validate the effectiveness of the proposed VSVM. The proposed strategy is performed by a Texas Instrument digital signal processor board (TI TMS320F28335). The parameters of the experiment are the same as in Table 1. Figure 15 shows the comparison of A-phase current, A-phase voltage, battery voltage, and DC current at 5 A. It can be seen that the proposed VSVM control strategy effectively reduces the DC current ripple of AC–DC MC compared with the C-SVM strategy in the range of high-modulation operation. The THDs of A-phase source current of both the C-SVM and the proposed VSVM methods are shown in Figure 16. The THD of the proposed VSVM method is slightly higher than the THD of the C-SVM method, this is a trade-off between ripple reduction and increasing current distortion.

The experimental waveforms of A-phase source current, A-phase source voltage, battery voltage, and DC current reference at 2 A of AC–DC MC under the C-SVM strategy and the proposed VSVM strategy are illustrated in Figure 17. Applying the optimized switching patterns for zero vector at low-modulation operation, the DC current ripple of the proposed VSVM is further reduced while maintaining the performance of AC–DC MC compared with the C-SVM. The transient state performances of AC–DC MC under the C-SVM and the proposed VSVM methods are illustrated in Figure 18. The proposed method successfully reduces the DC current ripple at both high and low power range compared with the C-SVM method. The experimental results show the effectiveness of the proposed method in the reduction of DC current ripple compared with the conventional methods in the whole range of operation. The assessment of the proposed VSVM method compared with the conventional methods in terms of the reduction of DC current ripples and the increase in the THD values of the input currents is shown in Table 2.

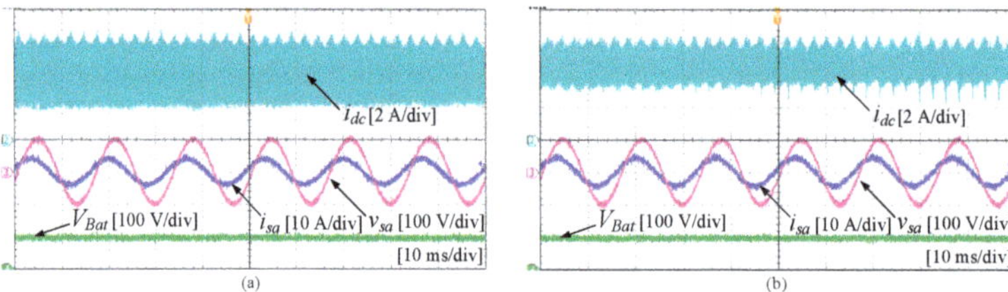

Figure 15. A-phase current, A-phase voltage, battery voltage, and DC current under different modulation strategies for AC–DC matrix converter at high-modulation operation. (**a**) C-SVM; (**b**) Proposed VSVM.

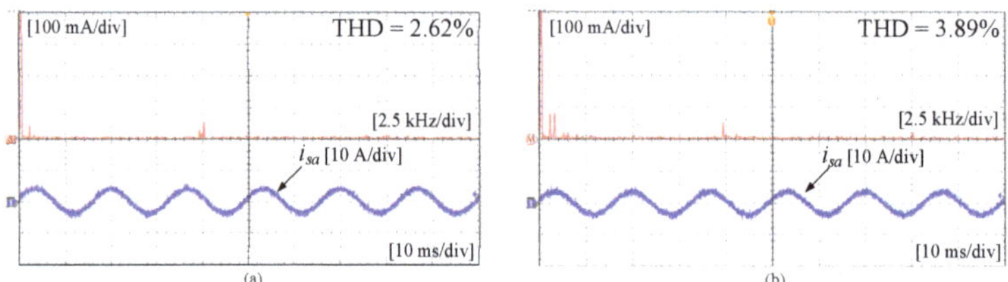

Figure 16. THD of A-phase source current under different modulation strategies for AC–DC matrix converter at high-modulation operation. (**a**) C-SVM; (**b**) Proposed VSVM.

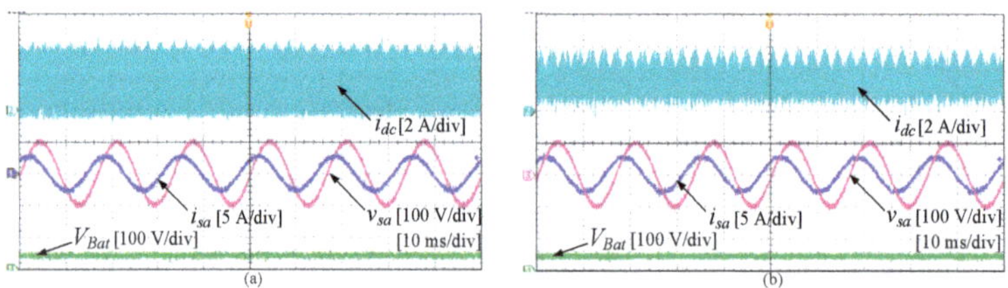

Figure 17. A-phase currents, A-phase voltage, battery voltage, and DC current under different modulation strategies for AC–DC matrix converter at low-modulation operation. (**a**) C-SVM; (**b**) Proposed VSVM.

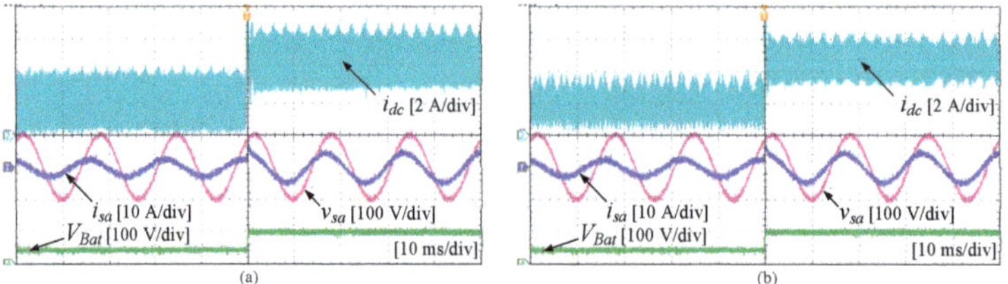

Figure 18. A-phase currents, A-phase voltage, battery voltage, and DC current under different modulation strategies for AC–DC matrix converter at transient state operation. (**a**) C-SVM; (**b**) Proposed VSVM.

Table 2. Percentage comparison of reducing DC current ripple and increasing THD value of the proposed method compared with the conventional methods.

Compared by	Proposed VSVM Method	
	Percentage of Reducing DC Current Ripple	Percentage of Increasing THD
C-SVM	43.1%	30.36%
C-VSVM	31.25%	14.06%

5. Conclusions

In this paper, the virtual space vector modulation control strategy is proposed for the AC–DC matrix converter to reduce the DC current ripple in the whole range of modulation. The proposed strategy successfully reduces the DC current ripples by reducing the longest switching period of the largest active vector and dividing one entire switching period with the five-segment optimized switching patterns. In addition, the optimized switching patterns for the zero vector are proposed to further reduce the DC current ripple at low-modulation operations. The THD value of the proposed VSVM method is slightly higher than that of the C-SVM method, this is a trade-off between the reduced DC current ripples and the increased input current distortion. The effectiveness of the proposed strategy was validated by the simulations and the experimental results.

Author Contributions: All authors contributed to this work by collaboration.

Funding: This research was supported by the National Research Foundation of Korea (NRF) grant funded by the Korean government (MSIP) (2017R1A2B4011444).

Acknowledgments: This research was supported by the National Research Foundation of Korea (NRF) grant funded by the Korean government (MSIP) (2017R1A2B4011444).

Conflicts of Interest: The authors declare no conflict of interest.

References

1. Casadei, D.; Serra, G.; Tani, A.; Zari, L. Matrix converter modulation strategies: A new general approach based on space-vector representation of the switching state. *IEEE Trans. Ind. Electron.* **2002**, *49*, 370–381. [CrossRef]
2. Nguyen, T.D.; Lee, H. Modulation Strategies to reduce common-mode voltage for indirect matrix converters. *IEEE Trans. Ind. Electron.* **2012**, *59*, 129–140. [CrossRef]
3. Rivera, M.; Wilson, A.; Rojas, C.A.; Rodrigez, J.; Espinoza, J.R.; Wheeler, P.W.; Empringham, L. A comparative assessment of model predictive current control and space vector modulation in a direct matrix converter. *IEEE Trans. Ind. Electron.* **2013**, *60*, 578–588. [CrossRef]
4. Rodriguez, J.; Rivera, M.; Kolar, J.W.; Wheeler, P.W. A review of control and modulation methods for matrix converters. *IEEE Trans. Ind. Electron.* **2012**, *59*, 58–70. [CrossRef]
5. Alesina, A.; Venturini, M.G.B. Analysis and design of optimim-amplitude nine-switch direct AC-AC converter *IEEE Trans. Power. Electron.* **1989**, *4*, 101–112. [CrossRef]
6. Vazquez, S.; Lukic, S.M.; Galvan, E.; Franquelo, L.G.; Carrasco, J.M. Energy storage systems for transport and grid applications. *IEEE Trans. Ind. Electron.* **2010**, *57*, 3881–3895. [CrossRef]
7. Yilmaz, M.; Krein, P. Review of battery charger topologies, charging power levels and infrastructure for plug-in electric and hybrid vehicles. *IEEE Trans. Power Electron.* **2013**, *28*, 2151–2169. [CrossRef]
8. Bhuiyan, F.A.; Yazdani, A. Energy storage technologies for grid connected and off-grid power system applications. In Proceedings of the 2012 IEEE Electrical Power and Energy Conference (EPEC), London, ON, Canada, 10–12 October 2012; pp. 303–310.
9. Roberts, B.P.; Sandberg, C. The role of energy storage in development of smart grids. *Proc. IEEE* **2011**, *99*, 1139–1144. [CrossRef]
10. Loh, P.C.; Rong, R.; Blaabjerg, F.; Wang, P. Digital carrier modulation and sampling issues of matrix converters. *IEEE Trans. Power Electron.* **2009**, *24*, 1690–1700. [CrossRef]

11. Yan, Z.; Jia, M.; Zhang, C.; Wu, W. An integration SPWM strategy for high-frequency link matrix converter with adaptive commutation in one step based on De-Re-coupling idea. *IEEE Trans. Power Electron.* **2012**, *59*, 116–128. [CrossRef]
12. Nguyen, T.; Lee, H. An enhanced control strategy for ac/dc matrix converters under unbalanced grid voltage. *IEEE Trans. Ind. Electron.* **2019**, 1718–1727. [CrossRef]
13. Nguyen, T.; Lee, H. Simplified model predictive control for ac/dc matrix converters with active damping function under unbalanced grid voltage. *IEEE J. Emerg. Sel. Top. Power Electron.* **2019**. [CrossRef]
14. Gokdag, M.; Gulbudak, O. Model predictive control of ac-dc matrix converter with unity input power factor. In Proceedings of the IEEE 12th International Conference on Compatibility Power Electronics and Power Engineering (CPE-POWERENG 2018), Doha, Qatar, 10–12 April 2018.
15. Metidji, R.; Metidji, B.; Mendil, B. Design and implementation of unity power factor fuzzy battery charger using ultrasparse matrix rectifier. *IEEE Trans. Power Electron.* **2013**, *28*, 2269–2276. [CrossRef]
16. You, K.; Xiao, D.; Rahman, M.F.; Uddin, M.N. Applying reduced general direct space vector modulation approach of AC-AC matrix converter theory to achieve direct power factor controlled three-phase AC-DC matrix rectifier. *IEEE Trans. Ind. Electron.* **2014**, *50*, 2243–2257. [CrossRef]
17. Feng, B.; Lin, H.; Wang, X. Modulation and control of ac/dc matrix converter for battery energy storage application. *IET Power Electron.* **2015**, *8*, 1583–1594. [CrossRef]
18. Feng, B.; Lin, H.; Hu, S.; An, X.; Wang, X. Control strategy of AC-DC matrix converter in battery energy storage system. In Proceedings of the IEEE Energy Conversion Congress and Exposition (ECCE), Raleigh, NC, USA, 15–20 September 2012; pp. 2128–2134.
19. Feng, B.; Lin, H.; Wang, X.; An, X.; Liu, B. Optimal zero-vector configuration for space vector modulated AC-DC matrix converter. In Proceedings of the IEEE Energy Conversion Congress and Exposition (ECCE), Raleigh, NC, USA, 15–20 September 2012; pp. 291–297.
20. Su, M.; Wang, H.; Sun, Y.; Yang, J.; Xiong, W.; Liu, Y. AC/DC matrix converter with an optimized modulation strategy for V2G applications. *IEEE Trans. Power Electron.* **2013**, *28*, 5736–5745. [CrossRef]
21. Kim, J.; Kwak, S.; Kim, T. Power factor control method based on virtual capacitor for three-phase matrix rectifiers. *IEEE Access* **2019**, *7*, 12484–12494. [CrossRef]
22. Holmes, D.G.; Lipo, T.A. Implementation of a controlled rectifier using AC–AC matrix converter theory. *IEEE Trans. Power Electron.* **1992**, *7*, 240–250. [CrossRef]
23. Ratanapanachote, S.; Cha, H.J.; Enjeti, P.N. A digitally controlled switch mode power supply based on matrix converter. *IEEE Trans. Power Electron.* **2006**, *21*, 124–130. [CrossRef]
24. Yang, X.J.; Cai, W.; Ye, P.S.; Gong, Y.M. Research on dynamic characteristics of matrix rectifier. In Proceedings of the IEEE Conference on Industrial Electronics and Applications, Singapore, 24–26 May 2006; pp. 1–6.
25. Guo, X.; Yang, Y.; Wang, X. Optimal space vector modulation of current source converter for DC link current ripple reduction. *IEEE Trans. Ind. Electron.* **2019**, *66*, 1671–1680. [CrossRef]
26. Tian, K.; Wang, J.; Wu, B.; Xu, D.; Cheng, Z.; Zargari, N.R. A virtual space vector modulation technique for the reduction of common-mode voltage in both magnitude and third-order component. *IEEE Trans. Power Electron.* **2016**, *31*, 839–848. [CrossRef]
27. Hu, C.; Yu, X.; Holmes, D.G.; Shen, W.; Wang, Q.; Luo, F.; Liu, N. An improved virtual space vector modulation scheme for three-level active neutral-point-clamped inverter. *IEEE Trans. Power Electron.* **2017**, *32*, 7419–7434. [CrossRef]
28. Gang, L.; Dafang, W.; Miaoran, W.; Cheng, Z.; Mingyu, W. Neutral-point-voltage balancing in three-level inverters using an optimized virtual space vector PWM with reduced commutations. *IEEE Trans. Ind. Electron.* **2018**, *65*, 6959–6969.

© 2019 by the authors. Licensee MDPI, Basel, Switzerland. This article is an open access article distributed under the terms and conditions of the Creative Commons Attribution (CC BY) license (http://creativecommons.org/licenses/by/4.0/).

Article

Temporary Fault Ride-Through Method in Power Distribution Systems with Distributed Generations Based on PCS

Jung-Hun Lee [1], Seung-Gyu Jeon [1], Dong-Kyu Kim [1], Joon-Seok Oh [2] and Jae-Eon Kim [1,*]

1 School of Electric Engineering, Chungbuk National University, Cheongju 28644, Korea; leejh6800@naver.com (J.-H.L.); sgjeon@cbnu.ac.kr (S.-G.J.); kimdk8079@naver.com (D.-K.K.)
2 Department of Distribution Planning, Korea Electric Power Corporation (KEPCO), Naju 58322, Korea; simeter@naver.com
* Correspondence: jekim@cbnu.ac.kr; Tel.: +82-10-9028-2423

Received: 30 December 2019; Accepted: 27 February 2020; Published: 2 March 2020

Abstract: The current practice of Distributed Generation (DG) disconnection for every fault in distribution systems has an adverse effect on utility and stable power trading when the penetration level of DGs is high. That is, in the process of fault detecting and Circuit Breaker (CB) reclosing when a temporary fault occurs, DGs should be disconnected from the Point of Common Coupling (PCC) before CB reclosing. Then all DGs should wait at least 5 minutes after restoration for reconnection and cannot supply the pre-bid power in power market during that period. To solve this problem, this paper proposes a control method that can keep operating without disconnection of DG. This control method is verified through modeling and simulation by the PSCAD/EMTDC software package for distribution systems with DGs based on PCS (Power Conditioning Systems) and CB reclosing protection.

Keywords: fault restoration; distribution generation; temporary fault ride-through; voltage control; inrush current control

1. Introduction

At present, the exhaustion of resources and environmental problems are continuously intensifying due to the indiscriminate use of fossil fuels. As a result, the Paris Convention, which will replace the Kyoto Protocol, which is scheduled to expire in 2020, was adopted at the Paris Climate Change Conference in France in 2015 with a focus on converting renewable energy sources into alternative energy sources for fossil fuels worldwide. Accordingly, the deployment of Distributed Generation (DG) based on Power Conditioning Systems (PCS) such as photovoltaic generation and wind power generation are rapidly increasing worldwide [1,2].

In this way, PCS-based DG without inertia may have some negative effects on reliability and power quality in power systems if it is introduced on a large scale. In particular, when a temporary fault occurs in distribution systems with many PCS-based DGs, they should be disconnected from the Point of Common Coupling (PCC) before Circuit Breaker (CB) reclosing. Then the voltage becomes unstable. Moreover, all DGs interconnected to distribution systems should wait more than 5 minutes for reconnection after restoration and cannot supply the pre-bid power in power market during that period [3,4]. In such a case, power trading cannot be performed, thereby leading to not only economic damages, but also unstable system operation such as power quality, etc. [5,6].

In this regard, the only existing studies for rapid restoration from faults are as follows: a multi-stage/multiple micro-grid technique for dividing the islanding operation section [7,8], an islanding operation method engaging the droop control of the DG when the fault section is isolated [9–13], a voltage control strategy that involves switching between the grid-connected mode and islanding operation mode for the VSC (Voltage-Sourced Converters) type of DGs [14], a Multi Agent System (MAS) technique for dividing switches, DGs, and load into agents for quick fault restoration [15–20], and a fault restoration method for minimizing the number of switch operations for load transfer [21]. However, these papers have focused on permanent faults and temporary faults have not been considered. That is, there are no measures against the voltage problems caused by the opening of CBs or recloser upon fault occurrence, disconnection of all DGs, and temporary inrush current problems caused by their reclosing.

On the other hand, in the conventional FRT(Fault Ride Through) regulations for grid code, when a fault occurs in transmission systems, the time required for fault detection and isolation is around 150 ms to 160 ms, and after that it takes about 2 s to return to the steady-state voltage. Taking this into consideration, DG should continue to operate within this range [22–26]. Control methods for complying with FRT operating conditions in which DG keeps operating during an accident are currently proposed and applied [27,28]. However, when a temporary fault occurs in distribution systems, CB operates within 5–6 cycles (about 0.1 s) immediately after a temporary fault occurs and then returns to the steady-state operation by performing CB reclosing within 0.5 s. As this abnormal condition is severe compared with the fault situation in transmission systems, FRT for temporary faults in distribution systems could not be considered until now and there were no suggestions to solve it. In addition, there is a problem with determining whether the fault is temporary or permanent before CB reclosing.

Therefore, in this paper, a temporary fault ride-through method in distribution systems is proposed. Section 2 describes the method of restoration from a fault in conventional distribution systems, and Section 3 describes some problems during restoration processes. Section 4 proposes a control method where all DGs connected to distribution systems can continue to operate even if a temporary fault occurs. In Section 5, the proposed method is verified through simulation by PSCAD/EMTDC and analysis.

2. Conventional Restoration Method for Temporary Faults

In distribution systems, the reclosing operation of a CB or RC(Recloser) is applied in order to determine whether there has been a fault, temporary or permanent [29–31]. The general protection method in distribution systems is described below.

The protection devices to break fault current in distribution systems are composed of a CB at the substation outlet and reclosers on distribution lines, which can detect a fault and reclose. In the following the characteristic operation of a CB and recloser is described.

2.1. CB

The operation of a CB at the substation outlet is carried out and ensured with some relays like Over Current Relay (OCR), Over Current Ground Relay (OCGR), and reclosing relay. There are various types of setting methods depending on the configuration of distribution lines. Basically, the CB acts as back-up protection for recloser downstream on distribution lines and reclosing operation to decide whether a fault is temporary or permanent. Then the CB retains lock-out and open state for a permanent fault.

2.2. Recloser

Recloser is generally installed on distribution lines and detects a fault, breaks the fault's current, and automatically performs the reclosing operation within a specified time. Reclosing is also applied because more than 80% of faults are temporary faults that are eliminated by themselves in a temporary period [32].

If the fault persists, the last reclosing function should be performed, followed by lock-out and open state. When a fault occurs on distribution lines, the upstream recloser nearest the fault point should perform the reclosing operation in accordance with specified operation obligation and satisfy the main or back-up protection relationship with CB at upstream substation. Figure 1 shows an operation example of the recloser under the occurrence of a fault between RC1 and RC2—RC1 is opened to break the fault current supplied from substation.

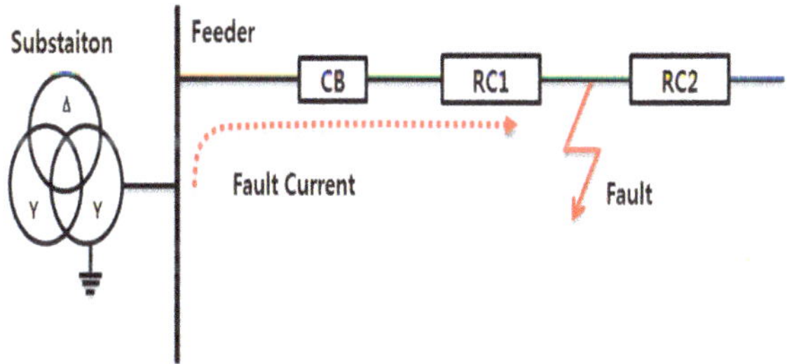

Figure 1. Configuration of the Circuit Breaker (CB) and Recloser (RC) in distribution systems.

The operating characteristics of a recloser generally operate with Two-Fast Two-Delay (2F2D). This can be accomplished by two fast trips and two delayed trips in order to eliminate fault or determine permanent lock-out. Figure 2 illustrates the 2F2D reclosing processes.

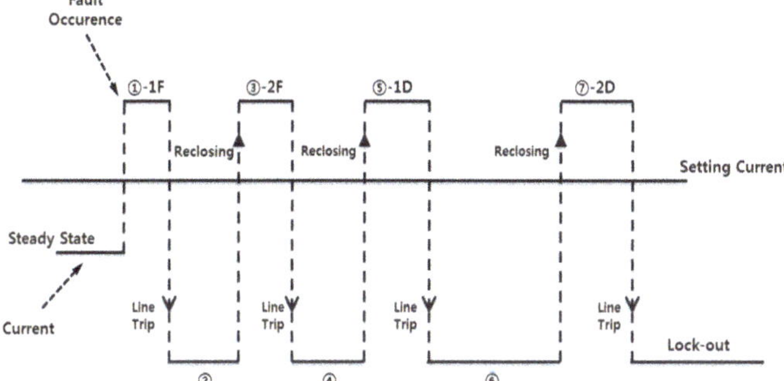

Figure 2. Typical Two-Fast Two-Delay (2F2D) operating scheme example of recloser.

3. Critical Problems in Conventional Fault Restoration Method

3.1. Discontinuity of DG Supply by Anti-Islanding Function

One of critical problems in the conventional fault restoration method is that the CB or recloser is opened when a fault occurs, and all DGs and loads on the stream line of the CB or recloser are islanding. At this time, for the safety of the human body and the protection of over-voltage due to the ratio of the power generation to the load, all DGs should detect the islanding state and be disconnected. In addition, after the fault is eliminated, distribution systems will be restored, and they will be reconnected after 5 minutes for normal operation based on grid code [32]. Then, DG cannot supply the pre-bid power to power market.

3.2. Power Quality Problems during Islanding Operation

When a fault occurs in distribution systems, the CB and recloser on distribution lines operate as shown in Figure 3. In order to prevent the islanding operation of the DG due to the opening of this recloser, there are several national grid codes as seen in Table 1. Most grid codes require that the islanding state should be detected and DG disconnected within around 0.5–2 s, considering the reclosing operation of a CB or recloser. This means that islanding operation may occur for 0.5–2 s after detecting a fault, which may cause power quality problems such as frequency and voltage fluctuation beyond permissible ranges and result in equipment damage and human body danger. As shown in

Figure 3, when the output of aggregated DGs is larger than aggregated loads in the islanding region, the over-voltage phenomenon occurs in a just few milliseconds after islanding operation, which may adversely affect loads, equipment, and the human body. In addition, in the case of an under-voltage, even if the fault is eliminated, a situation in which DG is disconnected occurs.

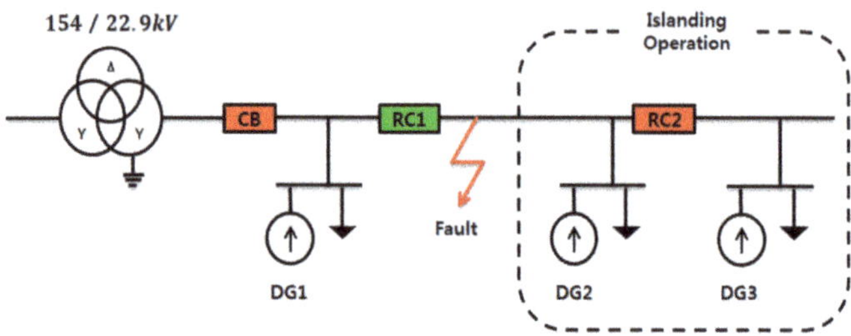

Figure 3. Configuration of AC grid with distributed generators. DG: Distributed Generation.

Table 1. Anti-islanding requirements in grid codes.

Name	Requirements	Detection Method
IEEE 1547 IEC 61727 [22,33]	Cease to energize within 2 s of the formation of the island	Active or Passive
VDE-AR-N 4105 [34]	Disconnect in 5 s	Active or Passive
BDEW 2008 [35]	Network operator may have special requirements	Not specified
JEAC 9701-2012 [36]	Detect within 0.5–1.0 s	Active
	Detect within 0.5 s	Passive
KEPCO Guideline [3]	Cease to energize within 0.5 s of the formation of the island	Active or Passive

IEEE: Institute of Electrical and Electronics Engineers, IEC: International Electrotechnical Commission, VDE: Verband Der Elektrotechnik, BDEW: Bundesverband der Energie-und Wasserwirtschaft, JEAC: Japan Electric Association Code, KEPCO: Korea Electric Power Corporation.

The ratio of DG output to load in the islanding region has the largest influence on voltage problems. This is an important factor, because the voltage V_{is} during islanding is determined by the ratio as follows [37]:

$$V_{is} = \frac{Total\ amount\ of\ power\ generation}{Total\ amount\ of\ load\ demand} \tag{1}$$

The load characteristic can be classified into three types: constant power, constant current, and constant impedance. Generally, power systems have all three types of load. The voltage V_{is} of the islanding condition can be expressed as shown in Equation (2) when DG with constant power is connected to a load with constant impedance [38,39].

$$V_{is} = V_L \frac{\sqrt{\sqrt{P_{DG}^2 + Q_{DG}^2}}}{\sqrt{\sqrt{P_L^2 + Q_L^2}}} \tag{2}$$

V_{is} Voltage amplitude during islanding.
V_L Voltage amplitude under steady-state.
P_{DG} Output active power of DG.
Q_{DG} Output reactive power of DG.

P_L Active power of load.
Q_L Reactive power of load.

Equation (2) means that the voltage during islanding is proportional to the square root of the apparent power ratio of DG to load. The research to solve this problem is currently underway on the output control of DG when islanding operation occurs. Several papers have applied the output control through the droop control of DG [11–15]. However, these controls have a slow response to prevent over-voltage, which rises rapidly due to the output control by the droop characteristic curve. Therefore, this paper considers that the voltage magnitude is controlled to be 0.5 p.u., which is the permissible voltage operation range of VRT (Voltage Ride Through), for the voltage problem during islanding operation generated by Equation (2).

4. Temporary Fault Ride-Through Method

Temporary fault most commonly occurs in distribution systems, and is eliminated before the first reclosing of the recloser. However, due to recloser opening, the anti-islanding function of DG is triggered, and all DGs are disconnected. Therefore, this paper proposes a control method to prevent unnecessary disconnection of DG in the case of a temporary fault. This method is defined as the Temporary Fault Ride-Through (TFRT), and consists of two controllers—Low/High Voltage Ride Through (L/HVRT) and Inrush Current Suppressing controller. The former suppresses over/under-voltages during temporary fault and the latter inrush current due to voltage amplitude and phase difference between both sides of a CB or recloser at their reclosing time. This control is described below.

4.1. L/HVRT Controller

The L/HVRT Controller controls the voltage to keep it within a permissible VRT range in an over- or under-voltage situation, which may occur by the ratio of generation to load after recloser opening. If a temporary fault occurs while operating with active and reactive power reference ($P_{reference}$, $Q_{reference}$) in steady-state, the recloser is open and the fault condition, which is unbalanced voltage, is maintained and after the temporary fault is eliminated, it changes to islanding operation state under balanced voltage. At this time, as shown in Figure 4a, the output reference d-axis current value of the controller is controlled through the PI (Proportional-Integral) controller so that the PCC voltage remains within VRT range. In this paper, the rms voltage V_{RMS} at the PCC is considered to be controlled to maintain $V_{RMS\text{-}reference}$, namely 0.5 p.u., which is the permissible range of VRT operation of IEEE 1547, regardless of the ratio of generation to load. At this time, as shown in Figure 4b, active power output is controlled by the q-axis current value $I_{q\_control}$, which is the square root of the square of the d-axis current value $I_{d\_control}$ subtracted from the square of the rated maximum current I_M of DG to be controlled within the rated range of the DG output.

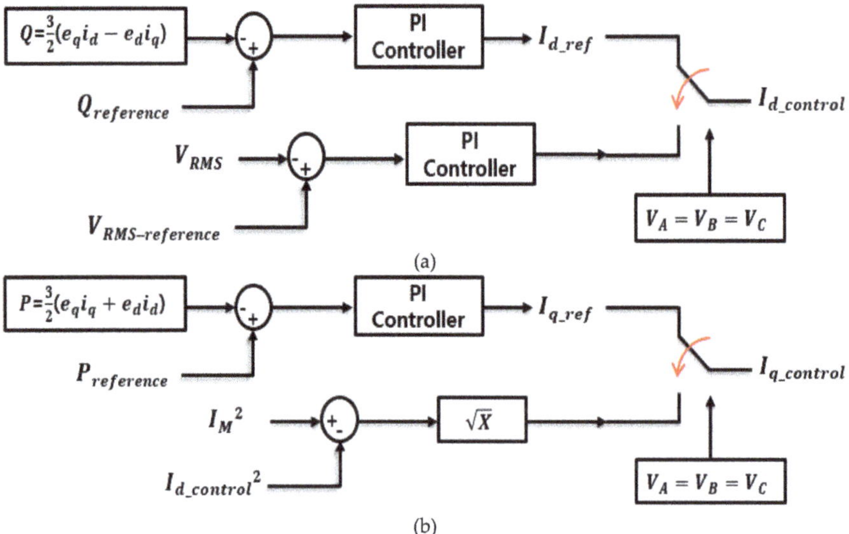

Figure 4. Block diagram of Low/High Voltage Ride Through (L/HVRT) controller. (**a**) I_d Current Controller; (**b**) I_q Current Controller.

4.2. Inrush Current Suppressing Controller

When the CB or recloser recloses, inrush current occurs due to the magnitude and phase difference of voltage between both their terminals. Since the inrush current has a negative effect on the PCS equipment and feeders, a control method of suppressing the inrush current is necessary. The magnitude of the inrush current I_{inrush} is derived as Equation (3) in the equivalent circuit of Figure 5, which is fed into DG and load. At this time, at the PCC of DG, the derivative value of current $I_{derivative\_max}$ flowing into DG is measured and multiplied to voltage reference $V_{d,q\_reference}$ to be controlled as shown in Figure 6. Then the output current I_{DG} from DG is shown in the blue line and offsets the inrush current I_{inrush} shown in the red line in Figure 5. This control is implemented during a few cycles just after reclosing.

$$I_{inrush} = \frac{V_S\angle 0 \ - \ V_R\angle\theta}{R \ + \ jX} \tag{3}$$

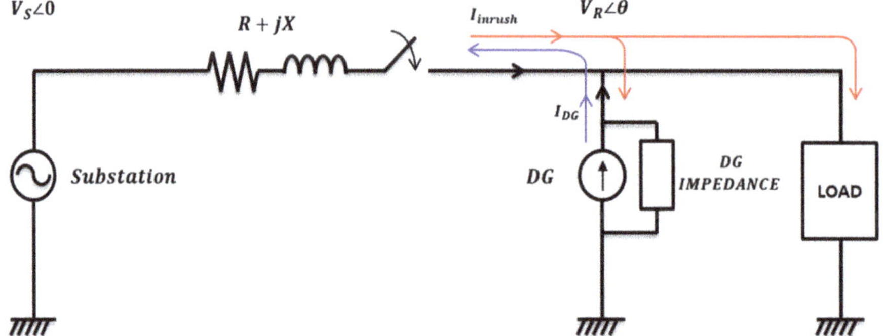

Figure 5. Equivalent circuit of current flowing in distribution systems with DG and loads.

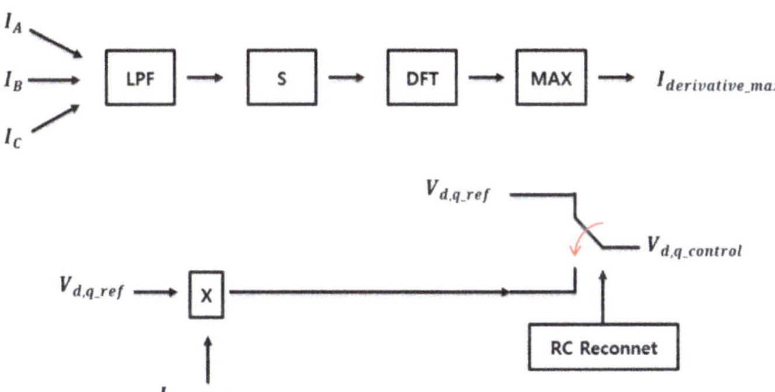

Figure 6. Block diagram of inrush current suppressing controller. LPF: Low Pass Filer, S; Differentiator, DFT: Discrete Fourier Transform.

I_{inrush} Current amplitude when recloser reclosing.
I_{DG} Current amplitude of DG.
V_S Voltage amplitude of substation.
V_R Voltage amplitude of DG.
θ Phase difference of substation and DG.

4.3. TFRT Algorithm

The FRT in the conventional grid code allows DG to continue operation without disconnection in the event of a fault in transmission systems, but under this condition, it cannot be applied in the event of a temporary fault in distribution systems. This is because the fault conditions of transmission systems and distribution systems are different, as described in the introduction. Therefore, this paper proposes a Temporary Fault Ride-Through (TFRT) algorithm so that DG can continue to operate in the temporary fault condition. The TFRT algorithm using the controller described in Sections 4.1 and 4.2 is shown in Figure 7. If a fault is detected on a feeder, the CB or recloser opens with their first fast operation, and then DG will run in islanding operation. If the unbalanced voltage continues until the reclosing operation of the CB or recloser, DG perceives the fault as permanent fault and is disconnected from the PCC. If the voltage is balanced, DG perceives the fault as a temporary fault and conducts control to maintain the voltage within a permissible operating range of the VRT. Here, the reference voltage value V_{RMS} was determined as 0.5 p.u., considering the VRT operating range of IEEE 1547. When the fault is eliminated and the CB or recloser is reclosed, the proposed controller suppresses inrush current to minimize severe or critical damages to DG and distribution systems.

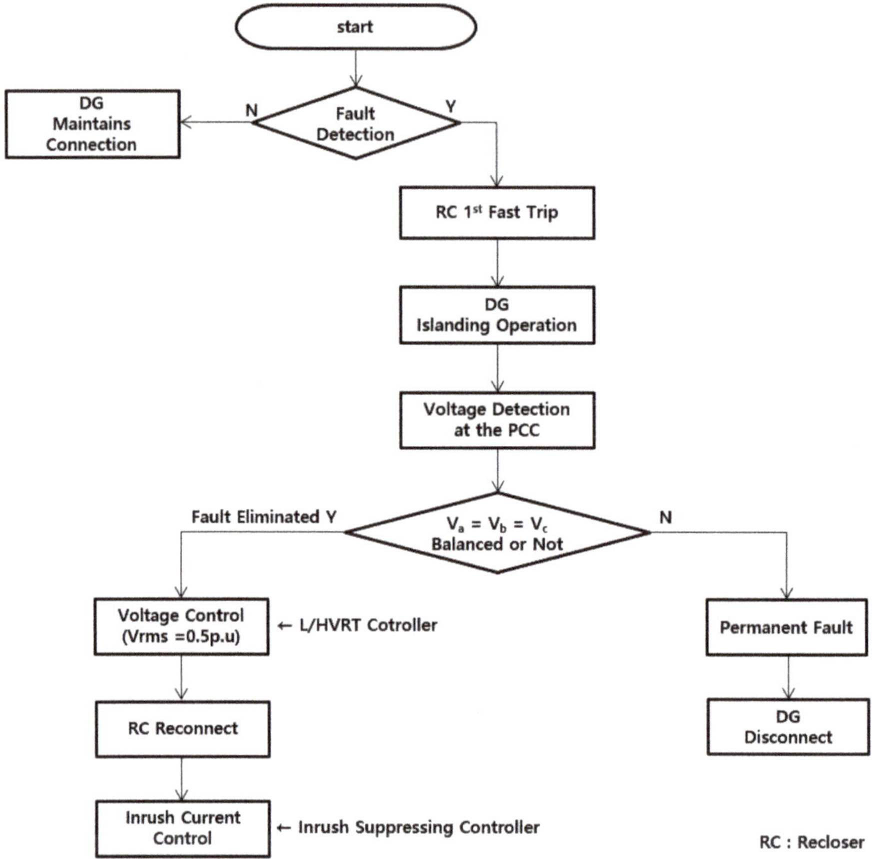

Figure 7. Proposed fault ride-through processing for a temporary fault. PCC: Point of Common Coupling.

5. Simulation and Analysis

In this section, we prove the proposed TFRT algorithm and control method by conducting a simulation and analyzing the results. Substation, PCS-based DG, load, and controllers were modeled through the PSCAD/EMTDC software package and the simulation was conducted according to the scenario set up to verify the continuous operation of DGs during temporary fault and restoration.

5.1. Configuration of Distribution Systems with DGs and Loads

The system configuration is shown in Figure 8, with the substation, distribution line, PCS-based DG, and loads. In addition, specific information on each model is summarized in Table 2. In the case of a renewable energy source, a model composed of a current controller is applied using the d–q-axis current value for MPPT (Maximum Power Point Tracking). The load is considered as constant impedance and is distributed equally in consideration of the length of the distribution line.

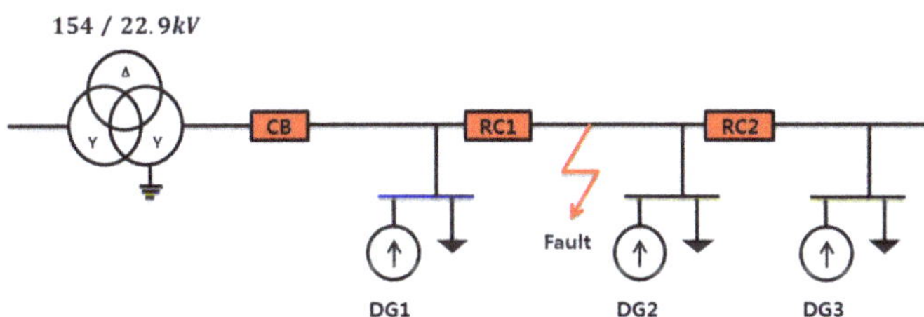

Figure 8. System model configuration.

Table 2. Specifications of the system model components.

Index	Value	Remark
154 kV Grid Source		
Positive Sequence %Z	0.08 + j 0.99	100 MVA Based
Zero Sequence %Z	0.34 + j 1.69	
3-Winding Transformer (154 kV/22.9 kV/6.6 kV)		
Rated Power	45/60 MVA	45 MVA Based
Positive Sequence %$X_{1\text{-}2}$	j 16.16	
Positive Sequence %$X_{2\text{-}3}$	j 6.69	
Positive Sequence %$X_{3\text{-}1}$	j 25.38	
Type	Y–Y_g –Δ	
Distribution Generation (22.9 kV)		
Rated Power of DG1	1 MVA (0.5 M × 2)	
Rated Power of DG2	2 MVA (0.5 M × 4)	
Rated Power of DG3	1 MVA (0.5 M × 2)	
Transformer Connection	Y_g –Δ	
Positive Sequence %X	j 0.05	

5.2. Configuration of Scenario

In the temporary fault scenario, a single-line ground fault on a-phase occurs in 3.0 s and the recloser on the distribution line trips in 3.1 s. In the case of a temporary fault, the fault lasts for 0.2 s and at 3.2 s the fault is eliminated. Therefore, the islanding operation period is divided into the one during 3.1 s to 3.2 s under fault condition and the other during 3.2 s to 3.6 s when the fault is eliminated. When a fault is cleared and DGs are ready to connect to the system, the recloser recloses in 3.6 s. Figure 9 shows a temporary fault scenario. In this section, the TFRT method in the case of a temporary fault is proved but also, in order to analyze the voltage at the moment of a permanent fault of the control algorithm, we added one scenario for a permanent fault separately from the temporary fault.

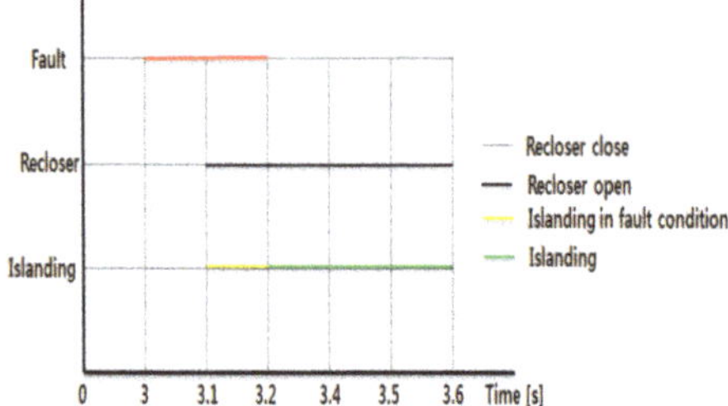

Figure 9. Simulation scenario for a temporary fault.

5.3. Simulation Results and Analysis

This section simulates the controller designed in the previous Section using the programming tool and analyzes the result. When a permanent fault occurs in a feeder with DG, the voltage waveforms at the PCC are shown in Figure 10. Therefore, DG perceives a permanent fault and disconnects.

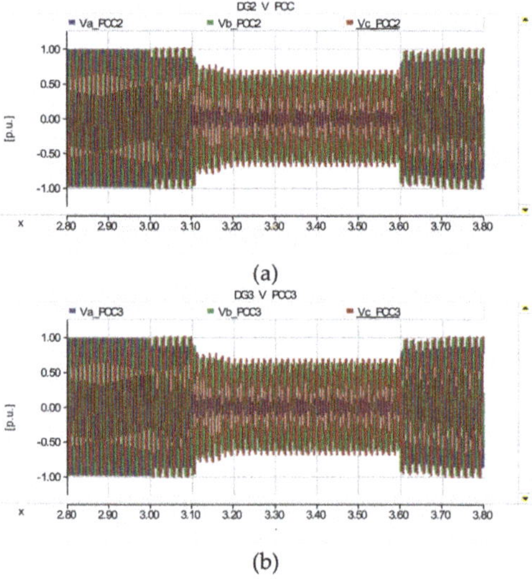

Figure 10. Phase Voltages at the PCC of DGs in the case of a permanent fault, (**a**) phase voltages at the PCC of DG2, (**b**) phase voltages at the PCC of DG3.

Firstly, the simulation waveforms for voltage and current when DG exceeds load are shown in Figure 10. Figure 10a shows the voltage of the PCC stage of DG2 and DG3. In Figure 11a, voltages are unbalanced between 3.1 s and 3.2 s. Conversely, as the is fault eliminated after 3.2 s, voltage is balanced. In this case, over-voltage occurred because the aggregated output of DGs is larger than the aggregated loads. However, when the proposed TFRT control method is applied, the voltage at the PCC was maintained at 0.5 p.u. In addition, over-voltage during reclosing of the recloser did not occur. Figure 11b shows the output current of DG in islanding operation. After 3.2 s, the output of DG1 upstream of the RC1 recloser does not change. The output currents in DG2 and DG3 remain constant after the fault is eliminated. However, inrush current occurred temporarily at 3.6 s when the RC1 recloser is reclosed. To solve this problem, when the proposed inrush suppressing control method was applied, the output currents of DGs increased and reduced the inrush current flowing from upstream. As a result, it is confirmed that the method can suppress the inrush current occurring during the reclosing period.

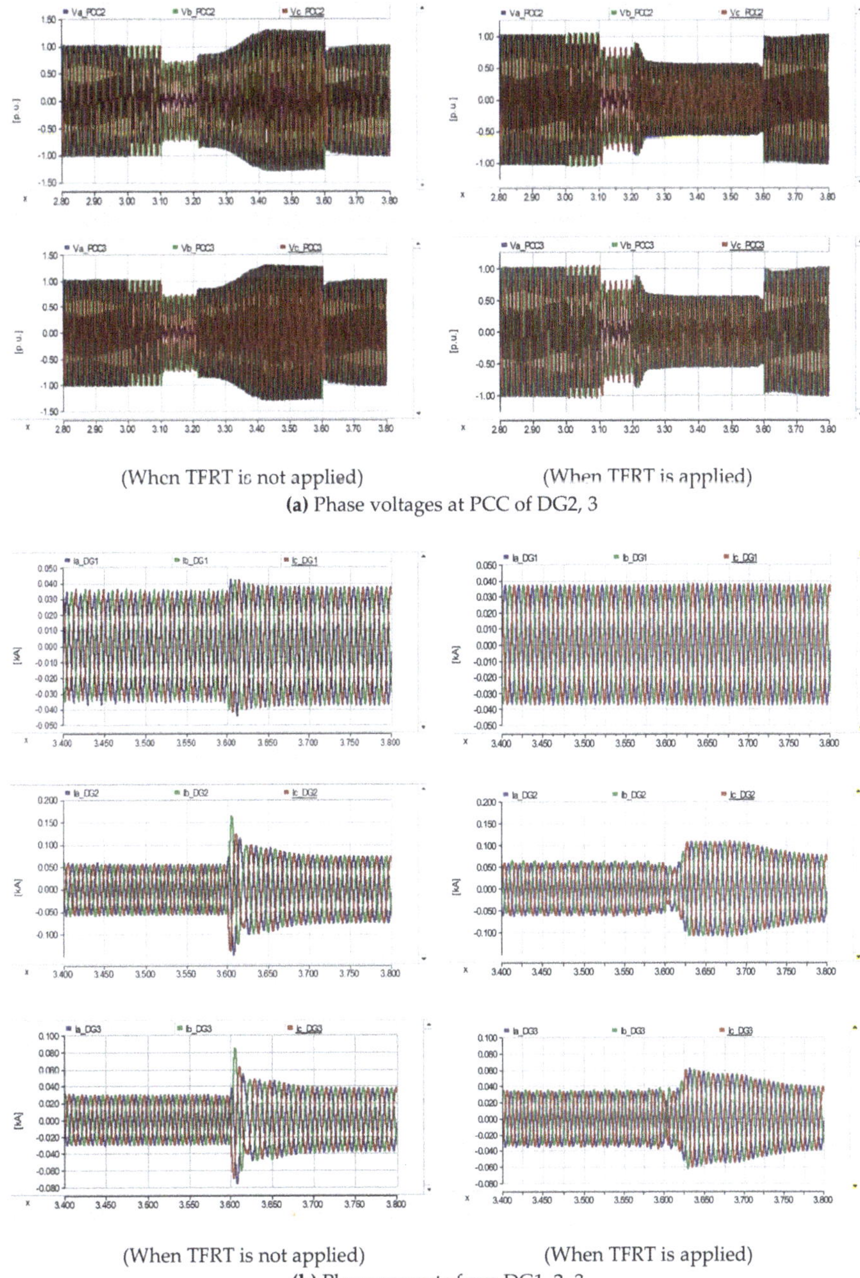

(When TFRT is not applied) (When TFRT is applied)

(a) Phase voltages at PCC of DG2, 3

(When TFRT is not applied) (When TFRT is applied)

(b) Phase currents from DG1, 2, 3

Figure 11. Comparison of the voltage and current with and without Temporary Fault Ride-Through (TFRT) control methods when the aggregated output of DGs is greater than the aggregated loads.

Secondly, voltage and current waveforms are shown in Figure 12 when the aggregated loads are greater than the aggregated output of DGs. As above, after the fault is eliminated, control was implemented when the voltage is balanced. If control is not implemented, the voltage is less than 0.5 p.u. In this case, a fault is eliminated, but DG is disconnected as it is outside the permissible operating range of the VRT. However, when the voltage is controlled at 0.5 p.u. using the proposed voltage control algorithm, after the fault is eliminated, the voltage kept a constant within the permissible range so that DG2 and DG3 can continue to operate. In the case of current, the inrush current is reduced as shown in Figure 12b using the proposed control method. Since the voltage is raised to maintain a voltage within the permissible range, the current was also higher than before the control.

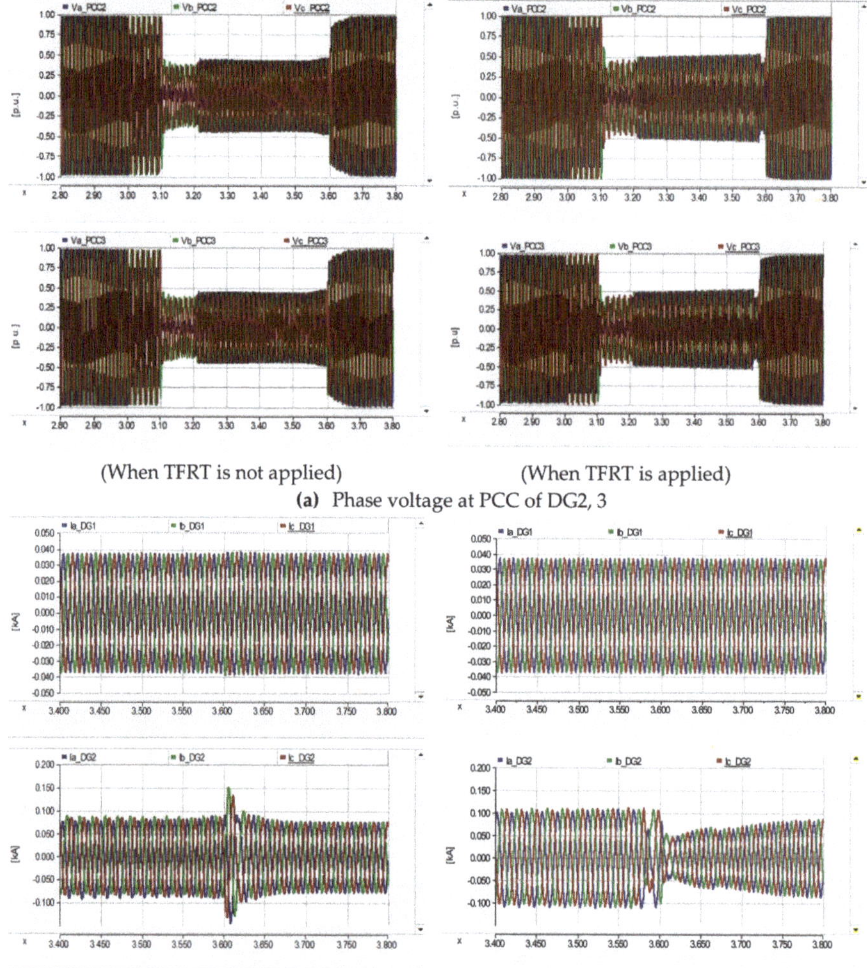

(When TFRT is not applied) (When TFRT is applied)

(a) Phase voltage at PCC of DG2, 3

Figure 12. *Cont.*

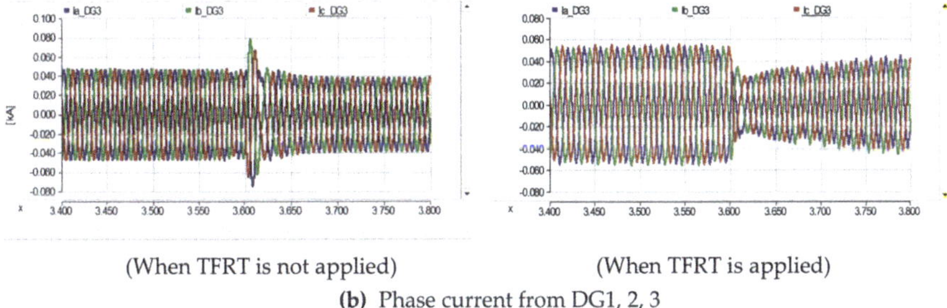

(When TFRT is not applied) (When TFRT is applied)

(b) Phase current from DG1, 2, 3

Figure 12. Comparison of the voltage and current with and without TFRT control methods when the aggregated loads are greater than the aggregated output of DGs.

As seen from the above simulation results, if a fault is permanent or temporary can be decided according to whether the output voltage of DG is balanced or not. In the case of a permanent fault where the output of voltage of DG is not balanced, DG is disconnected. In the event of a temporary fault where the output voltage is balanced, it was proofed that DG can maintain the voltage within the VRT permissible voltage range by maintaining a constant voltage regardless of the amount generation and load. It was also proofed that the inrush current during the reclosing of the CB or recloser is lower than without TFRT. Simulation results show that the proposed TFRT method can work effectively for temporary faults in distribution systems.

6. Conclusions

The hosting capacity of power distribution systems for introducing DGs based on PCS such as renewable energy sources is increasing, and the number of rotor-based power plants is being reduced. This tendency makes the stability and power quality in power systems critical, especially in the event of a temporary fault in distribution systems. As a result, the large increase of DG deployment may lead to not only economic damage, but also system operation problems.

To solve these problems, the TFRT method is proposed. That is, at first, it is determined whether a fault in distribution systems is temporary or permanent. Then DG is disconnected for a permanent fault and continues to operate for a temporary fault by applying the proposed TFRT method. It was verified using the PSCAD/EMTDC software package that the TFRT method can control DGs to keep operating even if a temporary fault occurs. Also, it can control current and voltage to prevent damage due to over-/under-voltage and over-current occurred by the reclosing of the CB or recloser.TDC programming tool. proposed e as under/over- voltage occurred.

Finally, it is expected that the proposed TFRT method can improve the reliability and stability of distribution systems and is included in current grid code worldwide through further research.

Author Contributions: Validation, J.-S.O.; Investigation, S.-G.J. and D.-K.K.; Writing—original draft preparation, J.-H.L.; Writing—review and editing, J.-E.K.; All authors have read and agreed to the published version of the manuscript.

Funding: This research received no external funding.

Acknowledgments: This work was supported by the Korea Institute of Energy Technology Evaluation and Planning (KETEP) grant funded by the Korean government (MOTIE) (20181210301470, Development and Proof on the Commercialization of the Broker Service based on Small-scale Distributed Energy Resources).

Conflicts of Interest: The authors declare no conflict of interest.

References

1. Teodorescu, R.; Liserre, M.; Rodriguez, P. *Grid Converters for Photovoltaic and Wind Power Systems*; John Wiley & Sons: New York, NY, USA, 2011.

2. *Global Trends in Renewable Energy Investment 2014*; Bloomberg New Energy Finance: Washington, DC, USA, 2014.
3. Korea Electric Power Corporation. *Guideline of Interconnection Technology of Distributed Generation in Distribution System*; Korea Electric Power Corporation: Naju, Korea, 2015.
4. Zidan, A.; El-Saadany, E.F. Incorporating load variation and variable wind generation in service restoration plans for distribution systems. *Energy* **2013**, *57*, 682–691. [CrossRef]
5. McDermott, T.E.; Dugan, R.C. PQ, reliability and DG. *IEEE Ind. Appl. Mag.* **2003**, *9*, 17–23. [CrossRef]
6. Kamel, R.M.; Chaouachi, A.; Nagasaka, K. RETRACTED: Wind power smoothing using fuzzy logic pitch controller and energy capacitor system for improvement Micro-Grid performance in islanding mode. *Energy* **2010**, *35*, 2119–2129. [CrossRef]
7. Zidan, A.; Khairalla, M.; Abdrabou, A.M.; Khalifa, T.; Shaban, K.; Abdrabou, A.; El Shatshat, R.; Gaouda, A.M. Fault Detection, Isolation, and Service Restoration in Distribution Systems: State-of-the-Art and Future Trends. *IEEE Trans. Smart Grid* **2017**, *8*, 2170–2185. [CrossRef]
8. Wang, F.; Chen, C.; Li, C.; Cao, Y.; Li, Y.; Zhou, B.; Dong, X. A Multi-Stage Restoration Method for Medium-Voltage Distribution System with DGs. *IEEE Trans. Smart Grid* **2017**, *8*, 2627–2636. [CrossRef]
9. Arefifar, S.A.; Mohamed, Y.A.-R.I.; El-Fouly, T.H. Optimized Multiple Microgrid-Based Clustering of Active Distribution Systems Considering Communication and Control Requirements. *IEEE Trans. Ind. Electron.* **2014**, *62*, 711–723. [CrossRef]
10. Dewadasa, M.; Ghosh, A.; Ledwich, G.F. Islanded operation and system restoration with converter interfaced distributed generation. In Proceedings of the 2011 IEEE PES Innovative Smart Grid Technologies, Perth, WA, Australia, 13–16 November 2011; pp. 1–8.
11. Ghoddami, H.; Yazdani, A. A Mitigation Strategy for Temporary Overvoltages Caused by Grid-Connected Photovoltaic Systems. *IEEE Trans. Energy Convers.* **2014**, *30*, 413–420. [CrossRef]
12. Guo, F.; Wen, C.; Mao, J.; Song, Y.-D. Distributed Secondary Voltage and Frequency Restoration Control of Droop-Controlled Inverter-Based Microgrids. *IEEE Trans. Ind. Electron.* **2014**, *62*, 4355–4364. [CrossRef]
13. Lissandron, S.; Sgarbossa, R.; Santa, L.D. DeltaP—deltaQ Area Assessment of Temporary Unintentional Islanding with Pf and QV Droop Controlled PV Generators in Distribution Networks. In Proceedings of the IEEE Energy Conversion Congress and Exposition, Montreal, QC, Canada, 20–24 September 2015.
14. Dietmannsberger, M.; Wang, X.; Blaabjerg, F. Restoration of Low-Voltage Distribution Systems with Inverter-Interfaced DG Units. *IEEE Trans. Ind. Appl.* **2018**, *54*, 5377–5386. [CrossRef]
15. Gao, F.; Iravani, M. A Control Strategy for a Distributed Generation Unit in Grid-Connected and Autonomous Modes of Operation. *IEEE Trans. Power Deliv.* **2008**, *23*, 850–859.
16. Nguyen, C.P.; Flueck, A.J. Agent Based Restoration with Distributed Energy Storage Support in Smart Grids. *IEEE Trans. Smart Grid* **2012**, *3*, 1029–1038. [CrossRef]
17. Lo, Y.L.; Wang, C.H.; Lu, C.N. A Multi-Agent Based Service Restoration in Distribution Network with Distributed Generations. In Proceedings of the 2009 15th International Conference on Intelligent System Applications to Power Systems, Curitiba, Brazil, 8–12 November 2009; pp. 1–5.
18. Elmitwally, A.; Elsaid, M.; Elgamal, M.; Chen, Z. A Fuzzy-Multiagent Service Restoration Scheme for Distribution System with Distributed Generation. *IEEE Trans. Sustain. Energy* **2015**, *6*, 1–12. [CrossRef]
19. Sharma, A.; Srinivasan, D.; Trivedi, A. A Decentralized Multiagent System Approach for Service Restoration Using DG Islanding. *IEEE Trans. Smart Grid* **2015**, *6*, 2784–2793. [CrossRef]
20. McArthur, S.D.J.; Davidson, E.M.; Catterson, V.M.; Dimeas, A.L.; Hatziargyriou, N.D.; Ponci, F.; Funabashi, T. Multi-agent systems for power engineering applications—Part I: Concepts, approaches, and technical challenges. *IEEE Trans. Power Syst.* **2007**, *22*, 1743–1752. [CrossRef]
21. McArthur, S.D.J.; Davidson, E.M.; Catterson, V.M.; Dimeas, A.L.; Hatziargyriou, N.D.; Ponci, F.; Funabashi, T. Multi-agent systems for power engineering applications—Part II: Technologies, standards, and tools for building multi-agent systems. *IEEE Trans. Power Syst.* **2007**, *22*, 1753–1759. [CrossRef]
22. Shirmohammadi, D. Service restoration in distribution networks via network reconfiguration. *IEEE Trans. Power Deliv.* **1992**, *7*, 952–958. [CrossRef]
23. IEEE Standards Association. *IEEE 1547 Standard for Interconnection and Interoperability of Distributed Resources with Assoiated Electric Power Systems Interface*; IEEE Standards Coordinating Committee: Washington, DC, USA, 2018.

24. International Electrotechnical Commission. *IEC TS 62786 Distributed Energy Resources Connection with the Grid*; International Electrotechnical Commission: Geneva, Switzerland, 2017.
25. CENELEC. *EN 50549-1,2 Requirements for Generating Plants to be Connected in Parallel with Distribution Networks*; CENELEC: Brussels, Belgium, 2019.
26. ENTSO-E AISBL. *ENTSO-E Netwrk Code for Requirements for Grid Connection Applicable to all Generators*; ENTSO-E: Brussels Belgium, 2013.
27. Bründlinger, R.; Schaupp, T.; Arnold, G.; Schäfer, N.; Graditi, G.; Adinolfi, G. Implentation of the European Network Code on Requirements for Generators on the European nation level. In Proceedings of the Solar Integration Workshop 2018: 8th International Workshop on Integration of Solar Power into Power Systems, Stockholm, Sweden, 16–17 October 2018.
28. Lopez, M.A.G.; De Vicuña, L.G.; Miret, J.; Castilla, M.; Guzmán, R. Control Strategy for Grid-Connected Three-Phase Inverters During Voltage Sags to Meet Grid Codes and to Maximize Power Delivery Capability. *IEEE Trans. Power Electron.* **2018**, *33*, 9360–9374. [CrossRef]
29. Zarei, S.F.; Mokhtari, H.; Ghasemi, M.A.; Blaabjerg, F. Reinforcing Fault Ride Through Capability of Grid Forming Voltage Source Converters Using an Enhanced Voltage Control Scheme. *IEEE Trans. Power Deliv.* **2019**, *34*, 1827–1842. [CrossRef]
30. Zidan, A.; El-Saadany, E.F. A Cooperative Multiagent Framework for Self-Healing Mechanisms in Distribution Systems. *IEEE Trans. Smart Grid* **2012**, *3*, 1525–1539. [CrossRef]
31. Ha, B.-N.; Lee, S.-W.; Shin, C.-H. Current practice of distribution automation technology and prospect. In Proceedings of the Summer Conference of Korean Institute of Electrical Engineers, Gangwon, Korea, 16–18 June 2008; pp. 313–314.
32. Oh, J.-H.; Yun, S.-Y.; Kim, J.-C.; Kim, E.-S. Particular characteristics associated with temporary and permanent fault on the multi-shot reclosing scheme. In Proceedings of the IEEE Power Engineering Society Summer Meeting, Seattle, WA, USA, 16–20 July 2003; pp. 421–424.
33. Liu, C.-C.; Lee, S.; Venkata, S. An expert system operational aid for restoration and loss reduction of distribution systems. *IEEE Trans. Power Syst.* **1988**, *3*, 619–626. [CrossRef]
34. International Electrotechnical Commission. *IEC 61727 ed2.0 Photovoltaic(PV) Systems Characteristics of the Utility Interface*; International Electrotechnical Commission: Geneva, The Netherlands, 2004.
35. VDE Association for Electrical, Electronic and Information Technologies. *VDE-AR-N 4105:2011-08 Power Generation Systems Connected to the Low Voltage Distribution Network*; VDE Association for Electrical, Electronic and Information Technologies: Frankfurt, Germany, 2011.
36. BDEW German Association of Energy and Water Industries. *BDEW Generating Plants Connected to the Medium-Voltage Network*; BDEW German Association of Energy and Water Industries: Berlin, Germany, 2008.
37. Japan Electric Association. *Grid Interconnection Code (JEAC 9701-2012)*; Japan Electric Association: Chiyoda, Japan, 2012.
38. Barker, P. Overvoltage considerations in applying distributed resources on power systems. In Proceedings of the 2002 IEEE Power Engineering Society Summer Meeting, Chicago, IL, USA, 21–25 July 2002; Volume 1, pp. 109–114.
39. Shin, S.-S.; Oh, J.-S.; Jang, S.-H.; Chae, W.-K.; Park, J.; Kim, J.-E. A Fault Analysis on AC Microgrid with Distributed Generations. *J. Electr. Eng. Technol.* **2016**, *11*, 1600–1609. [CrossRef]

© 2020 by the authors. Licensee MDPI, Basel, Switzerland. This article is an open access article distributed under the terms and conditions of the Creative Commons Attribution (CC BY) license (http://creativecommons.org/licenses/by/4.0/).

Article

Coordinated Frequency and State-of-Charge Control with Multi-Battery Energy Storage Systems and Diesel Generators in an Isolated Microgrid

Jae-Won Chang [1], Gyu-Sub Lee [1,*], Hyeon-Jin Moon [1], Mark B. Glick [2] and Seung-Il Moon [1]

1 Department of Electrical and Computer Engineering, Seoul National University, 1 Gwanak-ro, Seoul 08826, Korea; jwchang91@gmail.com (J.-W.C.); mhj7527@gmail.com (H.-J.M.); moonsi@plaza.snu.ac.kr (S.-I.M.)
2 Hawaii Natural Energy Institute, University of Hawaii at Manoa, 1680 East-West Road, Honolulu, HI 67822, USA; mbglick@hawaii.edu
* Correspondence: lgs1106@snu.ac.kr; Tel.: +82-2-886-3103

Received: 4 April 2019; Accepted: 23 April 2019; Published: 28 April 2019

Abstract: Recently, isolated microgrids have been operated using renewable energy sources (RESs), diesel generators, and battery energy storage systems (BESSs) for an economical and reliable power supply to loads. The concept of the complementary control, in which power imbalances are managed by diesel generators in the long time scale and BESSs in the short time scale, is widely adopted in isolated microgrids for efficient and stable operation. This paper proposes a new complementary control strategy for regulating the frequency and state of charge (SOC) when the system has multiple diesel generators and BESSs. In contrast to conventional complementary control, the proposed control strategy enables the parallel operation of diesel generators and BESSs, as well as SOC management. Furthermore, diesel generators regulate the equivalent SOC of BESSs with hierarchical control. Additionally, BESSs regulate the frequency of the system with hierarchical control and manage their individual SOCs. We conducted a case study by using Simulink/MATLAB to verify the effectiveness of the proposed control strategy in comparison with conventional complementary control.

Keywords: isolated microgrid; renewable energy source; diesel generator; battery energy storage system; hierarchical control

---

1. Introduction

Currently, with abundant natural resources in remote areas, numerous renewable energy sources (RESs) including photovoltaic (PV) and wind power are integrated into small isolated grids [1,2]. RESs have advantages in terms of cost effectiveness and environmental impact, but they can degrade the system stability through, for example, frequency fluctuations. Furthermore, since isolated grids generally have small inertia compared to large transmission networks, the intermittent outputs of RESs induce large frequency fluctuations [3]. These problems evoke the transition from conventional isolated grids, which have been operating with diesel-powered generators alone, to isolated microgrids which include battery energy storage systems (BESSs) in addition to the thermal generators [4]. In contrast to grids supported solely by slow diesel generators, the frequency can be regulated much more tightly with BESSs because they have short response times [4–6]. However, for utilizing BESSs as frequency-supporting resources in an isolated microgrid, the state of charge (SOC) of the BESSs must be managed efficiently because the capacity of batteries is limited [7].

To overcome the slow response of diesel generators and limited capacity of BESSs, complementary control schemes with diesel generators and BESSs for signals of different time scales were proposed [7–10]. With complementary control, diesel generators compensate for long-time-scale energy imbalances, while BESSs compensate for short-time-scale energy imbalance. As BESSs only compensate for frequent active power imbalances in the complementary control scheme, batteries with a relatively small capacity are required for frequency regulation in an isolated microgrid. Additionally, the system operator can efficiently take advantage of the attributes of diesel-powered thermal generators efficiently because diesel generators are used to compensate for long time-scale fluctuations [11]. As shown in Figure 1, complementary control strategies can be classified into three categories: (1) coordinated droop control (CD) [8], (2) control based on frequency distribution techniques (FD) [9,10], and (3) coordinated SOC and frequency control (CSF) [7].

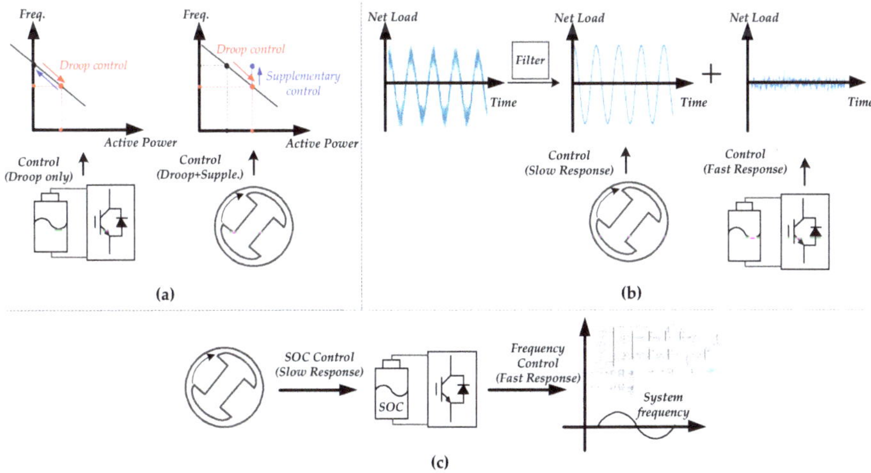

Figure 1. Complementary control methods with diesel generator and battery energy storage system (BESS) for system imbalance: (**a**) coordinated droop control (CD); (**b**) control based on frequency distribution techniques (FD); (**c**) coordinated state of charge (SOC) and frequency control (CSF).

In the CD method, diesel generators and BESSs regulate frequency with droop control. Additionally, for eliminating steady-state error resulting from droop characteristics, supplementary control is implemented in diesel generators [8], which implies that BESSs are only responsible for short-time-scale active power imbalances and diesel generators are responsible for imbalances of both time scales. In the FD method, filters, such as the wavelet transform [9] and discrete Fourier transform [10], are used for clearly dividing power imbalances of long and short time scales. After dividing power imbalances of different time scales, long and short time-scale imbalances are controlled by diesel generators and BESSs, respectively [9,10]. However, although imbalances of long and short time scales can be regulated complementarily by using the CD and FD methods, the SOC of BESSs is difficult to be managed because the energy stored in BESSs is not considered in these methods.

In the CSF method, which was proposed in [7], the grid frequency is rapidly regulated by a BESS, while a diesel generator controls the SOC of the BESS for capacity management which can lower the required capacity of batteries. With information only about the SOC of the BESS, the system can not only regulate short and long time-scale imbalances complementarily, but also manage the SOC of the BESS. However, the authors of [7] only focused on isolated microgrids featuring only one BESS and one diesel generator. For expanding the scale and enhancing the reliability, a system with several BESSs and diesel generators should be considered. However, it is difficult to apply the CSF method for multiple BESSs and diesel generators because the parallel operation of such devices and the SOC management of individual BESSs were not considered in the previous research.

In the present paper, we propose a new CSF control strategy for multiple BESSs and diesel generators in an isolated microgrid. In the proposed method, diesel generators manage an equivalent SOC, which represents the SOCs of all BESSs, with a hierarchical control scheme. BESSs control the frequency of the system with a hierarchical control structure, and a self SOC control mechanism of each BESS is proposed. Finally, a case study with the data of a real isolated microgrid in South Korea demonstrates the effectiveness of the proposed control method compared to conventional complementary control. The case study verifies that the proposed control method can regulate the frequency and individual SOCs of BESSs and enable the parallel operation of diesel generators and BESSs.

2. System Configuration and Control Strategy

2.1. *Configuration of the Test System*

As a test system, we utilize the Geocha Island network, which will be constructed as an actual isolated microgrid in South Korea. Geocha Island has been organized for an isolated microgrid with PV units, wind generation units, three diesel generators, and two BESSs. Figure 2 shows the planned structure of the Geocha Island microgrid system, including detailed information. All information about the system was obtained from the Korean Electric Power Corporation (KEPCO) and [12]. The nominal system frequency and voltage are 60 Hz and 6.9 kV, respectively.

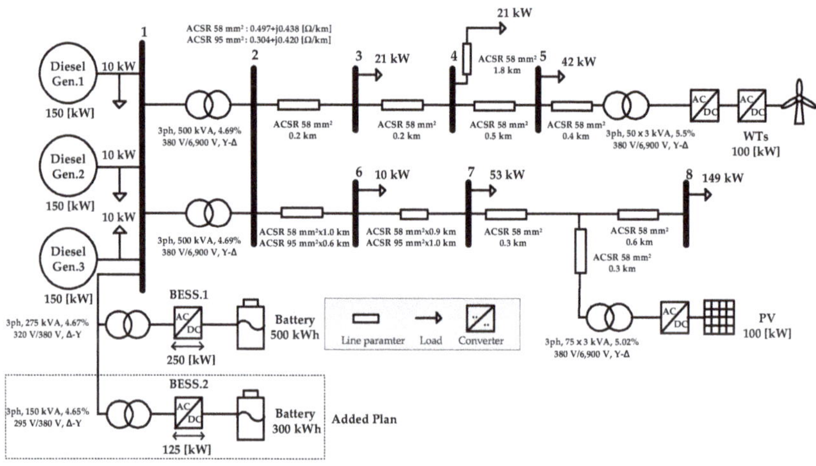

Figure 2. System configuration of the Geocha Island microgrid.

2.2. *Control Strategy of the System*

Although the conventional CSF has many advantages for controlling frequency and SOC, it cannot be adopted in the system with multiple diesel generators and BESSs, similarly to isochronous mode in a conventional power system. For applying the concept of previous CSF to multiple generators and BESSs, we suggest a new CSF method for parallel operation. Table 1 shows the comparison between conventional and proposed CSF methods.

Table 1. Comparison between conventional and proposed CSF methods.

Control Method	Parallel Operation of Diesel Generators	Parallel Operation of BESSs	Frequency Control	SOC Control
Conventional CSF	X	X	O	△ (Single)
Proposed CSF	O	O	O	O (Multiple)

The proposed control strategy is summarized as follows. (1) RESs are operated by the maximum power point tracking (MPPT) algorithm for maximizing the use of RESs. (2) Diesel generators are operated to manage the energy stored in BESSs by utilizing the proposed hierarchical control scheme. (3) BESSs operate in a hierarchical manner to control the grid frequency. (4) The SOCs of individual BESSs are managed by the proposed self SOC controller (SSC). Figure 3 shows the entire control strategy of the proposed method applied to the Geocha Island microgrid system. The frequency can be regulated from the frequency controllers of BESSs. While the frequency is regulated by BESSs, the individual SOCs of BESSs deviate from their reference values. For restoring the SOCs of all BESSs, the equivalent SOC, SOC_{eq}, which represents the SOCs of all BESSs, is controlled by diesel generators. Because only the total energy of all BESSs is considered in diesel generators, an additional controller is adopted in BESSs for restoring the individual SOCs. The detailed scheme of the proposed controllers is presented in the next sections.

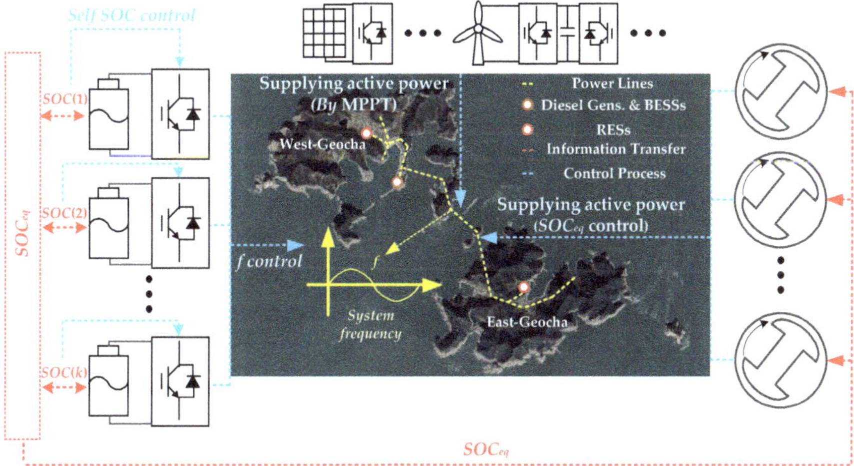

Figure 3. Proposed control strategy for the Geocha Island microgrid.

3. Proposed Control Strategy of Diesel Generators

In the previous CSF method of [7], a diesel generator regulates the SOC of a single BESS. However, there is no target variable for multiple BESSs because each BESS has a different SOC and capacity. To manage the SOCs of all BESSs, we propose the concept of SOC_{eq}, which can be defined from the definition of SOC [13] as follows:

$$SOC_{eq} = \frac{\text{Current Energy}}{\text{Rated Energy Capacity}} = \frac{\sum_{k=1}^{n} E(k)}{\sum_{k=1}^{n} E_{rate}(k)} = \frac{\sum_{k=1}^{n} C_{rate}(k)V_{dc}(k)SOC(k)}{\sum_{k=1}^{n} C_{rate}(k)V_{dc,rate}(k)}, \tag{1}$$

where n is the total number of BESSs, $E(k)$ is the stored energy (Wh), $E_{rate}(k)$ is the rated energy (Wh), $C_{rate}(k)$ is the rated capacity (Ah), $V_{dc}(k)$ is the dc voltage of the battery (V), $V_{dc,rate}(k)$ is the rated dc voltage of the battery (V), and $SOC(k)$ is the SOC of the k-th BESS. By regulating SOC_{eq} with respect to its reference value, SOC^*_{eq}, the energy stored in all BESSs can be maintained, and energy imbalance is regulated by diesel generators in the long time scale. Similar to the SOC_{eq} in (1), SOC^*_{eq} can be derived from the reference SOC values of individual BESSs.

3.1. SOC_{eq} Control Scheme for Single Diesel Generator

Firstly, we verify that SOC_{eq} can be regulated by a single diesel generator in an isolated microgrid with multiple BESSs. By differentiating (1), we obtain

$$(\sum_{k=1}^{n} C_{rate}(k)V_{dc}(k))\frac{d(SOC_{eq})}{dt} = \sum_{k=1}^{n} C_{rate}(k)V_{dc}(k)\frac{d(SOC(k))}{dt}. \quad (2)$$

From [13], the SOC of the k-th BESS can be expressed as follows:

$$SOC(k) = SOC^{0}(k) - \int \frac{P_{BESS}(k)}{V_{dc}(k)C_{rate}(k)}dt, \quad (3)$$

where $SOC^{0}(k)$ is the initial value of SOC and $P_{BESS}(k)$ is the active power of the k-th BESS. By taking the derivative of (3), we obtain

$$\frac{dSOC(k)}{dt} = -\frac{P_{BESS}(k)}{C_{rate}(k)V_{dc}(k)}. \quad (4)$$

By substituting (4) into (2) and under the assumption that $V_{dc}(k)$ is almost constant and equal to the rated value within the normal SOC region [7], the derivative of SOC_{eq} can be expressed as follows:

$$\frac{d(SOC_{eq})}{dt} = -\frac{(P_{BESS}(1) + \ldots + P_{BESS}(n))}{(\sum\limits_{k=1}^{n} C_{rate}(k)V_{dc,rate}(k))}. \quad (5)$$

To satisfy the power-balance equation of an isolated microgrid, the summation of the total active power outputs of the BESSs and diesel generator should be equal to the active power of net load.

$$P_{BESS}(1) + \ldots + P_{BESS}(n) + P_{d} = P_{net,load}, \quad (6)$$

where P_{d} is the active power output of the diesel generator and $P_{net,load}$ is the net load including the uncontrollable outputs of RESs, loads, and system losses.

For a single diesel generator, the isochronous control mode can be adopted for frequency regulation in a conventional power system [14,15]. Likewise, in the proposed method, a single diesel generator operates in the isochronous mode for SOC_{eq} control. Figure 4 represents the proposed SOC_{eq} control structure for a single diesel generator, including a corresponding plant model from (5) and (6). Since the plant model between the diesel output and SOC_{eq} is a first-order system as shown in Figure 4, a proportional integral (PI) controller can be adopted for SOC_{eq} control [16].

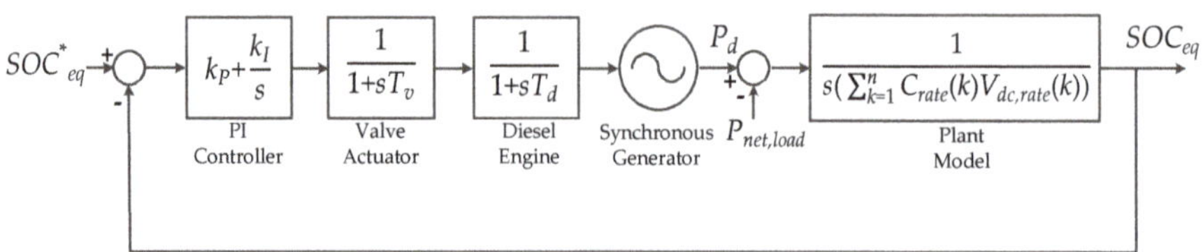

Figure 4. Plant model of the isochronous mode of the diesel generator.

3.2. SOC_{eq} Control Scheme for Multiple Diesel Generators

In a power system with multiple synchronous generators, a hierarchical frequency-control structure is utilized to prevent hunting effects on the frequency and inaccurate power sharing between generators at the steady-state [15]. Similarly, we propose a hierarchical control structure for multiple diesel generators to regulate SOC_{eq}. Owing to the proposed hierarchical control, SOC_{eq} can be regulated stably, and accurate power sharing between diesel generators at the steady-state is possible. Figure 5 shows the concept of the proposed hierarchical control structure for multiple diesel generators.

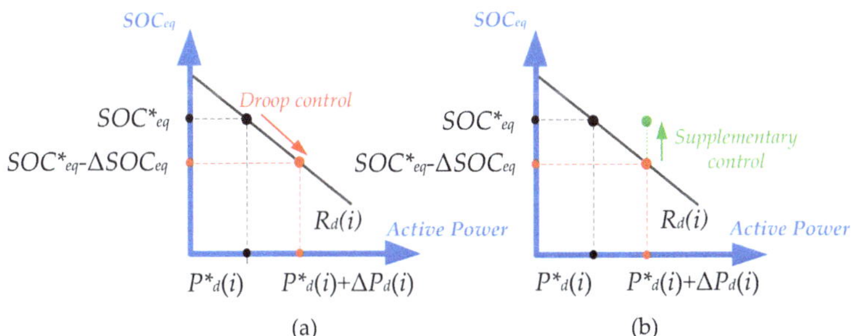

Figure 5. Hierarchical SOC_{eq} control of diesel generators: (**a**) droop; (**b**) supplementary control.

Firstly, we introduce the SOC_{eq}–P droop control strategy for primary responses of SOC_{eq} regulation and power sharing between diesel generators. As shown in Figure 5a, when an unexpected system variation occurs, SOC_{eq} deviates from the reference value, SOC^*_{eq}. Based on the droop characteristic, the active power outputs of diesel generators increase (or decrease) from the reference value of the active power output of the i-th diesel generator, $P^*_d(i)$. Eventually, SOC_{eq} can be saturated at the point where the active power is balanced (red dot in Figure 5), and the BESSs make zero active power. The droop coefficient of the i-th diesel generator can be defined based on the slope in Figure 5a and represented by $R_d(i)$ (i = 1, 2, and 3 for the target network).

Owing to the innate characteristics of the droop controller, steady-state error exists between the saturated value and the target reference value, ΔSOC_{eq}. To restore SOC_{eq} to its corresponding reference SOC^*_{eq}, supplementary control for secondary responses is provided as shown in Figure 5b. By integrating the concepts of the droop controller in Figure 5a and the supplementary controller in Figure 5b, we develop the proposed hierarchical control structure including the plant model between the active power of diesel generators and SOC_{eq}, as shown in Figure 6, where $Pf(i)$ is the participation factor, $T_v(i)$ is the time constant of the valve actuator, and $T_d(i)$ is the time constant of the diesel engine for the i-th diesel generator. The supplementary controllers are implemented with a PI controller for eliminating the steady-state error and participation factor for determining the sharing ratio. Similar to the conventional load frequency control structure, the supplementary controller must be operated to respond slower than the droop controllers [15] in diesel generators.

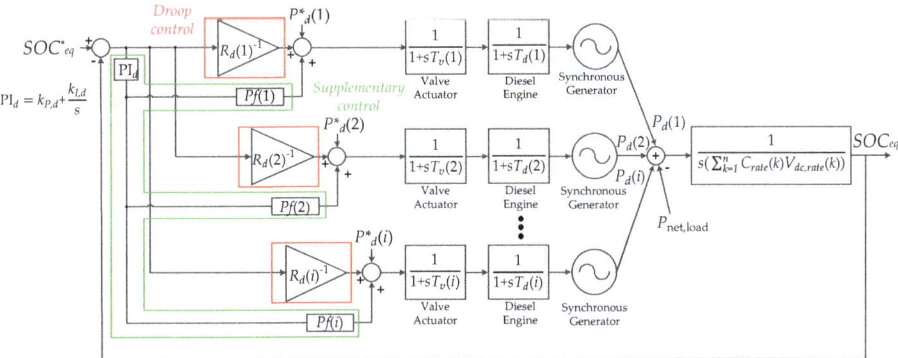

Figure 6. Plant model for the hierarchical control scheme of multiple diesel generators.

4. Control Strategy of Battery Energy Storage Systems (BESSs)

4.1. Frequency Control Scheme for Multiple BESSs

When a single BESS controls the frequency of the system, a converter is operated in the grid-forming mode to control the frequency directly. The detailed scheme of the grid-forming mode is shown in [16,17]. To regulate the grid frequency when using multiple BESSs with grid-forming converters, hierarchical control for frequency regulation can be utilized [18,19]. For the primary response, the *P-f* droop control method is adopted for the parallel operation of BESSs, as shown in Figure 7a. Additionally, for secondary response, as shown in Figure 7b, a supplementary controller including an integrator is exploited to eliminate the steady-state error of frequency generated from droop control.

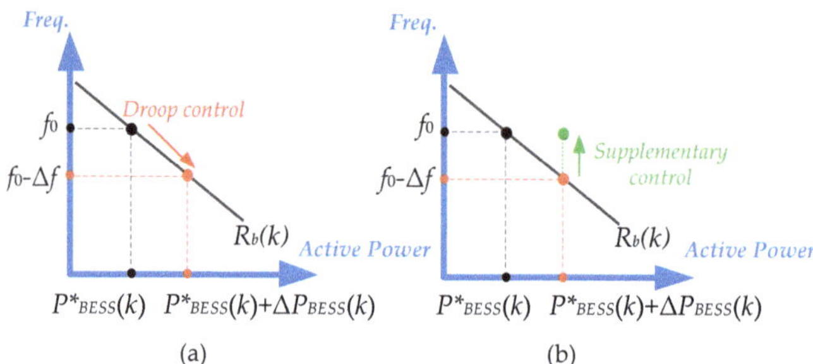

Figure 7. Hierarchical frequency control of BESSs. (**a**) Droop and (**b**) supplementary control.

In Figure 7, f_0 is the nominal frequency, $P^*_{BESS}(k)$ is the reference active power, and $R_b(k)$ is the droop constant of k-th BESS.

4.2. Self State of Charge (SOC) Controller for Individual BESSs

Frequency can be regulated almost perfectly by utilizing BESSs, and SOC_{eq} can be maintained at the desired reference value with diesel generators. However, the management of individual SOCs cannot be guaranteed by exploiting the controllers presented in the previous sections, because only the total energy of BESSs is considered in diesel generators. In addition, from the integral of supplementary frequency controllers, the power sharing between BESSs is not guaranteed from the desired value [18,19]. Therefore, the individual SOCs of BESSs should be controlled at their reference values by themselves with an additional controller. To regulate the individual SOCs of BESSs at the reference values, we develop SSC for restoring the individual SOCs. Since the BESSs are operated by grid-forming converters, which control the frequency of their terminals, BESSs should regulate their SOCs by adjusting their terminal frequencies. To validate the necessity of SSC, a simplified circuit for BESSs, as shown in Figure 8, is first investigated.

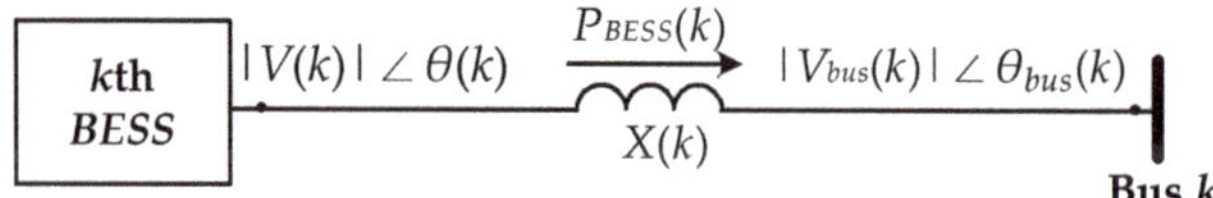

Figure 8. Simplified single line diagram between a terminal node and the main bus of the k-th BESS.

In Figure 8, $X(k)$ is the reactance component of the transformer, $\angle\theta(k)$ and $|V(k)|$ are the phase angle and magnitude of ac voltage at the terminal node of the k-th BESS, and $\angle\theta_{bus}(k)$ and $|V_{bus}(k)|$ are the phase angle and magnitude of ac voltage at the bus connected to the k-th BESS. As reactance is much larger than the resistance in the transformer [20], the transformer is modeled by a single reactance. From Figure 8, the active power output of the k-th BESS can be approximated as follows [21]:

$$P_{BESS}(k) = \frac{|V(k)||V_{bus}(k)|(\angle\theta(k) - \angle\theta_{bus}(k))}{X(k)}. \tag{7}$$

From (4) and (7), the SOC of the k-th BESS can be expressed as follows:

$$\frac{dSOC(k)}{dt} = -\frac{1}{C_{rate}(k)V_{dc,rate}(k)} \frac{|V(k)||V_{bus}(k)|(\angle\theta(k) - \angle\theta_{bus}(k))}{X(k)}, \tag{8}$$

with the assumption that the entire system except the k-th BESS is constant, the derivative of the phase angle for bus k is zero. By differentiating (8), we can obtain:

$$\frac{d^2SOC(k)}{dt^2} = -\frac{2\pi|V(k)||V_{bus}(k)|}{C_{rate}(k)V_{dc,rate}(k)X(k)} f(k), \tag{9}$$

where $f(k)$ is the frequency output of the k-th BESS. As the plant model between the frequency and SOC of the k-th BESS is a second-order system, each BESS can regulate its SOC by itself with a PI controller [16].

The purpose of SSC is to restore the SOC of the k-th BESS to its corresponding reference value, $SOC^*(k)$. However, as diesel generators respond slowly compared to BESSs, the transient difference between SOC_{eq} and its reference value, SOC^*_{eq}, should be considered in SSC. Therefore, in SSC, the reference value for the SOC of the k-th BESS must be modified as follows:

$$SOC_{ref}(k) = SOC^*(k) + SOC_{eq} - SOC^*_{eq}. \tag{10}$$

where $SOC_{ref}(k)$ is the adjusted reference value of $SOC(k)$ considering the transient state. Note that $SOC_{ref}(k)$ and $SOC^*(k)$ eventually become equal in the steady state because SOC_{eq} is regulated to SOC^*_{eq} by diesel generators in the long time scale. Figure 9 shows the total control structure of the k-th BESS including inner control loops [14,17], the hierarchical frequency controller, and SSC.

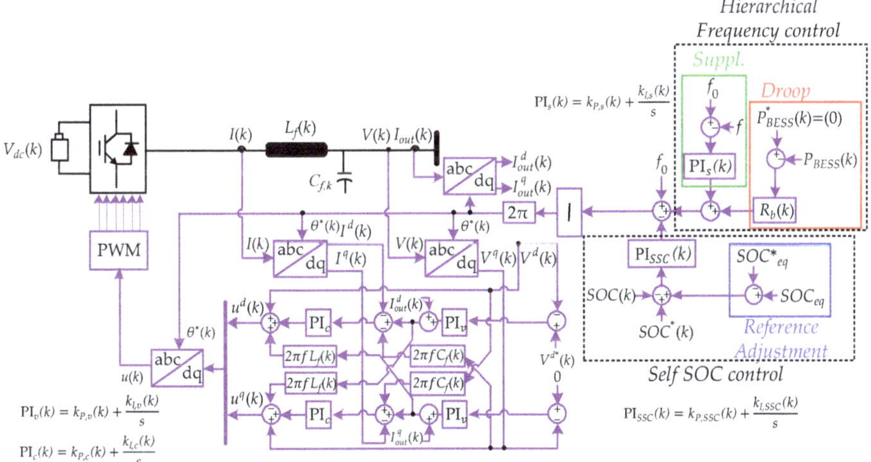

Figure 9. Proposed controller of BESS for frequency and self state of charge (SOC) recovery.

In Figure 9, $I(k)$ is the line current from the kth BESS, $V(k)$ is the terminal voltage, $I_{out}(k)$ is the line current to the terminal node, $L(k)$ and $C_f(k)$ are the filter components, $\theta^*(k)$ is the reference phase angle for the modified reference frequency value, and $u(k)$ is the modulating signal of the k-th BESS. Superscripts d and q indicate the dq components of corresponding variables.

The reference frequency value and $\theta^*(k)$ are determined from the hierarchical frequency controller and SSC. The dq components of the reference voltage are derived from the reference value of voltage magnitude, $V^{d*}(k)$, and $\theta^*(k)$. Finally, the terminal voltage of the k-th BESS is regulated to the reference value via a nested voltage and current control loop, as shown in Figure 9. Through the entire control loop, the frequency and SOC of each BESS are regulated.

4.3. Maintaining Desired Active Power Outputs of BESSs by the Linear Time-Varying SOC Control

By regulating SOCs as the linear time varying value, the BESSs can be controlled as the desired level of active power outputs [7]. In other words, while BESSs control frequency, they can maintain the scheduled or dispatched active power. From integrating (4), to make the active power of k-th BESS as the reference value $P^*_{BESS}(k)$, SOC reference value can be determined as:

$$SOC*(k) = \int -\frac{P*_{BESS}(k)}{C_{rate}(k)V_{dc}(k)}dt, \tag{11}$$

As $P^*_{BESS}(k)$ is desired constant value, SOC reference has the linear time varying value from (11). Through controlling SOCs as the linear time varying reference values as (11), BESSs can maintain the desired active power outputs.

5. Case Study

The isolated microgrid shown in Figure 2 was modeled and simulated by Simulink/MATLAB for the case study. All converters for RESs and BESSs in the system consisted of two-level half-bridge (HB) converters with switch models. Sine-pulse width modulation (SPWM) with switching frequency of 2 kHz was adopted to generate gate signals of switch model converters. The PV generator model consisted of photovoltaic source and converter [22]. The PV converter was controlled by the MPPT algorithm and inner dc voltage and current control loop. For MPPT, the perturbation and observation (P&O) algorithm in [23] was utilized and reference value of dc voltage was determined. To control the dc voltage of photovoltaic source as the reference value, the nested dc voltage and current control loops in [17] were used. On the other hand, the wind generator consisted of a permanent magnetic synchronous generator (PMSG), wind turbine, machine side converter, and grid side converter [22]. We utilized a permanent magnet synchronous generator model provided by Simulink/MATLAB. Wind turbine and controllers of the machine and grid side converters were modelled by the same procedure of [17]. The detailed parameters of the diesel generators and BESSs are listed in Tables 2 and 3. For voltage control, conventional reactive power (Q)—magnitude of the ac voltage (V_{ac}) droop method was adopted in diesel generators and BESSs [24]. Q–V_{ac} droop constants for three diesel generators were 0.05/20 ($V_{p.u}$/kVar) respectively and Q–V_{ac} droop constant for BESS 1 was 0.05/250 ($V_{p.u.}$/kVar) and for BESS 2 was 0.05/125 ($V_{p.u.}$/kVar). As the reactive power and magnitude of ac voltage were not the major concern of this paper, we just adopted the conventional method in [24].

The capacity of BESSs was scaled down from its real values, 500 and 300 kWh, by a factor of 1/100 in the simulation to illustrate the variation of SOCs clearly. The supplementary controllers of diesel generators have a sample time of 0.5 s to ensure slower response compared to the droop controller. Other components including transformers, lines, and loads were modeled with the parameters shown in Figure 2. Considering the fact that the efficiency of batteries is dependent on SOC level [25,26], we assumed that BESS 1 and 2 have the highest efficiency at 60% and 40% of SOCs, respectively.

Table 2. Control parameters in the conventional and proposed method.

Control Method	Type	Parameter	Symbol	Value
Conventional Method	Diesel Generators	Droop gain	$R_d(1)$, $R_d(2)$, $R_d(3)$	4.8/150 Hz/kW
		PI gain for suppl. control	$k_{P,sup}(1) + k_{I,sup}(1)/s$, $k_{P,sup}(2) + k_{I,sup}(2)/s$	60,000 + 45,000/s
		Participation factor	$Pf(1)$, $Pf(2)$, $Pf(3)$	1/3
	BESSs	Droop gain	$R_b(1)$, $R_b(2)$	0.6/250, 0.4/150 Hz/kW
Proposed Method	Diesel Generators	Droop gain	$R_d(1)$, $R_d(2)$, $R_d(3)$	5/150%/kW
		PI gain for suppl. control	$k_{P,d} + k_{I,d}/s$,	30,000 + 22,500/s
		Participation factor	$Pf(1)$, $Pf(2)$, $Pf(3)$	1/3
	BESSs	Droop gain	$R_b(1)$, $R_b(2)$	0.6/250, 0.4/150 Hz/kW
		PI gain for suppl. control	$k_{P,s}(1) + k_{I,s}(1)/s$, $k_{P,s}(2) + k_{I,s}(2)/s$	0 + 500/s, 0 + 100/s
		PI gain for self SOC control	$k_{P,SSC}(1) + k_{I,SSC}(1)/s$, $k_{P,SSC}(2) + k_{I,SSC}(2)/s$	20 + 5/s
		SOC reference value	$SOC_{ref}(1)$, $SOC_{ref}(2)$	60%, 40%

Table 3. Parameters of diesel generators and BESSs for the internal controllers.

Type	Parameter	Symbol	Value
Diesel Generators	Valve-actuator time constant	$T_v(1)$, $T_v(2)$, $T_v(3)$	0.05 s
	Diesel-engine time constant	$T_d(1)$, $T_d(2)$, $T_d(3)$	0.5 s
BESSs	Filter components	$L_f(1)$, $L_f(2)$, $C_f(1)$, $C_f(2)$	0.14 mH, 0.25 mH 4.8 mF, 3.4 mF
	Voltage-control PI gain	$k_{P,v}(1)+ k_{I,v}(1)/s$, $k_{P,v}(2)+ k_{I,v}(2)/s$	3 + 500/s, 3 + 1000/s
	Current-control PI gain	$k_{P,c}(1) + k_{I,c}(1)/s$, $k_{P,c}(2)+ k_{I,c}(2)/s$	32 + 300/s, 35 + 200/s
	Rated battery capacity	$C_{rate}(1)$, $C_{rate}(2)$	5000/710, 3000/650 Ah
	Nominal battery voltage	$V_{dc,rate}(1)$, $V_{dc,rate}(2)$	710, 650 V
	Initial SOC	$SOC^0(1)$, $SOC^0(2)$	60%, 40%

5.1. Case 1: Step Load Change with the Conventional and Proposed Control Method

In Case 1, to verify the effectiveness of the proposed strategy, the proposed method was compared to the CD method for the step load change. As CD method can be adopted in the system with multiple diesel generators and BESSs, has been widely adopted in many researches [27,28], and was the original control method in Geocha Island, we selected CD method for comparison. Before 10 s, all loads except the load at bus 7 were connected. After 10 s, the load at bus 7 was connected to the system. The active power consumption for loads was constant, as shown in Figure 2. The active power generated by the PV and wind generators was constant at 80 and 75 kW respectively. SOC reference values for BESS 1 and 2 were 60% and 40%, respectively.

Figure 10 shows the responses of frequency and SOC_{eq} in the conventional and proposed methods. As shown in Figure 10a, the frequency eventually maintains its rated value irrespective of the control method. However, while the conventional control method takes approximately 20 s to restore frequency, the proposed control takes less than 2 s, as shown in Figure 10a. BESSs are exploited for frequency restoration in the proposed scheme, while diesel generators are used for the conventional method. For this reason, the performance of frequency regulation is much better in the proposed method than the conventional method. Furthermore, as shown in Figure 10b, SOC_{eq} can be constant in both methods because long-time-scale energy imbalances are compensated for by diesel generators. However, as SOC management is not considered in the conventional method, SOC_{eq} deviates from the reference value. On the other hand, SOC_{eq} is regulated at the reference value by using the proposed strategy. This implies that BESSs with a relatively small capacity are required with the proposed method.

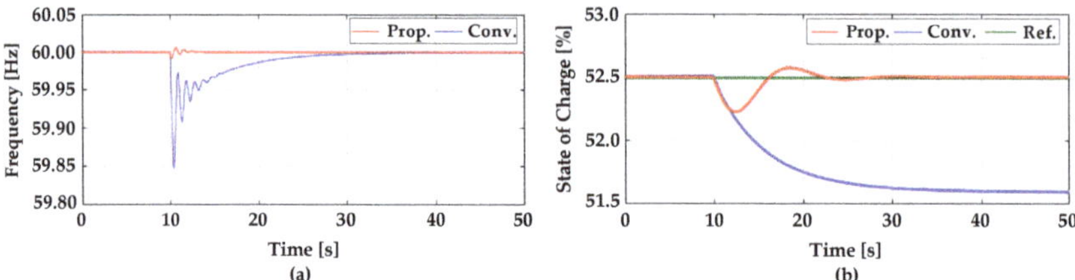

Figure 10. Frequency and SOC_{eq} in the conventional and proposed methods: (**a**) frequency; (**b**) SOC_{eq}.

Figure 11 shows the active power of diesel generators and BESSs in the conventional and proposed methods. In both methods, diesel generators share the load equally at the desired power-sharing ratio, as shown in Figure 11a, and BESSs make zero output equally as desired, as shown in Figure 11b. In the proposed method, parallel operation and power sharing of diesel generators and BESSs are possible, as desired.

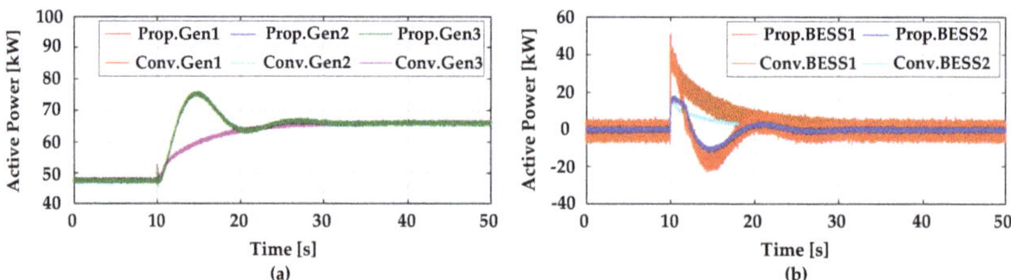

Figure 11. Active power in the conventional and proposed methods: (**a**) diesel generators; (**b**) BESSs.

5.2. Case 2: Step Load Change with and without Self SOC Controller (SSC)

In Case 2, the loads, PV generation, and wind generation are the same as in Case 1. To verify the necessity of SSC, the responses of BESSs with and without SSC were compared. SOC reference values for BESS 1 and 2 were 60% and 40%, respectively. Figure 12 shows the frequency and SOC_{eq} with and without SSC. In both cases, the responses of the frequency and SOC_{eq} are almost identical, which implies that, except for the individual SOCs of BESSs, the SSC hardly affects the performance of the system.

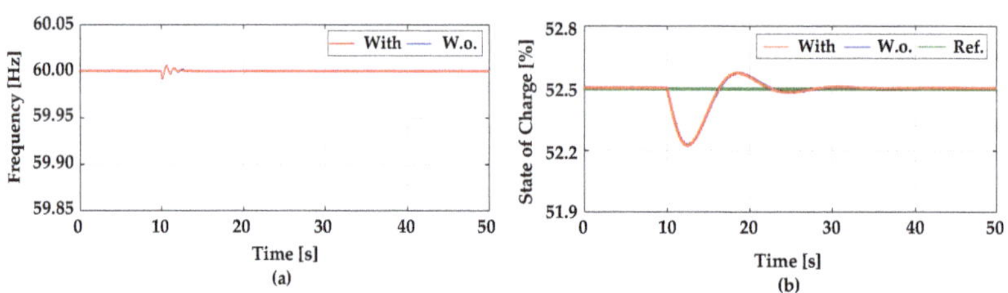

Figure 12. Frequency and SOC_{eq} of BESSs in Case 2: (**a**) frequency; (**b**) SOC_{eq}.

Figure 13 shows the individual SOCs of BESSs in the proposed control with and without SSC. Although SOC_{eq} is regulated at the reference value irrespective of whether SSC is applied, the SOC of each BESS in the proposed control with and without SSC has different responses. In the case without SSC, the SOC of each BESS deviates from its corresponding reference value. This phenomenon is similar to the occurrence of a circulating current, in which summation is equal to the desired value but individual components have different sign [29]. In the proposed control with SSC, the SOC of each

BESS can be restored to its reference value. Thus, SSC is necessary to maintain the individual SOCs of BESSs. Furthermore, as SOCs of BESS 1 and 2 are maintained as 60% and 40% respectively, we can utilize BESSs at their highest efficient points with SSC.

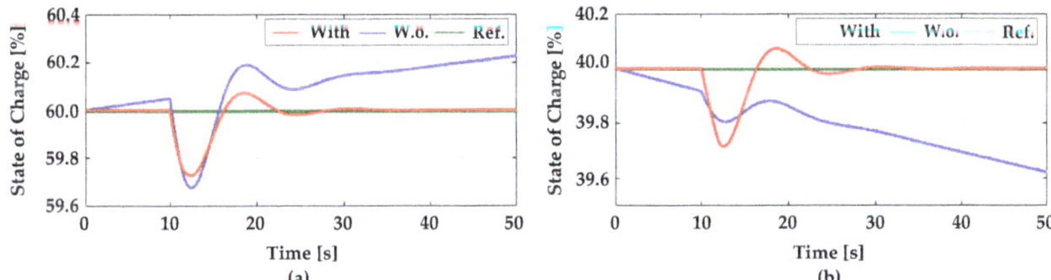

Figure 13. SOC of each BESS in Case 2: (**a**) BESS 1; (**b**) BESS 2.

5.3. Case 3: Fluctuation of Renewable Energy Sources (RESs) with the Conventional and Proposed Control Method

In Case 3, to verify the robustness of the system with the proposed control, fluctuations of the outputs of RESs were simulated when applying the proposed and conventional CD methods. The loads were constant, as shown in Figure 2. SOC reference values for BESS 1 and 2 were 60% and 40%, respectively. PV generation and wind generation were varied, as shown in Figure 14.

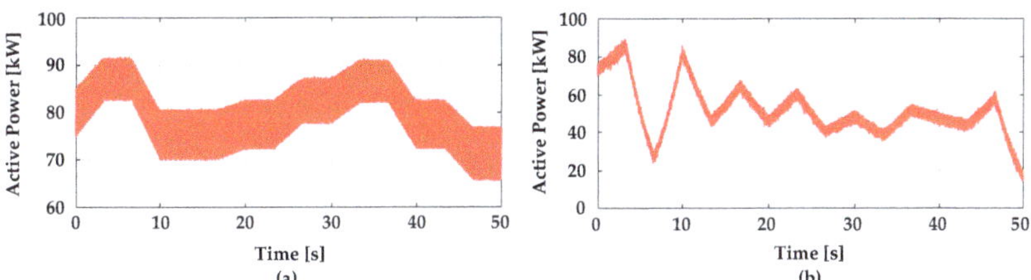

Figure 14. Variation of the active power of photovoltaic (PV) and wind generation in Case 3: (**a**) PV; (**b**) wind.

Figure 15 shows the frequency and SOC_{eq}. As shown in Figure 15a, the frequency is fluctuated by the intermittent outputs of RESs with the conventional method. However, with the proposed method, the frequency is near the nominal value and can be controlled quickly. In addition, as shown in Figure 15b, while SOC_{eq} deviates from the reference value with the conventional method, the proposed method maintains SOC_{eq} near the reference value with a small fluctuation, which was due to the slow response of diesel generators. These results imply that the proposed method can make the system robust and improve its reliability, even with a high penetration of RESs.

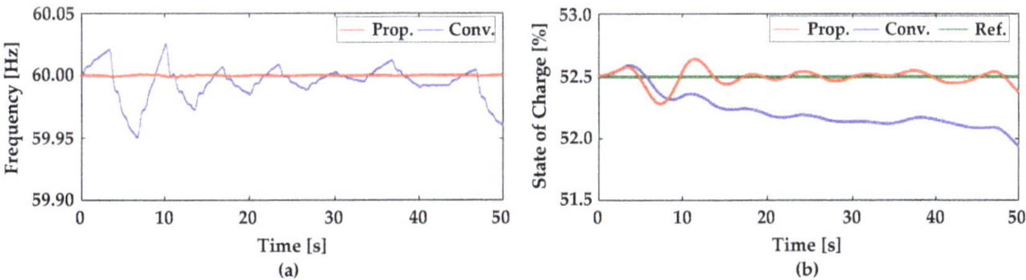

Figure 15. Frequency and SOC_{eq} in Case 3: (**a**) frequency; (**b**) SOC_{eq}.

Figure 16 shows the individual SOCs of BESSs. As shown in Figure 16, while the SOC of each BESS deviates from the reference value in the conventional method, the SOC of each BESS is near the reference value in the proposed method, similar to SOC_{eq} in Figure 15b. Thus, the proposed method can manage not only SOC_{eq} but also the SOC of each BESS. Even if a large power fluctuation occurred in the system, the individual SOCs of BESSs can be maintained and bounded tightly. This implies that batteries with a relatively small capacity are required, even with a high penetration of RESs, when the proposed method is adopted.

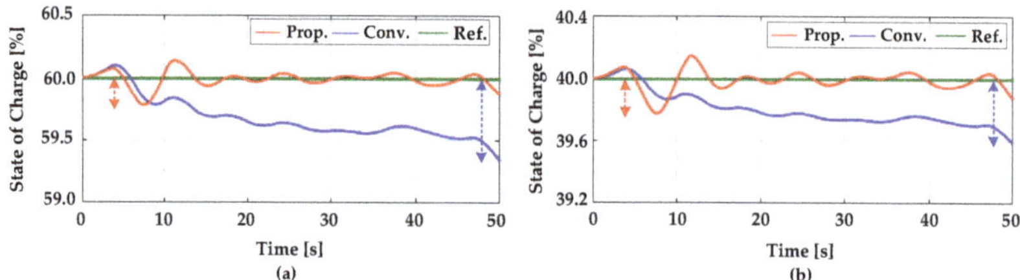

Figure 16. SOC of each BESS in Case 3: (**a**) BESS 1; (**b**) BESS 2.

5.4. Case 4: Consideration of Communication Delays

In Case 4, we verify the performance of the proposed method when communication delays are considered. When diesel generators control SOC_{eq} and SSCs of BESSs are operated, the information of SOC_{eq} is required. Because capacities of BESSs are constant, only simple communication lines for transmitting information of SOCs are required. In the case of Geocha Island microgrid, diesel generators and BESSs are located in the same station as shown in Figure 3. For this reason, communication delay can be ignored in Geocha Island microgrid. However, to adopt the proposed control method in other islanded microgrids where BESSs and diesel generators are far away from each other, we have to verify the performance of the proposed method when communication delays exist. The communication delays range from several milliseconds to about 100 ms in the wide area measurement/monitoring system for the large scale power system [30], hence, communication delays in islanded microgrids may be smaller than 100 ms. To verify the possibility for application of the proposed method, we tested the system when communication delays were 0, 50, and 100 ms. All other conditions are the same as Case 1. Figure 17 shows frequency and SOC_{eq} of the proposed method with 0, 50, and 100 ms delay when load change occurred at 10 s.

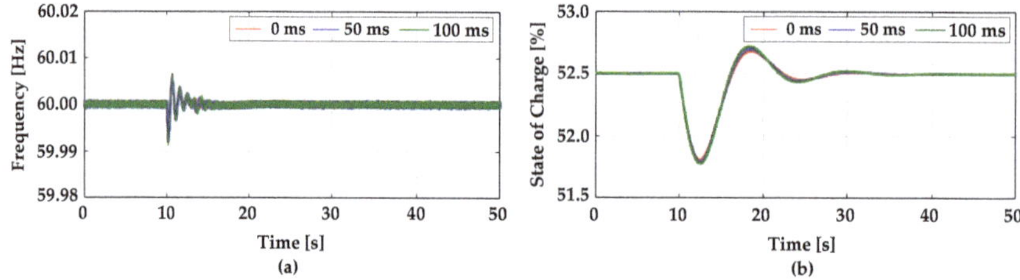

Figure 17. Frequency and SOC_{eq} in Case 4: (**a**) frequency; (**b**) SOC_{eq}.

As shown in Figure 17, if communication delays are prolonged, saturation time of frequency and SOC_{eq} takes a little longer. However, regardless of communication delays, frequency and SOC_{eq} are eventually saturated. To ensure that individual SOCs of BESSs can be controlled even with communication delays, Figure 18 shows individual SOCs of BESSs.

As shown in Figure 18, if communication delays are prolonged, SOCs of BESSs are a little more slowly saturated. But, SOCs of BESSs can be regulated as the reference values regardless of communication delays. Therefore, the proposed method can be implemented in other islanded microgrids where communication delays should be considered.

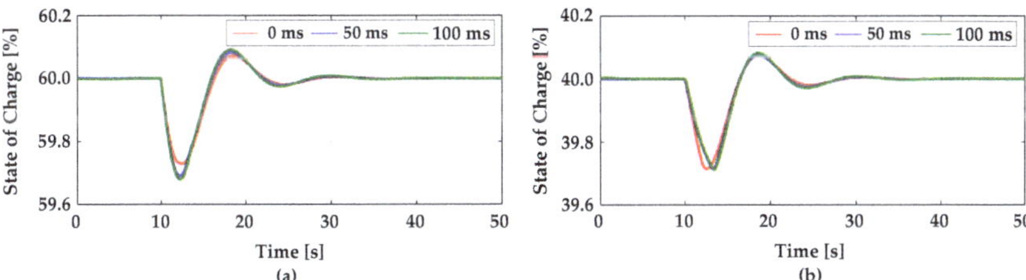

Figure 18. SOC of each BESS in Case 4: (**a**) BESS 1; (**b**) BESS 2.

5.5. Case 5: Regulating Active Power of BESSs by the Linear Time-Varying Control of SOCs

In Case 5, to verify that the active power outputs of BESSs can be controlled as the desired values, we provided continuous responses of BESSs and diesel generators for linear time-varying control of SOCs of which reference values were determined by the active power references. All loads and wind and PV generation are the same as Case 1. Before 30 s, active power references, $P^*_{BESS}(1)$ and $P^*_{BESS}(2)$ were 20 and −10 kW (negative sign means that BESS was charged), respectively. After 30 s, $P^*_{BESS}(1)$ and $P^*_{BESS}(2)$ were 30 and 10 kW. According to determining active power references, SOC reference values were decided as shown in (11). Figure 19 shows SOCs and SOC reference values of BESSs.

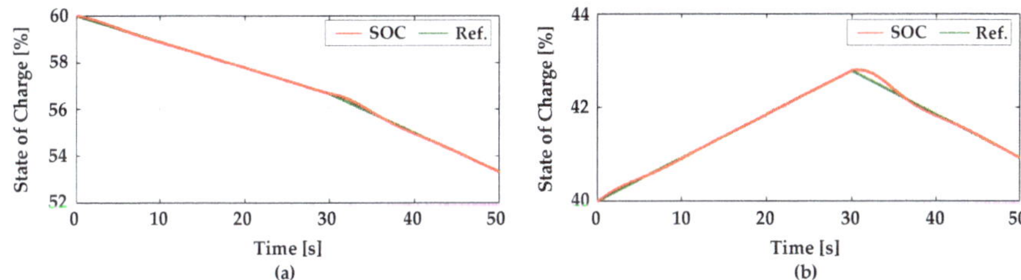

Figure 19. SOCs and SOC reference values of BESSs in Case 5: (**a**) BESS1; (**b**) BESS2.

As shown in Figure 19, by the active power reference values, the linear time-varying SOC reference values are determined. The slopes of SOC reference values are varied when the active power references are changed. Regardless of the slopes of SOC reference values, the SOCs of BESSs can be controlled as the reference values. To verify that frequency and active power outputs of BESS are controlled, Figure 20 shows the frequency and active power outputs of BESSs.

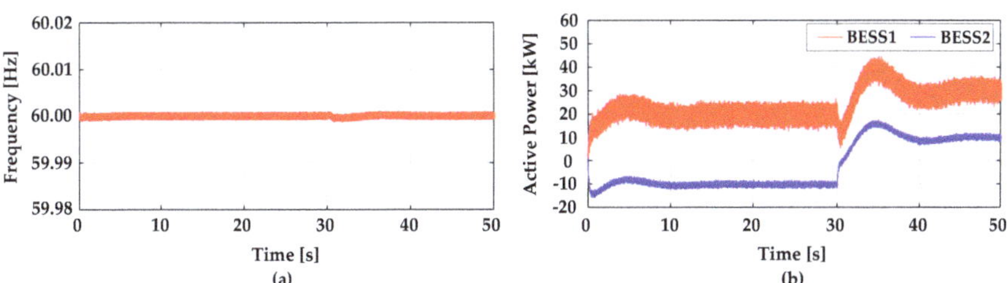

Figure 20. Frequency and active power outputs of BESSs in Case 5: (**a**) frequency; (**b**) active power outputs of BESSs.

Frequency can be regulated as the nominal value, as shown in Figure 20a. Although the active power references of BESSs are changed, frequency maintains the nominal value with just small fluctuation. As described in Figure 20b, active power outputs of BESSs are near the reference values. Case 5 implies that by adopting the proposed control method, frequency and active power outputs of BESSs can be controlled simultaneously as the desired values.

6. Conclusions and Future Work

This paper proposed a coordinated frequency and SOC control method for implementing isolated microgrids having a high penetration of RESs with multiple diesel generators and BESSs. The stored energy of all BESSs can be managed with the hierarchical controllers in diesel generators. In addition, frequency can be controlled quickly and individual SOCs of BESSs can be managed at their reference values with the hierarchical controllers and SSCs in BESSs. From Case Study, we validated that the proposed method can give a chance to (1) lower the requirement for the capacity of batteries, (2) utilize BESSs at their highest efficiency, (3) maintain the active power of BESSs for scheduled or dispatched values, and (4) operate the system robustly even with large fluctuation of RESs.

One of the future works is the magnitude of ac voltage and reactive power control. If effective voltage/reactive power control method is coordinated with our proposed method, more stable and efficient operation of islanded microgrids will be implemented.

Author Contributions: J.-W.C. organized the manuscript and wrote the manuscript; G.-S.L. conceived the main idea; H.-J.M. performed the simulations; M.B.G. gave some useful information for the manuscript; S.-I.M. provided the overall guidance for the manuscript.

Funding: This research received no external funding.

Acknowledgments: This work was supported by the Korea Institute of Energy Technology Evaluation and Planning (KETEP) and the Ministry of Trade, Industry and Energy (MOTIE) of the Republic of Korea (No. 20188550000410).

Conflicts of Interest: The authors declare no conflict of interest.

References

1. Pashajavid, E.; Ghosh, A. Frequency support for remote microgrid systems with intermittent distributed energy resources—A two-level hierarchical strategy. *IEEE Syst. J.* **2014**, *12*, 2760–2771. [CrossRef]
2. Mohamed, S.; Shaaban, M.F.; Ismail, M.; Serpedin, E.; Qaraqe, K. An efficient planning algorithm for hybrid remote microgrids. *IEEE Trans. Sustain. Energy* **2018**, *10*, 257–267. [CrossRef]
3. Etxegarai, A.; Equia, P.; Torres, E.; Iturregi, A.; Valverde, V. Review of grid connection requirements for generation assets in weak power grids. *Renew. Sustain. Energy Rev.* **2015**, *41*, 1501–1514. [CrossRef]
4. Kim, Y.S.; Hwang, C.S.; Kim, E.S.; Cho, C. State of charge based active power sharing method in a standalone microgrid with high penetration level of renewable energy sources. *Energies* **2016**, *9*, 480. [CrossRef]
5. Liu, Y.; Du, W.; Xiao, L.; Wang, H.; Cao, J. A method for sizing energy storage system to increase wind penetration as limited by grid frequency deviations. *IEEE Trans. Power Syst.* **2016**, *31*, 729–737. [CrossRef]
6. Diaz, N.L.; Luna, A.C.; Vasquez, J.C.; Guerrero, J.M. Centralized control architecture for coordination of distributed renewable generation and energy storage in islanded ac microgrids. *IEEE Trans. Power Electron.* **2017**, *32*, 5202–5213. [CrossRef]
7. Kim, Y.S.; Kim, E.S.; Moon, S.I. Frequency and voltage control strategy of standalone microgrids with high penetration of intermittent renewable generation systems. *IEEE Trans. Power Syst.* **2016**, *31*, 718–728. [CrossRef]
8. Delille, G.D.; Francois, B.; Malarange, G. Dynamic frequency control support by energy storage to reduce the impact of wind and solar generation on isolated power system's inertia. *IEEE Trans. Sustain. Energy* **2012**, *3*, 931–939. [CrossRef]
9. Jiang, Q.; Hong, H. Wavelet-based capacity configuration and coordinated control of hybrid energy storage system for smoothing out wind power fluctuations. *IEEE Trans. Power Syst.* **2013**, *28*, 1363–1372. [CrossRef]

10. Xiao, J.; Bai, L.; Li, F.; Liang, H.; Wang, C. Sizing of energy storage and diesel generators in an isolated microgrid using discrete fourier transform (DFT). *IEEE Trans. Sustain. Energy* **2014**, *5*, 907–916. [CrossRef]
11. Singh, B.; Solanki, J. Compensation for diesel generator-based isolated generation system employing DSTATCOM. *IEEE Trans. Ind. Appl.* **2011**, *5*, 907–916. [CrossRef]
12. Moon, H.J.; Kim, Y.J.; Chang, J.W.; Moon, S.I. Decentralised active power control strategy for real-time power balance in an isolated microgrid with an energy storage system and diesel generators. *Energies* **2019**, *12*, 511. [CrossRef]
13. Gao, L.; Liu, S.; Dougal, R.A. Dynamic lithium-ion battery model for system simulation. *IEEE Trans. Compon. Packag. Technol.* **2002**, *25*, 495–505.
14. Rocabert, J.R.; Luna, A.L.; Blaabjerg, F.; Rodriquez, P. Control of power converters in ac microgrids. *IEEE Trans. Power Electron.* **2012**, *27*, 4734–4749. [CrossRef]
15. Kundur, P. *Power System Stability and Control*; McGraw-Hill: New York, NY, USA, 1994.
16. Ogata, K. *Modern Control Engineering*; Prentice Hall: Upper Saddle River, NJ, USA, 2009.
17. Yazdami, A.; Iravani, R. *Voltage-Sourced Converters in Power System*; Willy Press: Hoboken, NY, USA, 2010.
18. Kim, Y.S.; Kim, E.S.; Moon, S.I. Distributed generation control method for active power sharing and self-frequency recovery in an islanded microgrid. *IEEE Trans. Power Syst.* **2017**, *32*, 544–551. [CrossRef]
19. Kosari, M.; Hosseinian, S.H. Decentralized reactive power sharing and frequency restoration in islanded microgrid. *IEEE Trans. Power Syst.* **2017**, *32*, 2901–2912. [CrossRef]
20. IEEE. *IEEE Recommended Practice for Protection and Coordination of Industrial and Commercial Power Systems*; IEEE: Piscataway, NJ, USA, 1986.
21. Bergen, A.R. *Power System Analysis*; Prentice-Hall: Englewood Cliffs, NJ, USA, 1986.
22. Teodorescu, R.; Liserre, M.; Rodriguez, P. *Grid Converters for Photovoltaic and Wind Power Systems*; John Wiley & Sons: Chichester, UK, 2011.
23. Femia, N.; Petrone, G.; Spagnuolo, G.; Vitelli, M. Optimization of perturb and observe maximum power point tracking method. *IEEE Trans. Power Electr.* **2005**, *20*, 963–973. [CrossRef]
24. Paquette, A.D.; Reno, M.J.; Harley, R.G.; Divan, D.M. Sharing transient loads: Causes of unequal transient load sharing in islanded microgrid operation. *IEEE Ind. Appl. Mag.* **2014**, *20*, 22–34. [CrossRef]
25. Badin, F.; Le Berr, F.; Briki, H.; Dabadie, J.-C.; Petit, M.; Magand, S.; Condemine, E. Evaluation of EVs energy consumption influencing factors, driving conditions, auxiliaries use, driver's aggressiveness. In Proceedings of the 2013 World Electric Vehicle Symposium and Exhibition (EVS27), Barcelona, Spain, 17–20 November 2013; pp. 1–12.
26. Ng, K.S.; Moo, C.S.; Chen, Y.P.; Hsieh, Y.C. Enhanced coulomb counting method for estimating state-of-charge and state-of-health of lithium-ion batteries. *Appl. Energy* **2009**, *86*, 1506–1511. [CrossRef]
27. Delille, G.; Francois, B.; Malarange, G. Dynamic frequency control support: A virtual inertia provided by distributed energy storage to isolated power systems. In Proceedings of the IEEE Innovative Smart Grid Technologies Europe Conference (ISGT), Gothenburg, Sweden, 11–13 October 2010.
28. Wang, Y.; Delille, G.; Guillaud, X.; Colas, F.; Francois, B. Real-time simulation: The missing link in the design process of advanced grid equipment. In Proceedings of the IEEE PES General Meeting, Minneapolis, MN, USA, 25–29 July 2010.
29. Pan, C.T.; Liao, Y.H. Modeling and coordinate control of circulating currents in parallel three-phase boost rectifiers. *IEEE Trans. Ind. Appl.* **2007**, *54*, 825–838. [CrossRef]
30. Xie, X.; Xin, Y.; Xiao, J.; Wu, J.; Han, Y. WAMS application in Chinese power system. *IEEE Power Energy Mag.* **2006**, *4*, 54–63.

© 2019 by the authors. Licensee MDPI, Basel, Switzerland. This article is an open access article distributed under the terms and conditions of the Creative Commons Attribution (CC BY) license (http://creativecommons.org/licenses/by/4.0/).

Article

A Control Strategy of Modular Multilevel Converter with Integrated Battery Energy Storage System Based on Battery Side Capacitor Voltage Control

Zhe Wang, Hua Lin * and Yajun Ma

State Key Laboratory of Advanced Electromagnetic Engineering and Technology, Huazhong University of Science and Technology, Wuhan 430074, China; zhe_wang@hust.edu.cn (Z.W.); mayajun@mail.hust.edu.cn (Y.M.)
* Correspondence: lhua@mail.hust.edu.cn; Tel.: +86-135-4423-772

Received: 17 April 2019; Accepted: 3 June 2019; Published: 5 June 2019

Abstract: A modular multilevel converter with an integrated battery energy storage system (MMC-BESS) has been proposed for high-voltage applications for large-scale renewable energy resources. As capacitor voltage balance is key to the normal operation of the system, the conventional control strategy for the MMC can be significantly simplified by controlling the individual capacitor voltage through a battery side converter in the MMC-BESS. However, the control strategy of the MMC-BESS under rectifier mode operation has not yet been addressed, where the conventional control strategy cannot be directly employed due to the additional power flow of batteries. For this defect, the rectifier mode operation of the MMC-BESS based on a battery side capacitor voltage control was analyzed in this paper, proposing a control strategy for this application scenario according to the equivalent circuit of MMC-BESS, avoiding passive impact on the state-of-charge (SOC) equalization of batteries. Furthermore, the implementation of a battery side converter control is proposed by simplifying the capacitor voltage filter scheme within phase arm, which enhances its performance and facilitates the realization of control strategy. Finally, simulation and experimental results validate the feasibility and effectiveness of the proposed control strategy.

Keywords: renewable energy; battery energy storage system (BESS); control strategy; modular multilevel converter; state-of-charge (SOC) equalization

1. Introduction

In recent years, the grid-connected applications of large-scale renewable energy resources have gradually become a trend, presenting new challenges to the power electronics converters applied in high-voltage and high-power fields [1]. Multilevel topologies can reduce the requirement of voltage rate of switching devices, achieving higher voltage levels [2]. Among existing multilevel topologies, a modular multilevel converter (MMC) is considered the most promising topology and has been extensively studied in the past decade [3–5]. Modular multilevel converters share the common advantages of multilevel converters, such as low harmonic content of output voltage and small voltage stress of switching devices. Compared with the cascaded H-bridge (CHB) topology with the modular structure, the MMC topology has only half of the arm current at the same power rate. There also exists a common DC-link in the MMC topology which is unavailable in CHB topology [6]. Based on this consideration, MMCs can be utilized as AC/DC interlinking converters in medium- and high-voltage renewable energy generation systems, e.g., offshore wind farm generation systems [7] and microgrids [8], as shown in Figure 1.

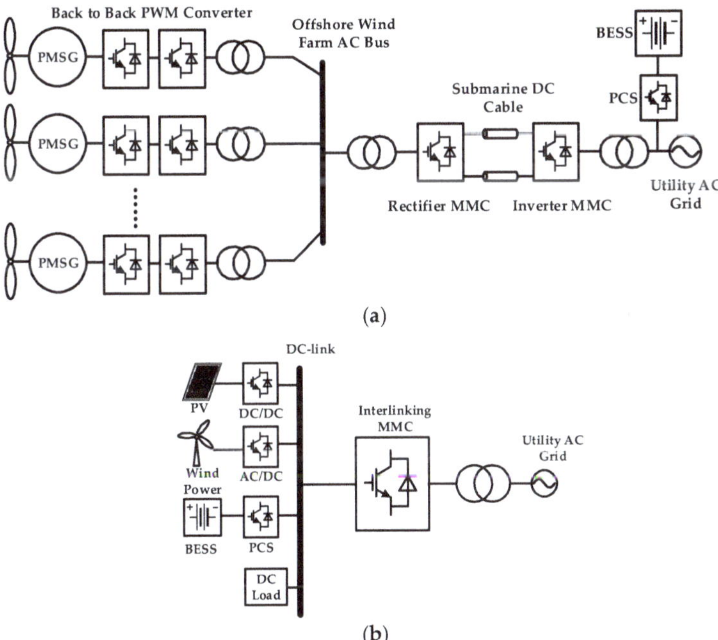

Figure 1. Applications of modular multilevel converter (MMC) topology with large-scaled renewable energy sources. (**a**) Offshore wind farm generation system based on two-terminal MMCs. (**b**) DC microgrid using MMC as interlinking converter.

Due to the characteristics of randomness and intermittency of renewable energy resources, the normal operation and power quality of the power system would be remarkably affected, reducing the voltage and frequency stability. To attenuate the passive impact caused by renewable energy resources, a battery energy storage system (BESS) is a reasonable and efficient solution for grid-connected renewable energy generation systems, as shown in Figure 1a,b. Recently, to simplify the configuration of BESS in the scenarios of MMC-based applications, the MMC with integrated BESS is proposed by combining the MMC and BESS together [9]. The extra power conversion system (PCS) of BESS can be saved by this combination. In this paper, this topology is abbreviated as the MMC-BESS. The integration was implemented by inserting battery cells into each submodule (SM) of MMC directly or through a DC/DC interface [10,11]. Due to this distributed integration mode of BESS, the state-of-charge (SOC) equalization of each battery was facilitated, improving the effective utilization rate of BESS compared with the centralized scheme at the DC-link of MMC [12].

In addition to the inherent advantages provided by BESS for the power system, the integration scheme of batteries through DC/DC interfaces provided and additional degree of freedom (DOF) to the system control strategy. As the capacitor voltage balancing control was significant to conventional MMCs, the cascaded control structure for balancing the capacitor voltage in the phase-, arm- and individual SM-level is reported in Reference [13], which was consequently employed in the MMC-BESS in Reference [14]. Nevertheless, with the additional DOF provided by batteries in the MMC-BESS, the capacitor voltage balancing control can be significantly simplified through battery side control, which makes individual SM capacitor behave as a voltage source to the MMC, avoiding the cascaded control structure and relatively complicated capacitor voltage balancing algorithms introduced in Reference [4]. It is worth noting that the premise of the battery side capacitor voltage control was that each SM must contain a battery module, otherwise the capacitor voltage would be incontrollable.

Based on the battery side capacitor voltage control, some studies have been reported. In Reference [11], the control strategy of a MMC-BESS operating as an inverter was researched, proposing the three-level SOC equalization control structure. In Reference [15], the SOC equalization of MMC-BESS considering the capacity inconsistency of batteries was proposed. In References [10,16], the MMC-BESS was applied in battery electric vehicles and the efficiency of this topology was assessed. Furthermore, the control strategy of MMC-BESS under AC and DC fault has been studies in Reference [17], realizing SOC equalization under fault conditions. Nevertheless, the current research is not deep enough, as the existing control strategies merely focus on the inverter mode operation of MMC-BESS. Usually, in MMC-based multi-terminal DC transmission system (e.g., Figure 1a), one of the MMCs must be responsible for DC-link voltage control while the other MMCs should be responsible for power control [18]. Similarly, in Figure 1b, the interlinking MMC should take the DC-link voltage as the control objective. Accordingly, the rectifier mode operation of MMC-BESS is required. However, the DC-link voltage control structure in conventional MMCs [18] cannot be directly employed in the MMC-BESS due to the additional power flow of BESS, which has not been investigated in detail in the current literature. As the battery side converter is employed to control the capacitor voltage, the SOC equalization must be achieved by MMC side control. Hence, the combination of SOC equalization and DC-link voltage control is mandatory in the MMC-BESS. Besides, as the capacitor voltage is controlled by battery side converter, the performance of the system would be greatly dependent on the battery side control strategy. As the individual capacitor voltage contains ripples at multiple frequencies while the battery current is expected to be pure DC component, the battery side control strategy should be investigated to ensure the performance both in a steady and dynamic state.

In this paper, based on battery side capacitor voltage control, the control strategy of rectifier mode operation was analyzed according to the equivalent circuit of the MMC-BESS. The DC-link voltage control was combined with the SOC equalization control, different from the control strategy applied in conventional MMCs. Furthermore, a simplified capacitor voltage filter scheme was proposed in this paper to facilitate the implementation of battery side control strategy.

The rest of this paper is organized as follows. Section 2 describes the basic configuration and the equivalent circuit of the MMC-BESS. Then the control strategy of rectifier mode operation of MMC-BESS is proposed in Section 3. To enhance the performance of battery side control strategy, a simplified capacitor voltage filter scheme is proposed in Section 4. Finally, the effectiveness and feasibility of the proposed control strategy of rectifier mode operation are validated by simulations and experimental results in Section 5.

2. Mathematic Model and Principles of MMC-BESS

The topology of three-phase MMC-BESS is shown in Figure 2. Each phase leg is comprised of $2N$ SMs in series, where the midpoint connected to the AC side divides a phase leg into two phase arms. Here, N represents the number of SMs in series per phase arm. Throughout this paper, the subscript k = (a, b, c) refers to the individual phase; j = (p, n) refers to the upper and lower arms; i = (1, 2 ... N) refers to the individual SM within phase arm. The DC-link voltage and grid phase-voltage are denoted as V_{dc} and v_{gk}, respectively. Each phase arm contains an arm inductor L_s with losses denoted by R_s (not shown in the figure). In general, the topology of the MMC-BESS is consistent with conventional MMCs. The main difference locates on the topology of SM. In addition to the half-bridge structure with the SM capacitor, BESS is integrated into each SM in different ways, which forms two types of SM as shown in Figure 2. For SM type A, the battery is directly connected to the terminals of individual SM capacitors; for SM type B, the battery is connected to individual SM capacitors through a DC/DC interface, while L_b is the inductor of this DC/DC converter. In the authors' view, SM type A compels the capacitor voltage to follow the battery voltage, inducing the power fluctuation caused by ripples in capacitor voltage into the battery which would impair the health or shorten the lifespan of an individual battery. On the contrary, the DC/DC interface in SM type B can decouple the battery side and MMC side through the SM capacitor, preventing the ripples from flowing into the battery through an appropriate control strategy, but it would comparatively degrade the power conversion efficiency due to the DC/DC interface. Nevertheless, the DC/DC interface provides a degree of freedom for the system control which SM type *A* does not, and the requirement of the terminal voltage of the battery module can be greatly lower than the rated capacitor voltage. Besides, it would not affect the original power conversion efficiency between AC and DC side of MMC itself. Hence, SM type *B* was employed in this paper. It should be noted that the topology of the DC/DC interface in SM type *B* was not restricted to a non-isolated buck–boost converter as shown in Figure 2, which was adopted in this paper for the convenience and simplicity.

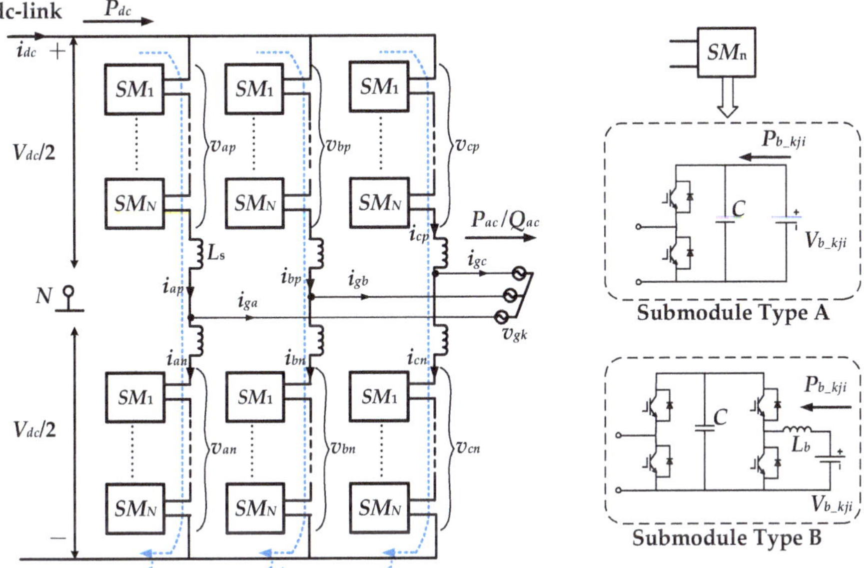

Figure 2. Topology of the three-phase MMC- battery energy storage system (BESS).

The MMC-BESS can be modelled identically to a conventional MMC as follows [19]:

$$\begin{cases} v_{sk} = \frac{R_s}{2} i_{gk} + \frac{L_s}{2} \frac{di_{gk}}{dt} + v_{gk} \\ v_{ck} = L_s \frac{di_{ck}}{dt} + R_s i_{ck} \end{cases} \quad (1)$$

where v_{sk} is the voltage required to drive the AC output current i_{gk}, and v_{ck} is the voltage required to drive the difference current i_{ck}. The difference current i_{ck} is a circulating current flowing through upper and lower arms within one phase simultaneously, which would not affect the AC side.

The upper and lower arm currents were composed of a circulating current and a half of AC output current as [19]:

$$\begin{cases} i_{kp} = \frac{i_{gk}}{2} + i_{ck} \\ i_{kn} = -\frac{i_{gk}}{2} + i_{ck} \end{cases} \quad (2)$$

The difference current can contain component at any frequency, but only DC and fundamental frequency components are necessary for power conversion in the MMC-BESS [14]. Thus, assuming:

$$i_{ck} = I_{dck} + I_{1k} \cos(\omega t + \varphi_{1k}) \quad (3)$$

where I_{dck} denotes the DC component; I_{1k} and φ_{1k} are the amplitude and angle of fundamental component in phase k; ω is the fundamental angular frequency of AC side.

By applying Kirchhoff's Voltage Laws(KVL) to Figure 2, the output voltages of the upper and lower arms can be yielded as

$$\begin{cases} v_{kp} = \frac{V_{dc}}{2} - v_{sk} - v_{ck} \\ v_{kn} = \frac{V_{dc}}{2} + v_{sk} - v_{ck} \end{cases} \tag{4}$$

Due to the battery power existing in the system, the following relationship is satisfied throughout neglecting losses:

$$P_{ac} = P_{dc} + \sum_{k}^{a,b,c} \sum_{j}^{p,n} \sum_{i=1}^{N} P_{b\_kji} \tag{5}$$

where P_{ac} and P_{dc} are total active powers of the AC and DC side, respectively, while P_{b_kji} is the battery power of individual SM. Note that the power directions are all in accordance with Figure 2 in this paper. To compensate or absorb power fluctuations of the AC or DC side of the MMC-BESS, the individual battery would discharge when $P_{b\_kji} > 0$, vice versa.

Assuming the grid voltage and arm current of phase k in time domain are expressed as:

$$\begin{cases} v_{gk} = V_{gk} \cos \omega t \\ i_{gk} = I_{gk} \cos(\omega t + \varphi_k) \end{cases} \tag{6}$$

where I_{gk} and φ_k are amplitude and angle of grid current; V_{gk} is the amplitude of the grid voltage. Substituting Equation (6) into the individual arm voltage (4), then multiplying it with the arm current (2) and taking battery powers into account, the total battery power of each arm over one fundamental period can be derived as

$$\begin{cases} \sum\limits_{i=1}^{N} P_{b\_kpi} = -P_{kp} = -\frac{\omega}{2\pi} \int_{t}^{t+\frac{2\pi}{\omega}} v_{kp} i_{kp} = \frac{1}{2} P_{dck} - \frac{1}{2} P_{ack} + P_{diffk} \\ \sum\limits_{i=1}^{N} P_{b\_kni} = -P_{kn} = -\frac{\omega}{2\pi} \int_{t}^{t+\frac{2\pi}{\omega}} v_{kn} i_{kn} = \frac{1}{2} P_{dck} - \frac{1}{2} P_{ack} - P_{diffk} \end{cases} \tag{7}$$

where

$$\begin{cases} P_{dck} = V_{dc} I_{dck} \\ P_{ack} = \frac{1}{2} V_{gk} I_{gk} \cos \varphi_k \\ P_{diffk} = \frac{1}{2} \left(\sum\limits_{i=1}^{N} P_{b\_kni} - \sum\limits_{i=1}^{N} P_{b\_kpi} \right) = \frac{1}{2} V_g I_{1k} \cos \varphi_{1k} \end{cases} \tag{8}$$

Here, P_{diffk} is defined as the difference power between upper and lower arms of phase k. According to Equation (7), the power flows of the three-phase MMC-BESS can be shown as Figure 3. It can be concluded that the DC circulating current is the carrier of DC power between the MMC-BESS and DC-link; the fundamental frequency circulating current transfers power between the upper and lower arms within one phase; the fundamental frequency grid current transfers active and reactive power between the MMC-BESS and the AC grid. Therefore, by regulating these components in corresponding currents, the power flows among batteries, DC-link, and AC grid can be controlled as the system requires.

If the individual capacitor voltage is controlled by MMC side, the cascaded capacitor voltage control strategy would be employed at the DC side [14], and the individual battery power would be directly controlled by battery side converter, which makes it behave as a constant power load (CPL) to the corresponding SM. On the contrary, if the battery side converter is utilized to control the individual capacitor voltage, the capacitor would behave as a voltage source to the MMC side. Consequently, the cascaded capacitor voltage balancing controls in conventional MMCs can be avoided. Nevertheless, due to the existence of the additional power flow of BESS, when the DC-link voltage is required to be controlled by the MMC-BESS itself in rectifier mode operation, the original control strategy in conventional MMCs (outer-loop DC-link voltage control and inner-loop active grid current control [18]) would be invalid because this manner would lead to the uncertainty of battery power, bringing instability to the system operation. In essence, the AC power and DC power are not strictly equal to each other in the MMC-BESS compared with conventional MMCs. The DC-link voltage is not only decided by the active grid current at AC side, but also the battery power each phase. Hence, to make the batter-side based capacitor voltage control effective in the rectifier mode, the control strategy will be proposed in the following section.

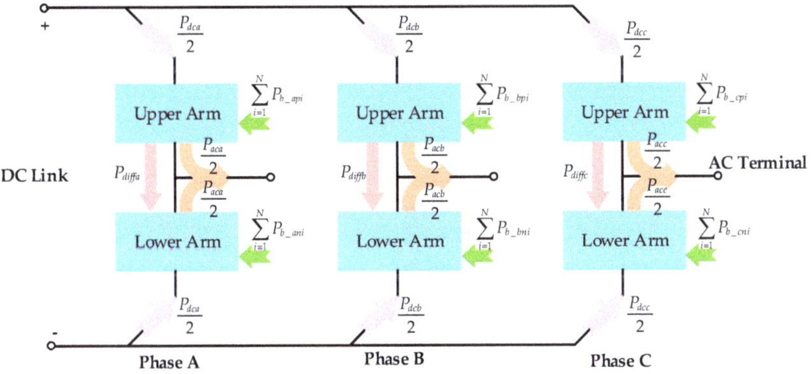

Figure 3. Power flows of the MMC-BESS.

3. Control Strategy of Rectifier Mode Operation of MMC-BESS

3.1. SOC Equalization of MMC-BESS

As an important function of a PCS, SOC equalization of batteries must be achieved during operation of the MMC-BESS. Based on the structure of the MMC-BESS, the SOC equalization is divided into three levels: among three phases, between upper and lower arms, and among SMs within one arm. The definition of SOC is expressed as

$$SOC = \frac{\text{Storaged Charges}}{\text{Nominal Capacity}} \times 100\% \tag{9}$$

The SOC of individual battery can be established as

$$SOC_{kji}(t) = SOC_{kji}(t_0) + \frac{1}{E_{b\_kji}} \int_{t_0}^{t} P_{b\_kji} dt \tag{10}$$

where P_{b_kji} is the individual battery power; E_{b_kji} is the nominal energy of individual battery given by the production of battery voltage and its capacity. According to Equation (10), the SOC of individual battery can be controlled through the direct regulation of corresponding battery power. Since E_{b_kji} may be different in individual SM due to the age of battery, the SOC equalization should take the capacity of individual battery into consideration. According to the research in Reference [15], battery capacity should be employed to produce the coefficient for corresponding SOC equalization control. For the sake of simplicity, it is assumed that all batteries are of the same capacity, which would not affect the following investigations in this paper.

The required power adjustments are generated through closed-loop controls of average SOC in corresponding levels as shown in Figure 4, where K_{ph}, K_{arm}, and K_{SM} are the coefficients of SOC equalization controller each level. Since the battery side converters are employed to control the capacitor voltages, the SOC equalization must be achieved by MMC side control. Hence, the SOC equalizations are realized by adjusting the modulation waveforms at MMC side according to the power differences ΔP_{b_k}, ΔP_{b_kj}, and ΔP_{b_kji} in Figure 4.

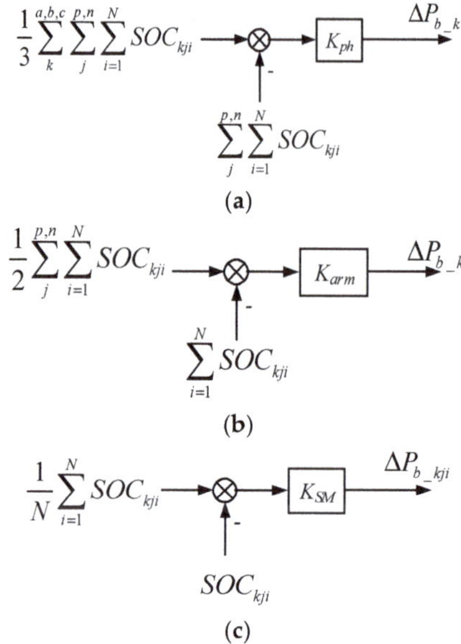

Figure 4. Three-level State-of-charge (SOC) equalization control. (**a**) SOC equalization among three phases. (**b**) SOC equalization between upper and lower arms. (**c**) SOC equalization among submodules (SMs) within phase arm.

3.1.1. SOC Equalization among Three Phases

Seen from Figure 3 and Equation (7), the total battery power within one phase is composed by DC power and AC power, so the battery power can be regulated by these two components. Nevertheless, due to the power control requirement at AC side, the AC powers must be symmetric in three phases under a balanced AC grid. Hence, to equalize the SOCs among phases, DC power should be regulated at each phase.

When the MMC-BESS operates as an inverter, the DC-link voltage is fixed to the system and the DC side would be controlled as the current sources each phase independently [17]. However, when the MMC-BESS operates as a rectifier, this manner would be invalid because the DC-link must be controlled as a voltage source by the DC side control of MMC-BESS. With the individual capacitor voltage controlled by battery side converter, the DC side equivalent circuit of MMC-BESS can be present as Figure 5. The capacitors behave as a controllable voltage source, then the DC-link voltage should be controlled by three-phase DC circulating current collectively. Therefore, the DC side control is proposed based on this DC side equivalent circuit in this paper as shown in Figure 6.

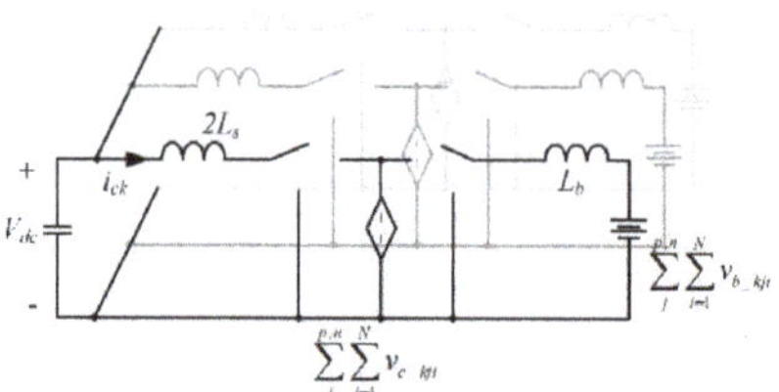

Figure 5. DC side equivalent circuit of the MMC-BESS.

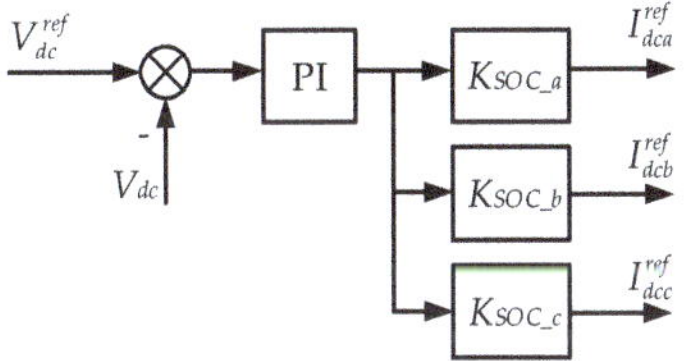

Figure 6. DC-link voltage control diagram.

In Figure 6, the DC-link voltage is regulated by a proportional–integral (PI) controller. The coefficient K_{soc_k} is employed to redistribute the DC power among the three phases according to the difference power in Figure 4a as:

$$K_{SOC\_k} = \frac{V_{dc}I_{dc}/3 + \Delta P_{b\_k}}{V_{dc}I_{dc}} \tag{11}$$

when the SOCs of three phases have been equalized, K_{soc_k} would be 1/3 consequently, which means DC power is equally divided by three phases. Through the proposed DC-link voltage control, the DC circulating current reference is generated by each phase.

3.1.2. SOC Equalization between Upper and Lower Arms

According to Equation (8), the fundamental frequency circulating current can transfer power between the upper and lower arms within one phase. With the difference power generated by SOC equalization controller in Figure 4b, the fundamental frequency circulating current reference can be obtained. Due to the demand of preventing fundamental frequency circulating current per phase from flowing into DC-link, the reactive components are injected into the other two phases as [11]:

$$\begin{bmatrix} i_{1a}^{ref} \\ i_{1b}^{ref} \\ i_{1c}^{ref} \end{bmatrix} = \frac{2}{V_g} \begin{bmatrix} \cos\varphi & -\frac{1}{\sqrt{3}}\sin\varphi & \frac{1}{\sqrt{3}}\sin\varphi \\ \frac{1}{\sqrt{3}}\sin(\varphi - \frac{2\pi}{3}) & \cos(\varphi - \frac{2\pi}{3}) & -\frac{1}{\sqrt{3}}\sin(\varphi - \frac{2\pi}{3}) \\ -\frac{1}{\sqrt{3}}\sin(\varphi + \frac{2\pi}{3}) & \frac{1}{\sqrt{3}}\sin(\varphi + \frac{2\pi}{3}) & \cos(\varphi + \frac{2\pi}{3}) \end{bmatrix} \begin{bmatrix} P_{diffa}^{ref} \\ P_{diffb}^{ref} \\ P_{diffc}^{ref} \end{bmatrix} \tag{12}$$

where

$$P_{diffk}^{ref} = \frac{\Delta P_{b\_kp} - \Delta P_{b\_kn}}{2} \tag{13}$$

Since DC and fundamental frequency components are required to be controlled in the circulating current each phase, a proportional–integral–resonant (PIR) controller tuned at fundamental frequency and double line-frequency was employed to regulate components at different frequencies simultaneously, as shown in Figure 7. In this paper, the reference of double line-frequency component i_{2k}^{ref} was set as zero to reduce the power losses [13].

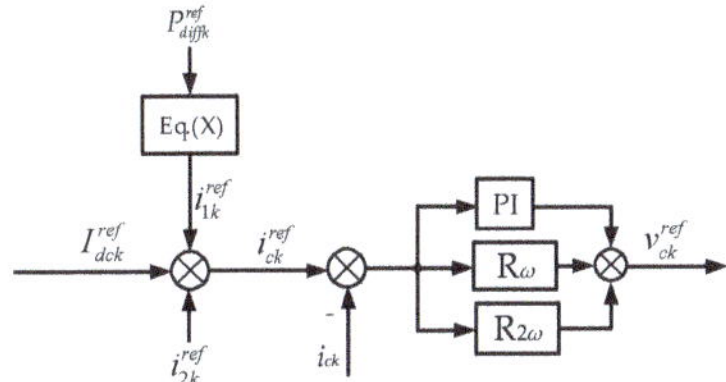

Figure 7. Circulating current control diagram.

3.1.3. SOC Equalization among SMs within Phase Arm

Since all SMs within the same phase arm share a common arm current, the individual SM power must be regulated by redistributing the terminal voltage without affecting the output voltage of the whole arm. Based on the former two SOC equalization controls, the total power of the phase arm can be generated as

$$P_{b\_kj}^{ref} = \frac{P_{ac} - P_{dc}}{6} - \frac{\Delta P_{b\_k}}{2} - \Delta P_{b\_kj} \tag{14}$$

Consequently, the individual battery power reference was obtained through the SOC controller in Figure 4c as:

$$P_{b\_kji}^{ref} = \frac{P_{b\_kj}^{ref}}{N} - \Delta P_{b\_kji} \tag{15}$$

Defining the power ratio of individual SM according to the battery power as:

$$m_{kji} = \frac{P_{b\_kji}^{ref}}{\sum_{i=1}^{N} P_{b\_kji}} \tag{16}$$

By multiplying m_{kji} with the arm voltage reference v_{kj}, the SM power can be regulated according to the SOC equalization control within phase arm. With the constraint as:

$$\sum_{i=1}^{N} m_{kji} = 1 \tag{17}$$

the output voltage of the whole arm would not be affected by the SOC equalization within phase arm.

3.2. *Control Strategy of MMC-BESS Based on Battery Side Capacitor Voltage Control*

According to the equivalent circuit in Figure 3, the control strategy of the MMC-BESS mainly consists of three parts: DC side control, AC side control, and battery side control. With the capacitor voltage controlled by the battery side converter, each SM, DC, and AC side control of the MMC are simplified in this paper. As per the analysis in Section 2, when the MMC-BESS operates as a rectifier in multi-terminal MMC-based applications, the output of the DC-link voltage controller cannot be employed as the reference of the active current at the AC side. Hence, the DC-link voltage control was proposed by regulating the DC circulating current at each phase according to the SOC equalization among three phases, as per Figure 6. On the other hand, the AC side should be controlled in a way in which the active and reactive powers are given directly according to the system's requirement [13]. By doing this, AC power is directly controlled by the reference, DC power is determined by external DC network, and battery power is controlled indirectly by capacitor voltage control.

The overall control strategy of the MMC-BESS based on battery side capacitor voltage control is shown in Figure 8.

The SOC equalization of all three levels are implemented by the MMC side control according to the proposed control strategies in Section 3.1. The common modulation waveform of the individual phase arm is generated by AC and DC side controls of the MMC, then the modulation waveform of individual SMs is generated by the power redistribution ratio according to Equation (16). Finally, carrier phase-shifted pulse-width modulation (CPS-PMW) is utilized to generate the pulse signal for MMC side-switching devices [20]. On the other hand, the implementation of battery side capacitor voltage control will be introduced in the next section.

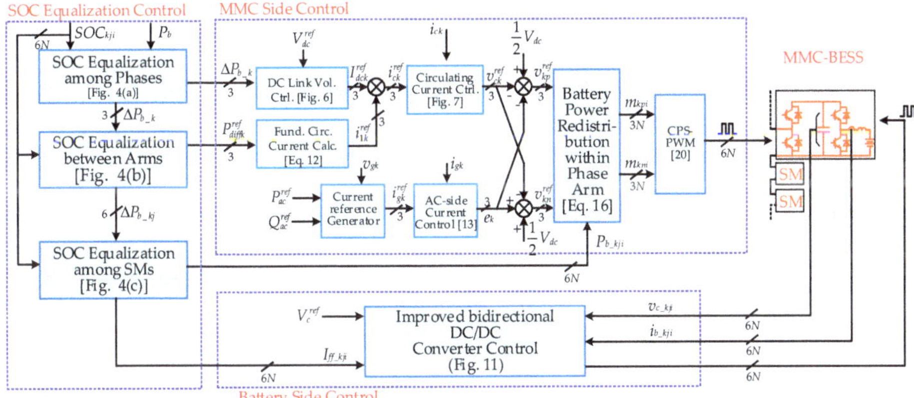

Figure 8. Overall control strategy of the MMC-BESS based on battery side capacitor voltage control.

4. Implementation of Battery Side Control Strategy

The local circuit of the individual bidirectional DC/DC converter with control diagram is shown in Figure 9, where the MMC side is taken as the load for the battery side represented as the current i_{M_kji}. The midpoint voltage of the battery side half-bridge $v_{T\text{-}kji}$ is taken as the input of the bidirectional DC/DC converter. In Figure 9, G_{VR} and G_{CR} are voltage and current regulators, respectively; The time delay caused by PWM is defined as G_d, which is treated as $1.5T_{sb}$ [21]. Here, T_{sb} is the switching period of the battery side DC/DC converter.

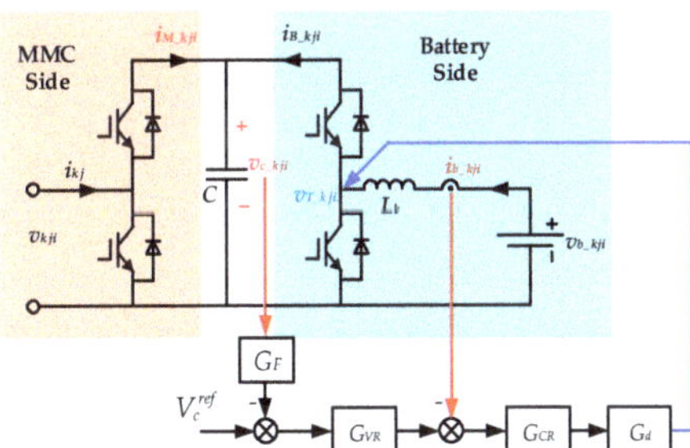

Figure 9. Local circuit of the bidirectional DC/DC converter with control diagram.

Although some research on the control strategy of bidirectional DC/DC converters has been published [21,22], the application of MMC-BESS still has special issues which should be investigated. The control targets are the individual capacitor voltage v_{c_kji} with a stable DC component and the battery current i_{b_kji} without a low-frequency AC component. However, the voltage ripples (mainly at fundamental and double-line frequency) caused by operation of the MMC side would bring negative influence to the DC/DC converter control. To eliminate these ripples, the filter block G_F is employed as shown in Figure 9, but the type of G_F should be discussed. Since moving average filter (MAF) can attenuate components in specific frequencies [23], it is especially suitable for filtering harmonics in capacitor voltage of MMC-BESS. The transfer function of MAF in discrete domain is

$$G_F = \frac{1}{N_F}\frac{1 - z^{-N_F}}{1 - z^{-1}} \tag{18}$$

where N_F is the number of sampling data stored in the digital processor. However, since the switching frequency of an individual DC/DC converter is usually well above the fundamental frequency at the MMC side, it will lead to the requirement of a large amount of data storage space in the DC/DC converter control. For example, for a sampling frequency (equal to the switching frequency of DC/DC converter) f_{sb} = 10 kHz and fundamental frequency of MMC side f = 50 Hz, N_F is given by f_{sb}/f = 200. With the increment of the number of SMs in the series per phase arm, the data storage space required by the MAFs would consequently become large-scale. To avoid the defect of large-scale data storage space in this application, a common MAF scheme shared by all SMs per arm is proposed in this paper.

The MMC side current i_{M_kji} in Figure 9 can be expressed as:

$$\begin{cases} i_{M\_kpi} = i_{kp}m_{kpi} = (\frac{I_g\cos(\omega t+\varphi)}{2} + I_{dck} + I_{1k}\cos(\omega t+\varphi_{1k}))(m_{dc\_kpi} - m_{ac\_kpi}\cos\omega t) \\ i_{M\_kni} = i_{kn}m_{kni} = (-\frac{I_g\cos(\omega t+\varphi)}{2} + I_{dck} + I_{1k}\cos(\omega t+\varphi_{1k}))(m_{dc\_kpi} + m_{ac\_kpi}\cos\omega t) \end{cases} \tag{19}$$

Then the individual capacitor voltage can be derived as:

$$\begin{cases} v_{c\_kpi} = \frac{1}{C}\int(i_{M\_kpi} + i_{B\_kpi})dt + v_{c\_kpi\_0} \\ v_{c\_kni} = \frac{1}{C}\int(i_{M\_kni} + i_{B\_kni})dt + v_{c\_kni\_0} \end{cases} \tag{20}$$

where the $v_{c\_kpi\_0}$ and $v_{c\_kpi\_0}$ are initial voltages of capacitors in the upper and lower arms, respectively. The DC component of the capacitor current must be zero, otherwise the capacitor voltage would be instable with time, which can be deduced as

$$\begin{cases} m_{dc\_kpi}I_{dck} - \frac{1}{4}m_{ac\_kpi}I_g\cos\varphi - \frac{1}{2}m_{ac\_kpi}I_{1k}\cos\varphi_{1k} + i_{B\_kpi} = 0 \\ m_{dc\_kni}I_{dck} - \frac{1}{4}m_{ac\_kni}I_g\cos\varphi + \frac{1}{2}m_{ac\_kni}I_{1k}\cos\varphi_{1k} + i_{B\_kni} = 0 \end{cases} \tag{21}$$

Substituting Equations (19) and (21) into (20), the average ripples of individual capacitors are derived as:

$$\begin{cases} \frac{1}{N}\sum_{i=1}^{N}\widetilde{v}_{c\_kpi} & = \frac{1}{\omega C}(-\frac{1}{4}I_{1k}m_{ac\_kp}\sin(2\omega t+\varphi_{1k}) + I_{1k}m_{dc\_kp}\sin(\omega t+\varphi_{1k}) \\ & -\frac{1}{8}I_g m_{ac\_kp}\sin(2\omega t+\varphi) + \frac{1}{2}I_g m_{dc\_kp}\sin(\omega t+\varphi) - I_{dck}m_{ac\_kp}\sin\omega t) \\ \frac{1}{N}\sum_{i=1}^{N}\widetilde{v}_{c\_kni} & = \frac{1}{\omega C}(\frac{1}{4}I_{1k}m_{ac\_kn}\sin(2\omega t+\varphi_{1k}) + I_{1k}m_{dc\_kn}\sin(\omega t+\varphi_{1k}) \\ & -\frac{1}{8}I_g m_{ac\_kn}\sin(2\omega t+\varphi) - \frac{1}{2}I_g m_{dc\_kn}\sin(\omega t+\varphi) + I_{dck}m_{ac\_kn}\sin\omega t) \end{cases} \tag{22}$$

Seen from Equations (20) and (22), the battery side current i_{B_kji} affects the DC component in capacitor voltage, while the individual capacitor voltage ripples are mainly decided by the common arm current flowing through all SMs per phase arm. It can be proved by the simulation waveforms in Figure 10, the individual battery power differentiated from each other up to ±40%, and the corresponding capacitor voltage ripples were almost identical due to the common arm current.

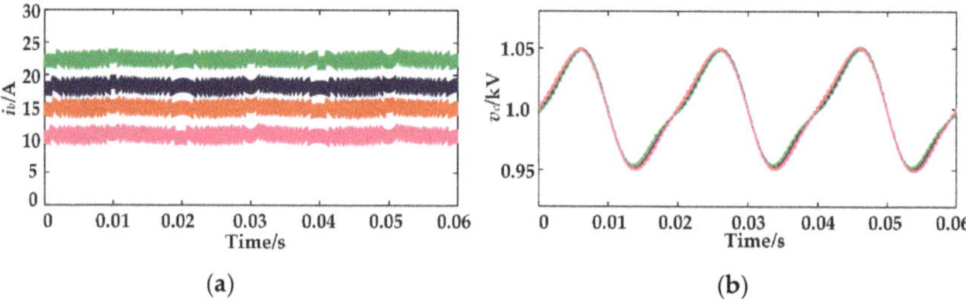

Figure 10. Capacitor voltages within one phase arm under unbalanced battery powers. (**a**) Unbalanced battery currents within phase arm. (**b**) Capacitor voltages within phase arm.

Based on the analysis above, an improved control strategy of a battery side DC/DC converter with the data storage space reduction of MAF is proposed in Figure 11. In this scheme, a common MAF was employed to extract the average capacitor voltage ripples of all SMs per phase arm, then the individual capacitor voltage ripples were counteracted by the average capacitor voltage ripples. Through the common MAF, the occupied data storage space can be reduced to $1/N$ compared with the original scheme implemented in each SM.

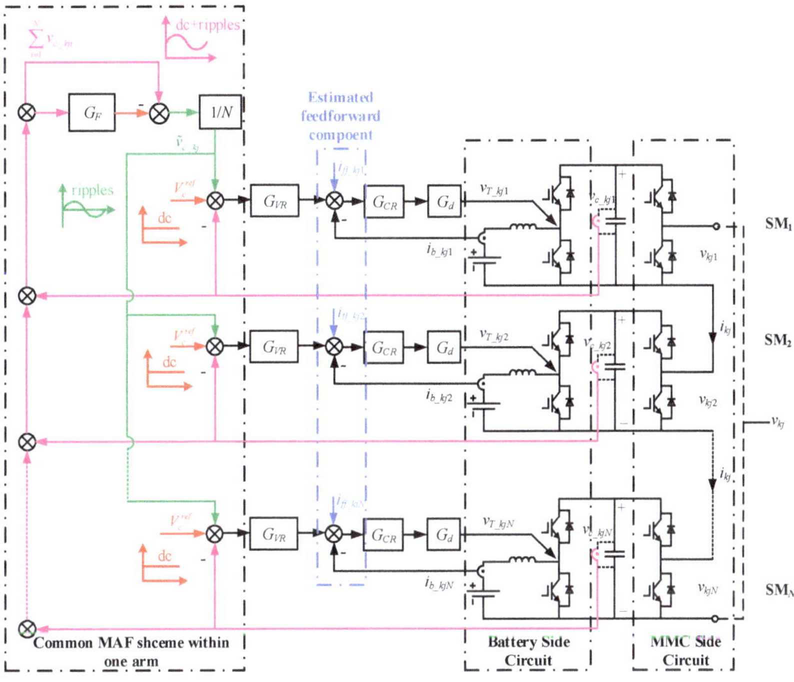

Figure 11. Improved control strategy based on moving average filter (MAF) for individual bidirectional DC/DC converter.

On the other hand, since the capacitor voltage was controlled by the individual DC/DC converter, the dynamic response was expected to be fast enough for MMC side power conversion. However, due to the existence of the MAF, the bandwidth of the voltage control loop cannot exceed the cut-off frequency of the MAF, which is unacceptable in application. Seen from Figure 9, the reference of the capacitor voltage was constant during normal operation of the MMC-BESS, hence, the disturbance from the MMC side i_{M_kji} is the exclusive factor that causes the fluctuation of the DC capacitor voltage. To eliminate the load disturbance, the load current feedforward can be implemented. However, the measurement of MMC side load in individual SM would greatly increase the cost of the system, and i_{M_kji} is a current in pulse form which is different from ordinary bidirectional DC/DC converter. Therefore, the load disturbance was mitigated by estimated battery current in this paper as:

$$I_{ff\_kji} = \frac{1}{v_{b\_kji}}\left(\frac{3v_{gd}i_{gd}/2 - V_{dc}I_{dc}}{6N} - \frac{\Delta P_{b\_k}}{2N} - \frac{\Delta P_{b\_kj}}{N} - \Delta P_{b\_kji}\right) \tag{23}$$

This estimated feedforward component was calculated based on the instantaneous power relationship in Equation (5), where the AC and DC powers were derived from measured values instead of the reference values, representing the dynamics of power flows in real time. Then the feedforward component was redistributed into each SM according to the SOC of individual battery. Consequently, the dynamic performance of the individual capacitor voltage control can be significantly improved.

Finally, the improved control strategy in Figure 11 can be simplified as Figure 12 through the block diagram transformation. D_{kji} is the ratio of capacitor voltage and individual battery voltage. The MAF block in Figure 11 is equivalent to the configuration in individual feedback loop in Figure 12. With this simplified control diagram, the design of controllers of capacitor voltage and battery current is convenient to be implemented using the control system theory.

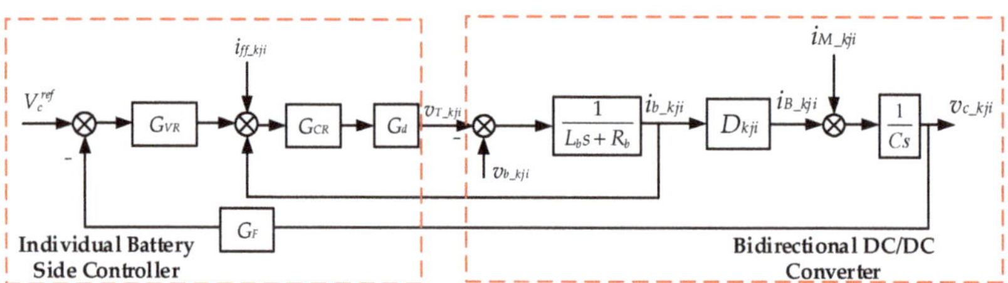

Figure 12. Simplified control block diagram of the individual DC/DC converter.

5. Simulations and Experimental Results

5.1. Simulation Results

The simulation model of the three-phase MMC-BESS is built in MATLAB/Simulink (R2016b, Mathworks, Natick, MA, USA). The main parameters are listed in Table 1. To verify the SOC equalization proposed in this paper, the capacity of individual battery is set as 0.3 Ah, so the SOCs can be equalized rapidly.

Table 1. Main parameters of the MMC-BESS simulation model.

Symbol	Quantity	Value
V_g	Grid voltage (line-to-ground, Root Mean Square (RMS) value)	1.13kV (M = 0.8)
V_{dc}	DC-link voltage	4 kV
S_{rated}	Rated apparent output power	1 MVA
f	AC line frequency	50 Hz
L_s	Arm inductance	5 mH
N	Submodule (SM) number per phase arm	4
C	SM capacitance	3000 μF
V_b	Rated battery voltage	600 V
L_b	Battery side DC/DC converter inductance	5 mH
Q_b	Rated battery capacity	0.3 Ah
f_{sm}	Carrier frequency at MMC side modulation	2 kHz
f_{sb}	Battery side converter switching frequency	10 kHz

5.1.1. Rectifier Mode Operation with Proposed Control Strategy Based on Battery Side Capacitor Voltage Control

The simulation waveforms of rectifier mode operation are shown in Figure 13 and the SOC equalization process is shown in Figure 14. The initial SOCs of all 24 SMs are arbitrarily distributed from 61% to 71%. The AC power increases from 0 to 480 kW through a slope reference, then it is transferred to 320 kW at t = 0.4 s and 640 kW at t = 0.6 s. A 480 kW external DC load is connected to the DC-link at t = 0.1 s. Seen from Figure 13, the AC power is indicated by grid currents in Figure 13a; the DC-link voltage is shown in Figure 13b. With the proposed control strategy for rectifier mode operation of MMC-BESS, the DC-link voltage is controlled to the reference value throughout. The capacitor voltages are shown in Figure 13c utilizing the proposed battery side-based capacitor voltage control strategy. Figure 13d shows the circulating currents of three phases. The DC circulating currents are controlled according to the coefficient in Equation (11) to equalize the SOCs among phases, while the fundamental frequency circulating currents are controlled according to Equation (12) to equalize the SOCs between the upper and lower arms each phase. With the SOCs gradually converging to the same value, the differences among circulating currents in three phases also decrease consequently. The battery currents of four SMs within upper arm of phase A are shown in Figure 13d, where the differences among SMs also decrease gradually during the SOC equalization process. In Figure 14, the SOC equalization of all three levels are illustrated. Figure 14a shows the SOC equalization of all 24 SMs. The SOC equalizations among three phases, between upper and lower arms, and among SMs are shown in Figure 14b–d, respectively. During the rectifier mode operation of MMC-BESS, the simulation results validate the effectiveness of the proposed control strategy based on battery side capacitor voltage control.

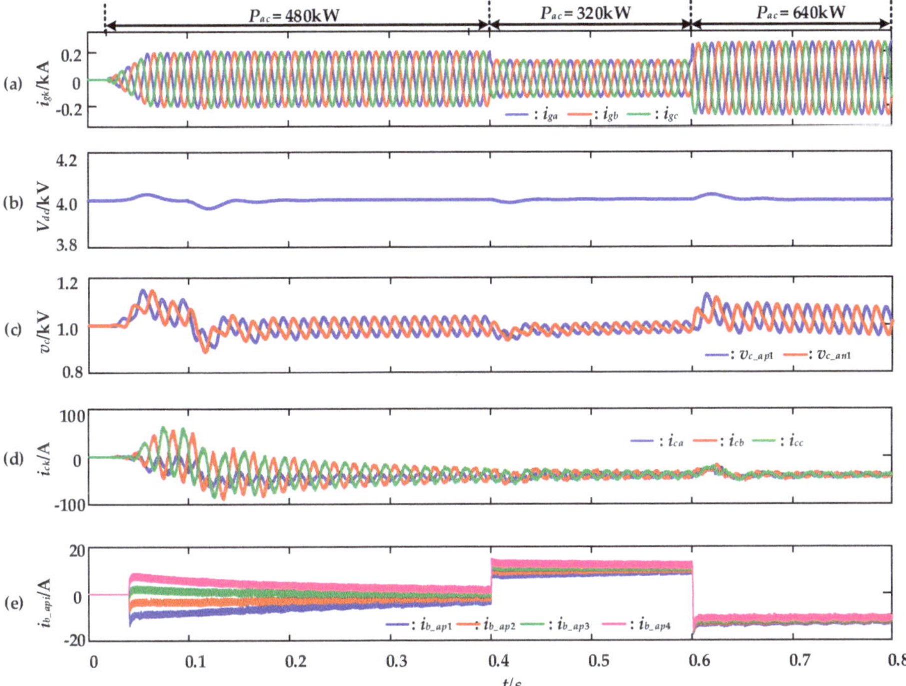

Figure 13. Simulation waveforms of rectifier mode operation of MMC BESS. (**a**) Grid currents. (**b**) DC-link voltage. (**c**) Capacitor voltages of upper and lower arm of phase A. (**d**) Circulating currents. (**e**) Battery currents of SMs within upper arm of phase A.

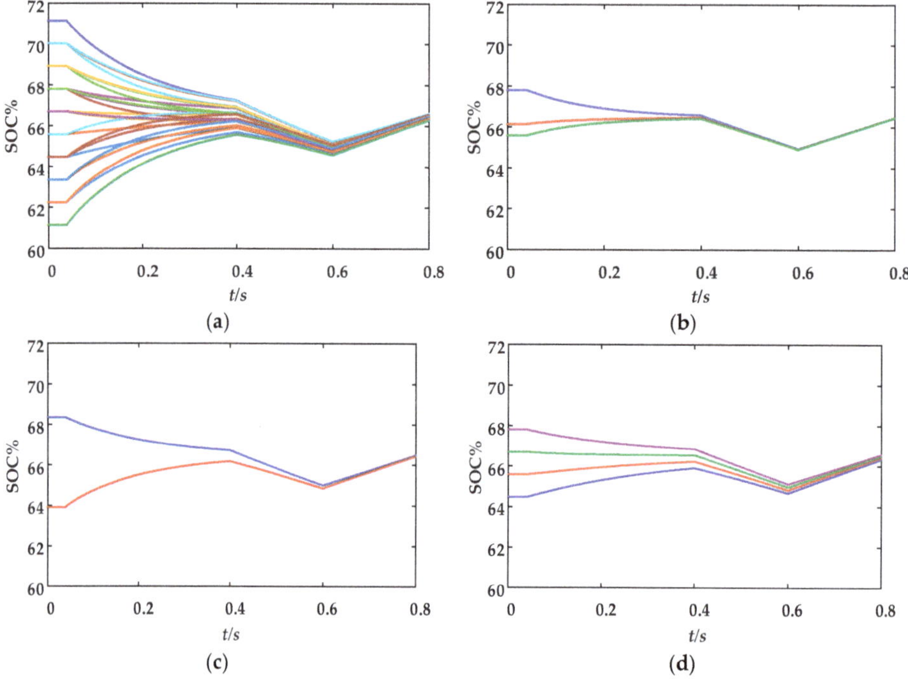

Figure 14. Three-level SOC equalization results. (**a**) SOC equalization of all 24 SMs. (**b**) SOC equalization among three phases. (**c**) SOC equalization between upper and lower arms of phase A. (**d**) SOC equalization among SMs within upper arm of phase A.

5.1.2. Verification of Proposed Battery Side Capacitor Voltage Control Strategy

The verifications of the proposed battery side capacitor voltage control strategy in steady and dynamic state are shown in Figure 15. Figure 15a is implemented without feedforward component of Equation (23), while this feedforward component is employed in Figure 15b,c. The difference of implementation between Figure 15b,c locates on the capacitor voltage filter scheme. The MAF is implemented in each SM capacitor voltage control in Figure 15b while the proposed common MAF scheme within one phase arm is employed in Figure 15c. At t = 0.3 s, the AC power reference is transferred from 480 kW to 960 kW, then at t = 0.6 s, the external DC load is transferred from 480 kW to 960 kW. In transient process, the capacitor voltage variation is up to $\Delta v_{c1} = -16\%$ and the regulation time of battery current is t_{r1} = 0.2 s after t = 0.3 s in Figure 15a. While in Figure 15b, with the estimated feedforward component at battery side control, the transient performance is significantly improved with $\Delta v_{c1} \approx -2\%$ and t_{r1} = 0.03 s. Similar improvement can be observed from the comparison of Figure 15a,b during 0.6–0.8 s. Hence, the effectiveness of the feedforward component is validated. One the other hand, in steady state in Figure 15b,c, the ripples in capacitor voltage are eliminated from battery current, leaving only DC component, which validate the feasibility of the proposed common MAF scheme within phase arm compared with utilizing MAF each SM.

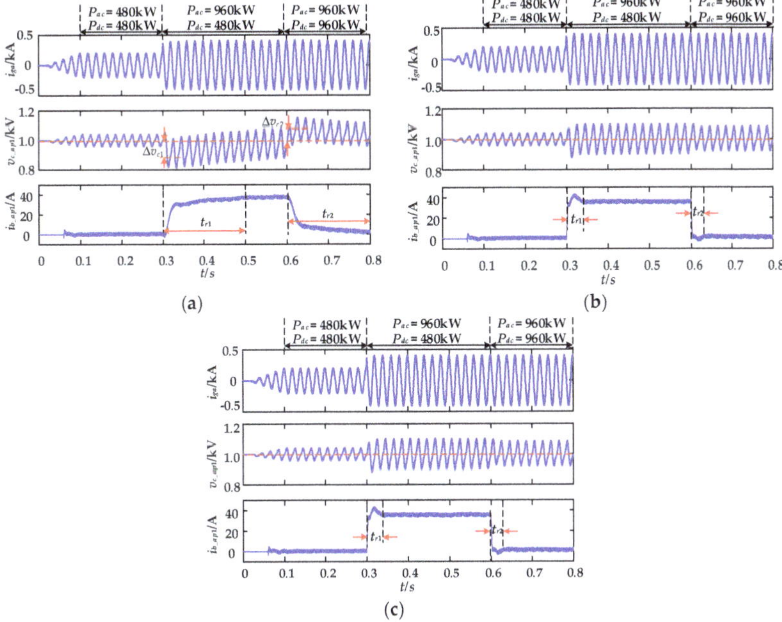

Figure 15. Verification of proposed battery side capacitor voltage control strategy. (**a**) Without estimated feedforward component. (**b**) Utilizing MAF each SM with estimated feedforward component. (**c**) Utilizing proposed common MAF each phase arm with estimated feedforward component.

5.2. Experimental Results

To verify the proposed control strategy of the MMC-BESS, a downscaled prototype was built in the laboratory [24]. The configuration of the prototype is shown in Figure 16. The main parameters are listed in Table 2. The controller structure of the prototype was designed with a single digital-signal processor (DSP) as the main controller and three field-programmable gate arrays (FPGAs) as the auxiliary controllers in each phase. The battery module in individual SMs is composed of three 12 V/24 Ah lead-acid battery cells due to the consideration of cost reduction compared with other type of batteries [25]. Note that the switching frequency of battery side DC/DC converters is different from MMC side converters. Given that the power rate of individual batteries is usually lower than the MMC side, semiconductor devices and switching frequency can be selected separately at the battery side DC/DC converter [26]. With the development of wide bandgap devices, the conversion efficiency between batteries and MMC can be enhanced at high-power levels [27].

Table 2. Main parameters of the prototype.

Symbol	Quantity	Value
V_g	Grid voltage (line-to-ground, RMS)	48V (M = 0.8)
V_{dc}	DC-link voltage	120 V
L_s	Arm inductance	5 mH
N	SM number per arm	2
C	SM capacitance	3000 μF
V_b	Rated battery voltage	36 V
L_b	Battery side DC/DC converter inductance	5 mH
Q_b	Rated battery capacity	24 Ah
f_{sm}	Carrier frequency at MMC side modulation	2 kHz
f_{sb}	Battery side converter switching frequency	10 kHz

(a)

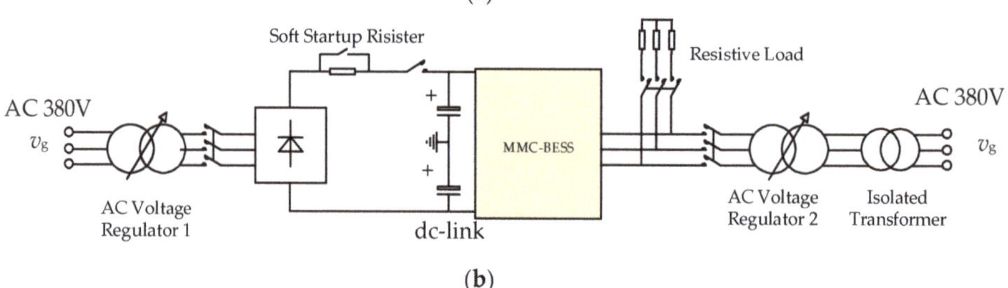

(b)

Figure 16. Configuration of the three-phase MMC-BESS prototype. (**a**) Picture. (**b**) Circuit.

5.2.1. Verification of SOC Equalization

The experimental waveforms of rectifier mode operation in steady state are shown in Figure 17. During the operation, the references of P_{ac} 800 W and an external 600 W DC load is connected to the DC-link throughout. In Figure 17a, the five-level modulation waveform of line voltage is presented with quite balanced capacitor voltages. Sinusoidal grid currents with low harmonic components are shown in Figure 17b. Circulating current of phase A is shown in Figure 17c with fundamental frequency component to equalize the SOC difference between upper and lower arms. Battery current with low harmonic components is also shown in Figure 17c to verify the effectiveness of the proposed capacitor voltage filter scheme. As the batteries operate in discharge mode, the SOC equalization process of the 12 battery modules are shown in Figure 17d. Due to the large capacity of the battery unit, the SOC equalization process is long, hence, the SOCs are sampled every ten minutes and presented by MATLAB.

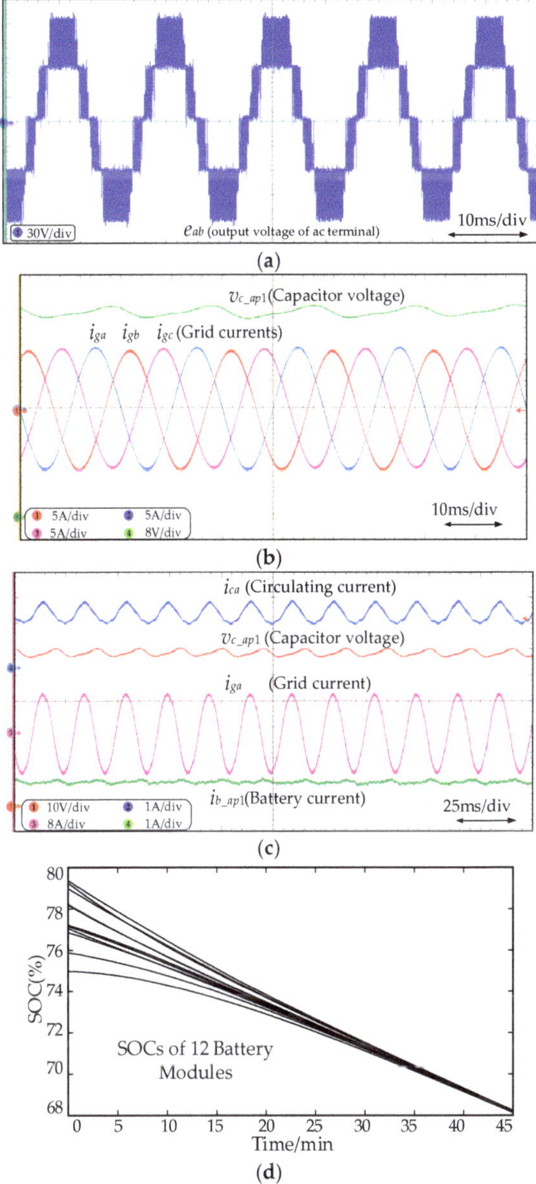

Figure 17. Experimental waveforms of steady-state operation of the MMC-BESS. (**a**) Modulated voltage waveform. (**b**) Grid currents and capacitor voltage. (**c**) Circulating current and battery current. (**d**) SOC equalization process.

5.2.2. Verification of Proposed Battery Side-Based Capacitor Voltage Control

The experimental waveforms of dynamic performance with and without the proposed control strategy at battery side are shown in Figure 18. The reference of AC power was 800 W throughout, and DC load was changed from 800 W to 500 W. Seen from Figure 18a, without the proposed control strategy at battery side, the battery current responded slowly, causing voltage fluctuation on capacitor voltage. Compared with Figure 18a, the battery current responded faster and capacitor voltage was stable during the power flow transfer process in Figure 18b. This comparison was in accordance with the simulation results in Figure 15.

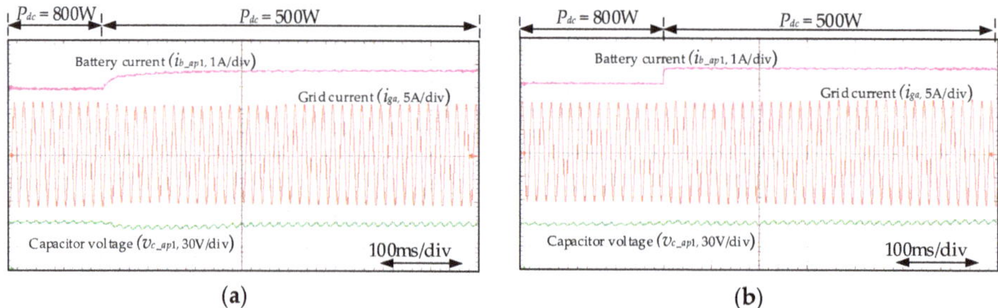

Figure 18. Experimental waveforms of dynamic performance of the MMC-BESS. (**a**) Without the proposed control strategy. (**b**) With the proposed control strategy.

5.2.3. Verification of Dynamics of Rectifier Mode Operation

The dynamic performance of rectifier mode operation is shown in Figure 19. Utilizing the proposed control strategy based on battery side capacitor voltage control, the DC-link voltage was controlled to 120 V throughout, AC power was regulated through the reference value directly according to the indications in Figure 19. The external 600 W DC load was connected to the DC-link during the operation, which is illustrated in the general view of the waveform. Seen from the experimental results, the AC power and DC-link voltage had no direct interactions due to the existence of additional power flow provided by the BESS, which is different from conventional MMCs. The dynamics of AC power and DC-link voltage were both able to satisfy the requirements of the system as a rectifier or a pure PCS.

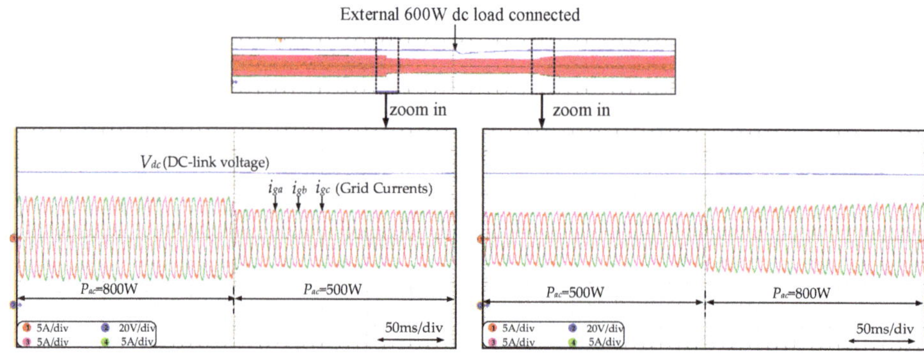

Figure 19. Experimental waveforms' dynamic performances of rectifier mode operation with proposed control strategy.

6. Conclusions

In this paper, the control strategy of the MMC-BESS for rectifier mode operation with capacitor voltage controlled by a battery side DC/DC converter was proposed. The main contributions can be summarized as follows:

- The reason why the rectifier mode control strategy in conventional MMCs invalidates in the MMC-BESS was analyzed in this paper. The ac power and dc power were decoupled by the battery side control, so it was impossible to control dc-link voltage through ac power. According to the analysis, the control strategy containing dc-link voltage control and SOC equalization was proposed in this paper to achieve the control objective based on battery side capacitor voltage control.

- At battery side, since the capacitor voltage was stabilized by a battery side DC/DC converter, the control strategy of the DC/DC converter was key to the performance of the whole system. Therefore, an improved control strategy for an individual capacitor voltage control was proposed in this paper, enhancing the dynamic performance of capacitor voltage, meanwhile greatly reducing the occupied storage space of the MAF by using a common filter per phase arm. Seen from the simulation and experimental results, the proposed battery side control strategy can satisfy the requirements of rapidity and accuracy, facilitating the application of the MMC-BESS in practice.

Author Contributions: Z.W., H.L. and Y.M. conceived and discussed the implementation of the control strategy; Z.W. and Y.M. performed the experiments; H.L. supervised the progress of the project; Z.W. and Y.M. wrote the paper.

Funding: This research was funded by National Natural Science Foundation of China grant number 51741703.

Conflicts of Interest: The authors declare no conflict of interest.

References

1. Carrasco, J.M.; Franquelo, L.G.; Bialasiewicz, J.; Galván, E.; Guisado, R.C.P.; Prats, M.Á.M.; León, J.I.; Moreno-Alfonso, N. Power-electronic Systems for the Grid Integration of Renewable Energy Sources: A Survey. *IEEE Trans. Ind. Electron.* **2006**, *53*, 1002–1016. [CrossRef]
2. Rodríguez, J.; Bernet, S.; Wu, B.; Pontt, J.; Kouro, S. Multilevel Voltage-Source-Converter Topologies for Industrial Medium-Voltage Drives. *IEEE Trans. Ind. Electron.* **2007**, *54*, 2930–2945. [CrossRef]
3. Lesnicar, A.; Marquardt, R. An Innovative Modular Multilevel Converter Topology Suitable for a Wide Power Range. In Proceedings of the 2003 IEEE Bologna Power Tech Conference Proceedings, Bologna, Italy, 23–26 June 2003; pp. 272–277.
4. Debnath, S.; Qin, J.; Bahrani, B.; Saeedifard, M.; Barbosa, P. Operation, Control, and Applications of the Modular Multilevel Converter: A Review. *IEEE Trans. Power Electron.* **2015**, *30*, 37–53. [CrossRef]
5. Perez, M.A.; Bernet, S.; Rodriguez, J.; Kouro, S.; Lizana, R. Circuit Topologies, Modeling, Control Schemes, and Applications of Modular Multilevel Converters. *IEEE Trans. Power Electron.* **2015**, *30*, 4–17. [CrossRef]
6. Akagi, H. Classification, Terminology, and Application of the Modular Multilevel Cascade Converter (MMCC). *IEEE Trans. Power Electron.* **2011**, *26*, 3119–3130. [CrossRef]
7. Vidal-Albalate, R.; Beltran, H.; Rolán, A.; Belenguer, E.; Peña, R.; Blasco-Gimenez, R. Analysis of the Performance of MMC under Fault Conditions in HVDC-Based Offshore Wind Farms. *IEEE Trans. Power Del.* **2016**, *31*, 839–847. [CrossRef]
8. Mishra, S.; Palu, I.; Madichetty, S.; Suresh Kumar, L.V. Modelling of Wind Energy-Based Microgrid System Implementing MMC. *Int. J. Energy Res.* **2016**, *40*, 952–962. [CrossRef]
9. Trintis, I.; Munk-Nielsen, S.; Teodorescu, R. A New Modular Multilevel Converter with Integrated Energy Storage. In Proceedings of the IECON 2011—37th Annual Conference of the IEEE Industrial Electronics Society, Melbourne, Australia, 7–10 November 2011; pp. 1075–1080.
10. Quraan, M.; Tricoli, P.; D'Arco, S.; Piegari, L. Efficiency Assessment of Modular Multilevel Converters for Battery Electric Vehicles. *IEEE Trans. Power Electron.* **2017**, *32*, 2041–2051. [CrossRef]
11. Vasiladiotis, M.; Rufer, A. Analysis and Control of Modular Multilevel Converters with Integrated Battery Energy Storage. *IEEE Trans. Power Electron.* **2015**, *30*, 163–175. [CrossRef]
12. Baruschka, L.; Mertens, A. Comparison of Cascaded H-Bridge and Modular Multilevel Converters for BESS Application. In Proceedings of the 2011 IEEE Energy Conversion Congress and Exposition, Phoenix, AZ, USA, 17–22 September 2011; pp. 909–916.
13. Hagiwara, M.; Maeda, R.; Akagi, H. Control and Analysis of the Modular Multilevel Cascade Converter Based on Double-Star Chopper-Cells (MMCC-DSCC). *IEEE Trans. Power Electron.* **2011**, *26*, 1649–1658. [CrossRef]

14. Soong, T.; Lehn, P.W. Internal Power Flow of a Modular Multilevel Converter with Distributed Energy Resources. *IEEE J. Emerg. Sel. Top. Power Electron.* **2014**, *2*, 1127–1138. [CrossRef]
15. Liang, H.; Guo, L.; Song, J.; Yang, Y.; Zhang, W.; Qi, H. State-of-Charge Balancing Control of a Modular Multilevel Converter with an Integrated Battery Energy Storage. *Energies* **2018**, *11*, 873. [CrossRef]
16. Quraan, M.; Yeo, T.; Tricoli, P. Design and Control of Modular Multilevel Converters for Battery Electric Vehicles. *IEEE Trans. Power Electron.* **2016**, *31*, 507–517. [CrossRef]
17. Chen, Q.; Li, R.; Cai, X. Analysis and Fault Control of Hybrid Modular Multilevel Converter with Integrated Battery Energy Storage System. *IEEE J. Emerg. Sel. Top. Power Electron.* **2017**, *5*, 64–78. [CrossRef]
18. Li, Y.; Shi, X.; Liu, B.; Lei, W.; Wang, F.; Tolbert, L.M. Development, Demonstration, and Control of a Testbed for Multiterminal HVDC System. *IEEE Trans. Power Electron.* **2017**, *32*, 6069–6078. [CrossRef]
19. Antonopoulos, A.; Angquist, L.; Nee, H.-P. On Dynamics and Voltage Control of the Modular Multilevel Converter. In Proceedings of the 2009 13th European Conference on Power Electronics and Applications, Barcelona, Spain, 8–10 November 2009; pp. 1–10.
20. Ilves, K.; Harnefors, L.; Norrga, S.; Nee, H.-P. Analysis and Operation of Modular Multilevel Converters with Phase-Shifted Carrier PWM. *IEEE Trans. Power Electron.* **2015**, *30*, 268–283. [CrossRef]
21. Zhang, J.; Lai, J.S.; Yu, W. Bidirectional DC-DC Converter Modeling and Unified Controller with Digital Implementation. In Proceedings of the 2008 23rd Annual IEEE Applied Power Electronics Conference and Exposition, Austin, TX, USA, 24–28 February 2008; pp. 1747–1753.
22. Zhu, G.; Ruan, X.; Zhang, L.; Wang, X. On the Reduction of Second Harmonic Current and Improvement of Dynamic Response for Two-Stage Single-Phase Inverter. *IEEE Trans. Power Electron.* **2015**, *30*, 1028–1041. [CrossRef]
23. Golestan, S.; Ramezani, M.; Guerrero, J.M.; Freijedo, F.D.; Monfared, M. Moving Average Filter Based Phase-Locked Loops: Performance Analysis and Design Guidelines. *IEEE Trans. Power Electron.* **2014**, *29*, 2750–2763. [CrossRef]
24. Wang, Z.; Lin, H.; Ma, Y.; Wang, T. A Prototype of Modular Multilevel Converter with Integrated Battery Energy Storage. In Proceedings of the 2017 Applied Power Electronics Conference and Exposition (APEC), Tampa, FL, USA, 26–30 March 2017; pp. 434–439.
25. Keshan, H.; Thornburg, J.; Ustun, T.S. Comparison of Lead-Acid and Lithium Ion Batteries for Stationary Storage in Off-Grid Energy Systems. In Proceedings of the 4th IET Clean Energy and Technology Conference (CEAT 2016), Kuala Lumpur, Malaysia, 14–15 November 2016; pp. 1–7.
26. Soong, T.; Lehn, P.W. Evaluation of Emerging Modular Multilevel Converters for BESS Applications. *IEEE Trans. Power Del.* **2014**, *29*, 2086–2094. [CrossRef]
27. Hudgins, J.L.; Simin, G.S.; Santi, E.; Khan, M.A. An Assessment of Wide Bandgap Semiconductors for Power Devices. *IEEE Trans. Power Electron.* **2003**, *18*, 907–914. [CrossRef]

© 2019 by the authors. Licensee MDPI, Basel, Switzerland. This article is an open access article distributed under the terms and conditions of the Creative Commons Attribution (CC BY) license (http://creativecommons.org/licenses/by/4.0/).

Article

Accuracy Improvement Method of Energy Storage Utilization with DC Voltage Estimation in Large-Scale Photovoltaic Power Plants

Yeuntae Yoo [1], Gilsoo Jang [1], Jeong-Hwan Kim [2], Iseul Nam [2], Minhan Yoon [3,*] and Seungmin Jung [2,*]

1 School of Electrical Engineering, Korea University, Seoul 136-713, Korea; yooynt@korea.ac.kr (Y.Y.); gjang@korea.ac.kr (G.J.)
2 Department of Electrical Engineering, Hanbat National University, Daejeon 305-719, Korea; kjh7384@hanbat.ac.kr (J.-H.K.); iseul9501@hanbat.ac.kr (I.N.)
3 Department of Electrical Engineering, Tongmyong University, Busan 48520, Korea
* Correspondence: minhan.yoon@gmail.com (M.Y.); seungminj@hanbat.ac.kr (S.J.); Tel.: +82-42-821-1096 (S.J.)

Received: 27 August 2019; Accepted: 15 October 2019; Published: 15 October 2019

Abstract: In regard to electric devices, currently designed large-scale distributed generation systems require a precise prediction strategy based on the composition of internal component owing to an environmental fluctuating condition and forecasted power variation. A number of renewable resources, such as solar or marine based energies, are made up of a low voltage direct current (DC) network. In addition to actively considering a power compensation plan, these generation systems have negative effects, which can be induced to a connected power system. When a storage is connected to a DC-based generation system on an inner network along with other generators, a precise state analysis plan should back the utilization process. This paper presents a cooperative operating condition, consisting of the shared DC section, which includes photovoltaic (PVs) and energy storage devices. An active storage management plan with voltage-expectation is introduced and compared via a commercialized electro-magnetic transient simulation tool with designed environmental conditions. Owing to their complexity, the case studies were sequentially advanced by dividing state analysis verification and storage device operation.

Keywords: PV diagnosis; ESS application; DC power flow; DC system dynamics; hybrid generation system

1. Introduction

Solar power still has huge potential and the rate of installation has been growing drastically. A report by the European Photovoltaic Industry Association (EPIA) revealed that the European cumulative photovoltaic (PV) capacity, had increase to more than 120 GW from around 29 GW in 2010 [1]. The report suggests that the PV installation rates will continue to grow through the next decade. A small-scale PV system for residential use is connected to the distribution network, and acts to reduce the electric fee by supplying local demand. An energy management system (EMS) including a building energy management system (BEMS) could be implemented for these kinds of sources, although grid operators concentrate mainly on farm-scale generation systems. Currently, PV plants with a capacity of over 1 GW have been connected worldwide as described in [2], and a number of megawatt (MW)-scaled PV arrays have been configured in those farm networks. These trends will continue to increase, and the production management which covers their stochastic characteristics will pose a challenge in the power system industry.

To stably and efficiently harness energy from various renewable resources, clustered distribution farms have pursued integrating compensation devices for supporting grid operation with enhanced controllability [3]. These circumstances have been presented by the reported variation resulting from unpredictable natural resources, which is a major issue in renewable energies [4]. Although a clustered farm exhibits better power profile with the in-built smoothing effect, a classic power system still requires advanced solutions by concentrating on predictable usage of storage devices. Currently, most storage applications are oriented to time shifting of real power supply rather than support the connected grid [5]. However, as the requirement in terms of response between cluster and operator expands, further concretion on the controller is required in the power system industry [6].

Recently, the renewable-storage application has been increased due to the increasing large-scale wind farm, something that could significantly affect an integrated power grid, [7]. As a fastidious requirement from grid operator is expected in future power grid, the authors in [8] highlight the importance of both real and reactive power compensation according to the power extracting condition. The reactive power compensation generally focuses on the voltage at local power system; it relies on installed auxiliary devices. On the other hand, an energy storage system (ESS) such as a battery storage could act as real power compensation at the distribution generator side. Both compensation options are based on power-conversion system (PCS) that normally follow an optimized signal generated based on the reference by the main system [9–11]. To achieve a close cooperation with distinct distribution units, conceptual approaches that integrate ESS at the direct current (DC) network have been considered through full converter-based wind turbines in Ref. [12]. Recently, typical DC-based sources including PV and tidal have been extended as an ESS clustered form with other resources as described in Ref. [13]. The main objective of those integration forms is to collaborate among the connected devices, which can enhance the controllability based on the imposed system order [14]. The PV system that is able to compose a hybrid network along with ESS at the DC section is usually composed of a number of solar panels for generating usable input voltage and reasonable power extraction. As described in Refs. [15,16], these specifications generally allow not only electric loss in a cable but also voltage drop below the standard testing condition. Therefore, a direct application of ESS in large-scale PV-based DC clusters is classified as being supported by current analysis at each point. In industrial application forms which are based on inbuilt complementary modes to respond to grid requirements, both limitation and compensation options have to match the harsh power fluctuation of renewable sources [17]. In order to continuously utilize these hybrid features, power flow analysis-based approaches and appropriate ESS design are required.

Meanwhile, the increased utilization of storage in the power system industry has resulted in demands of enhanced storage feature, since power swapping requires fast response capability and suitable energy/power capacity [18]. In the case of conventional ESS applications, utilization of a number of control schemes including hierarchical control were tried as described in Ref. [19]. The described works suggest that we can utilize an ESS configuration to achieve power balance in a power grid where there is presence of several renewable energy and variable loads. However, in previous research, the ESSs were separately composed of the renewable network, and the classified power conversion devices had to respond to the order from the operator based on their own topology. To perform a detailed control within a combined DC circuit that integrates ESS, an operation signal for charging/discharge must be generated reflecting an algorithm based on the expectation of power flow. In case the ESS solution is to be implemented in a PV-integrated DC system, it is deemed that a fast voltage analysis method should support the order decision process in order to take instantaneous voltage fluctuation into the controller [20]. In particular, given that renewable generators based on DC could impose unpredictable profiles on ESS operating processes, specific support strategies are required to manage a transient situation to cover these issues.

In this paper, the main objective is to come up with an ESS compensation scheme that considers exact voltage level so as to implement a power-management plan for a DC-combined system by focusing on the demand of the power system. What distinguishes it from previous research on the renewable-ESS integrated networks is that it focused on a method to minimize errors based on detailed circuit analysis rather than systemic utilization. The main analysis in this document focuses on the small power generated by the modules rather than simply focusing on the power extracted from each unit. The entire management process is formed based on power flow analysis for DC systems, which deals with controllable elements that can be reflected on connected equipment. A simulation for verification is designed using the electro-magnetic transient DC (EMTDC) tool. In order to cause a considerable voltage fluctuation according to the output of the connected distributed resources, and to carry out an experiment, a distribution network with a scale of 0.5 km was constructed considering the real radial system. The ESS model configurations, including control topologies, are utilized to implement the case studies. The case studies focus on the operational accuracy of the proposed control scheme in regards to imposed order from operator.

2. System Configuration

2.1. General Objective

The typical layout of a MW-class PV farm system is illustrated in Figure 1. A large-scale PV array contains distinct inverters and transformers. In regard to compensation, when an ancillary device for renewable system is considered, the installation location should be at the main bus or sub-station of the PV system. As shown in the figure, a common area is formed by linking the MW-scale solar power generation arrays with a storage device in the DC section. The main purpose is to define possible impact from current flows in the DC section, and preparing a control method for supporting the compensation process in integrated ESS.

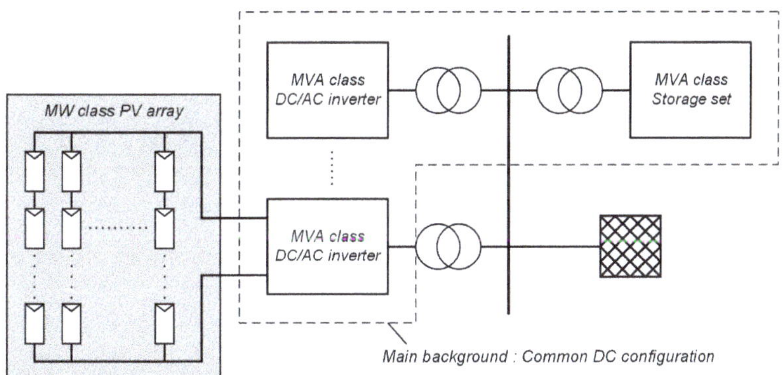

Figure 1. A megawatt (MW)-scale photovoltaic/energy storage system (PV/ESS) for common direct current (DC) configuration.

In various research, large-scale PV plants are modeled with a lumped single generator, by neglecting current flow from each PV module. In terms of power system analysis, this modeling method is fast and useful in identifying the influence of generated power from farm network. However, as mentioned in Ref. [21], a detailed model construction is necessary when deriving a concrete compensation plan is required through confirmation of an instantaneous production of a PV system. It is necessary to select the available topology for industrial use, when implementing DC power flow with a focus on large PV configurations. Generally, to secure the robustness in terms of power extraction, a single MW PV array is connected to the grid through a central inverter topology [22]. With the topology, a mega-voltage ampere (MVA) class DC/AC inverter can accommodate several thousands of PV panels. However, unlike other topologies, the central topology exhibits low levels of flexibility and high mismatching losses due to its huge configuration characteristics. New PV panels for the MW class even consider mismatching losses in PV array arrangement through the use of several re-configuration techniques, as described in Ref. [23].

In DC distribution network research, several studies have been conducted in regards to the effects of DC current owing to PV modules. Authors in Refs. [24,25] focused on large-scale integration of the PV system in terms of grid impact such as voltage drop and defined capacity range of a suitable PV generation system. There has been demand in PV module about a DC current analysis. Refs. [26,27] consider actual losses due to a realistic maximum power point tracking (MPPT) errors in large-scale PV system. Reference [23] pursues minimizing mismatching loss by adopting module sorting techniques. A more detailed interpretation is required to establish a more accurate ESS compensation plan in view of the instantaneous power of the internal module. It is likely that the common DC system mentioned in this section will be affected by the current generated by each module, and the resulting DC voltage variation affects the existing ESS control schemes. In the case of PV, since the power generated by each module varies continuously, the voltage calculation of each point through the power flow analysis can be undertaken instantly. By utilizing this values, effective control method for internal ESS can be derived. As a result, data from the external environment can be used to derive its impacts on the internal electric network to use at the main controller. These measures can be useful in the structure of modern PV management systems that use various sensors to obtain environmental information or utilize predictions.

2.2. Conceptual Design

Figure 2 represents a conceptual control plan of the proposed method. The entire power from DC-based source is transferred to the connected grid through a main PCS. The ESS is integrated into a common DC link using a DC/DC converter. Since the compensation devices in these structures should be located at the front of the main PCS, the power flow from DC generators (PV or wave) should pass through the junction of the ESS [28]. Through this flow, the integrated storage device is directly affected by the electrical conditions with respect to the voltage level which is among the main reference signal of the control process. The main purpose of the compensation devices is to store excess output power (above the reference signal) and utilize a bi-directional PCS to discharge the stored energy on the operators' request. This paper describes DC power flow analysis, including DC network characteristics for PV generators. Section 2 covers a basic system description.

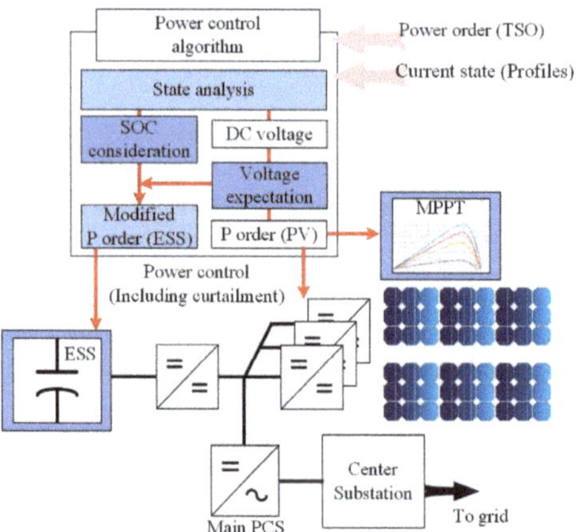

Figure 2. Power control concept of ESS-combined PV system.

2.3. Alternating Current/Direct Current (AC/DC) Hybrid System

General voltage-sourced converter (VSC)-based structures are established at the front of generation system, which means that the generated power must pass through a DC section through its own converter. Therefore, as described in Ref. [29], the power supply from combined DC network could experience more ohmic losses compared to other single generation system. There are several studies on the DC farm topologies to increase efficiency and mitigate these issues as a result [30]. The conventional DC-based farm have extended the available farm scale through the increase in number of the sources and boosting level of the classified voltage.

The model utilized in this paper focuses on coupled power resources structure constructed in DC. Once the ESS is planned to attach to the structure, it should be placed in front of the main sub-system and follow the operators' order. The ESS applications applied in the DC section are based on improved usability. However, the electrical flow in the low voltage section can result to unstable voltage conditions in terms of compensation device usage. As a compensation device in the common DC network is expected in several renewable energy sources, an advanced technique based on DC analysis is required to maintain the system's performance.

2.4. DC Network

2.4.1. Generation System

This paper presents a hybrid system that installs a common DC network for PV and ESS. If used in real-time conditions, then a fast computation method management should be composed to handle electronic based resources. Prior to defining the relationship between order and voltage level, a general power flow analysis have to compose focused on power production. The flow injected by PV modules can be given by the sum of the power produced from each module with incurred losses as follows:

$$P_{\mathrm{PV}} = \sum_{l=1}^{L}\sum_{n=1}^{N} P_n - \sum_{l=1}^{L}\sum_{n=1}^{N} P_{\mathrm{loss}}(n) \tag{1}$$

where, P_{PV} is total output power from PVs, P_n is output power from nth module, P_{loss} is generated loss in the cable, L is total number of arrays, and N is total number of modules.

The ohmic loss can be defined as follows:

$$P_{\mathrm{loss}} = r_{nn+1}\frac{P_{nn+1}^2}{V_n^2} \tag{2}$$

where, P_{nn+1} is real power flow from module n to n+1, R_{nn+1} is resistance value between module n and n+1, and V_n is voltage magnitude of module n.

A PV system generates real power according to the irradiation profile by utilizing applied converter based on the designed curve which is related to the power coefficient. The PCS selects the power point within the operational range to extract the maximum available power. This paper considers the ESS application in the DC link as a power compensation device. If the extracted power of the hybrid system exceeds the designated order by the operator, the ESS is activated to support the entire output power based on the operator's order. It can be composed when the total DC flow of the entire circuit is analyzed using appropriate formulas. The total DC flow with capacity constraint can be defined as in Equation (3), and DC current flow to utilize flow analysis as shown is as follows:

$$P_{\mathrm{dc}} = P_{PV} + P_{ESS}, \qquad |P_{ESS}| \le S_{dc-dc} \tag{3}$$

where, P_{dc} is real power injection from DC system, P_{ESS} charging/discharging quantity from ESS, and $S_{dc\text{-}dc}$ is the Power capacity of DC/DC converter for the ESS.

$$i_{\mathrm{dc}} = \sqrt{g_{\mathrm{eq}} \cdot P_{\mathrm{ref}}} \tag{4}$$

where, i_{dc} is current flow at the converter, g_{eq} is equivalent admittance of PCS, and P_{ref} is reference signal of real power for the PCS.

The constraint in the limit process can be defined as follows:

$$\max P_{dc} \leq \sqrt{S_{PCS}^2 - Q_{PCS}^2} \tag{5}$$

where, S_{pcs} is power capacity of main PCS, and Q_{PCS} is reactive power production from main PCS.

Since it is expected that several modules are integrated into a single array in a PV farm, a voltage variation of the DC side is dependent on the PVs' output power. When an operator wants to handle demand response through applied ESS (charging or discharge), the voltage variations induced in the DC network have to be reflected.

2.4.2. Storage System

The application of the ESS in renewable energy should consider both the own operational state and the generated signals from the grid operator. In general, there is a standardized form of the ESS in an industrial power system. (1) Usually, the ESS connection point provides the point of common coupling (PCC) for renewable energy. (2) To maintain the DC level of the battery, an ESS should apply a distinct DC/DC conversion device. (3) An ESS performs charge/discharge based on imposed limitation (voltage) or a direct order. In case of voltage focused solutions, previous research utilized the specified voltage level as the limit, which was utilized to decide whether the ESS was attached or detached to the grid. Figure 3 describes the concept with a flywheel solution, which employs a super-conducting rotating device [31]. This application relies on a bi-directional controller to control electrical flow and determine whether it charges or discharges by detecting the voltage limitation. In this topology, the controller can strictly manage the voltage levels within the specified range. However, the supplied power from the ESS cannot be defined as a constant value which is able to utilize power balancing. The proposed plan proposes storing and releasing real power with a strict control signal instead of voltage limitation. A bi-directional PCS focuses on designated order to handle entire output power from the DC network.

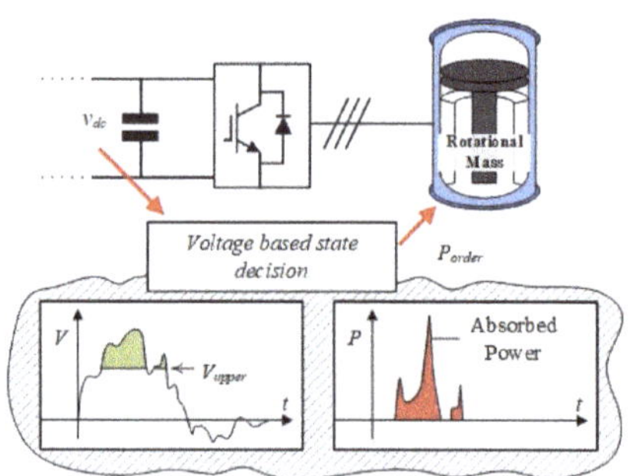

Figure 3. Voltage limitation-based ESS control scheme.

In case of order for ESS, both charging and discharging signals can be represented with the connected PV as follows:

$$P_{ESS}(t) = P_{dc} - \sum P_{PV}(t) \tag{6}$$

In order to correctly extract the power signal, the proposed ESS controller adopts additional DC analysis blocks to enhance the accuracy of charging/discharging signals.

3. DC Flow Analysis

To perform the current flow analysis about a target DC network, an electric based circuit model should first be developed. In order to estimate the voltage variation of each section based on the power extraction, an equivalent circuit was developed for power flow analysis.

In case of the general analysis of the PV circuit for current estimation, the resistance in the circuit is divided into two categories (shunt and series) as shown in Figure 4 to advance it in terms of simplification [32]. With this simplification, environmental factors such as irradiation (G) can be reflected in output current expression. On the PV system, however, since a current flows through the negative pole which is for grounding, for detailed current analysis, a method to reflect this circumstance should be derived. In Ref. [33], a method is derived to eliminate the leakage current in consideration of the earth impedance. In this regard, a method of applying the grounding components to the detailed analysis is continually being studied. In addition, although the general method introduces each environmental variable as a constant value to advance the prediction of current flow, it is difficult for these methods to reflect a real-time environment. In this paper, we focused on the method of deriving the amount of current based on real-time power output, and tried to improve the accuracy by minimizing the variables. Figure 5 shows a simplified circuit which includes PV, ESS and DC/AC inverter to connect the main grid. The resistance components between each PV module were imposed in the circuit for purposes of clearly describing voltage fluctuation on a low-voltage DC network. Reflection of the inner voltage variation condition for large-scale PV generation system is a demanded feature which is also considered in commercialized simulation tool as well. In this paper, the resistance components at the negative DC pole which is usually considered as ground section is utilized to improve the accuracy of voltage calculation.

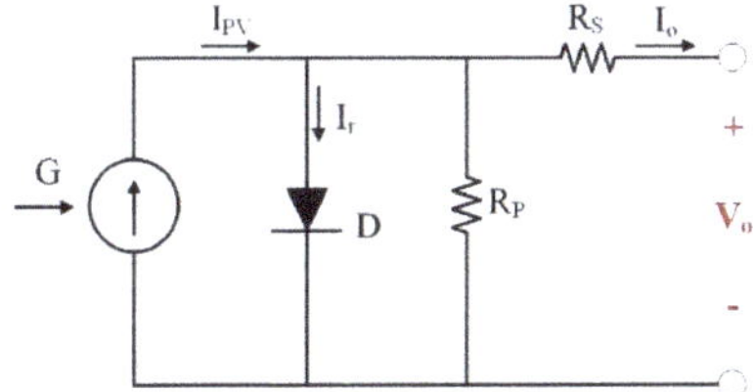

Figure 4. Equivalent circuit of single diode model for PV.

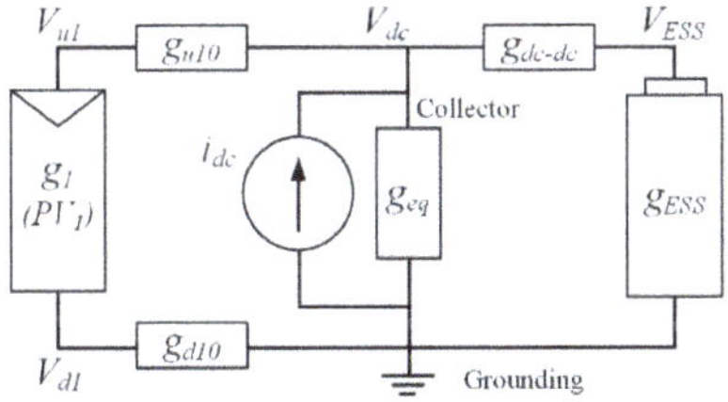

Figure 5. Power flow analysis model for DC.

Firstly, the equivalent component of the PV module in the figure is based on a single module; the basic objective of the described circuit is to derive a mathematical expression for voltage level of each section including the front of the inverter device. In the case of the PCS for a PV system, since the AC side is connected to stable grid network, it is capable of converging as an equivalent circuit. To reflect the voltage variation in the DC section, the Norton equivalent method is applied in this study. When DC network analysis is imposed into the PV system, there is a need to imply the power extracting condition for each module in current flow form. The output power of the PV module can be modified to obtain an equivalent circuit with negative values if each module is considered as an admittance component. In case of ESS component, both charging and discharging power can be implemented using an equation that has described modeling processes. Each component in the figure can be transformed using the components mentioned as follows:

$$\begin{vmatrix} g_1 + g_{u10} & -g_1 & -g_{u10} & 0 \\ -g_1 & g_1 + g_{d10} & 0 & 0 \\ -g_{u10} & 0 & g_{u10} + g_{eq} + g_{pcs} & -g_{eq} \\ 0 & 0 & -g_{eq} & g_{eq} + g_{ESS} \end{vmatrix} \begin{vmatrix} V_{u1} \\ V_{d1} \\ V_{dc} \\ V_{ESS} \end{vmatrix} = \begin{vmatrix} 0 \\ 0 \\ i_{dc} \\ 0 \end{vmatrix} \tag{7}$$

The DC section of single array would consist directly connected PV module, main PCS for subsystem, and DC/DC convertor for introduced storage devices. The three basic models could be transferred as an electrical component in order to proceed as an iterative calculation process. The iteration requires input parameters including a known quantity to derive each section's voltage levels which are considered as unknown values in the circuit elements. The system input parameters include admittance values, which are repeatedly updated with extracted power from series-connected PV modules, and derived based on Equation (8).

$$g_n = -\frac{P_n}{(V_{un} - V_{dn})^2} \tag{8}$$

The voltage for PV modules considers both the upper and lower side and are continuously modified throughout the iteration process as unknown values. The admittance component of ESS are as follows in Equation (9) by assuming shared grounding option.

$$g_{ESS} = -\frac{P_{ESS}}{V_{ESS}^2} \tag{9}$$

Since the voltage level of each section is dependent on electrical current via power flows, the modified values would continuously affect the admittance component until every input parameter is converged. The corresponding current of PCS would hold the relevant DC power flow as a strict component in the system matrix. To perform the iteration, which is considered as a main calculation, an inverse matrix is utilized to progress the iteration as described in Equation (10).

$$[V] = [g]^{-1} \times [I] \tag{10}$$

The contents of Figure 5 are based on single PV module; hence, large-scaled PV generation system requires further dimensional matrix. If n modules are added, the circuit will be expanded as mentioned above, which results in a change in the basic equation as follows in (11).

When adding n modules, an $n \times 2$ matrix size expansion is progressed on an existing equation. If this equation is analyzed in detail, it is possible to consider a difference of production by large modules and organize an equation about the profile of each module with the required data.

$$\begin{vmatrix} g_1 + g_{u1n} & -g_1 & -g_{u1n} & 0 & 0 & 0 \\ -g_1 & g_1 + g_{d1n} & 0 & -g_{d1n} & 0 & 0 \\ -g_{u1n} & 0 & g_{u1n} + g_n + g_{un0} & -g_n & -g_{un0} & 0 \\ 0 & -g_{d1n} & -g_n & g_{d1n} + g_n + g_{dn0} & 0 & 0 \\ 0 & 0 & -g_{un0} & 0 & g_{un0} + g_{pcs} + g_{eq} & -g_{eq} \\ 0 & 0 & 0 & 0 & -g_{eq} & g_{eq} + g_{ESS} \end{vmatrix} \begin{vmatrix} V_{u1} \\ V_{d1} \\ V_{un} \\ V_{dn} \\ V_{dc} \\ V_{ess} \end{vmatrix} = \begin{vmatrix} 0 \\ 0 \\ 0 \\ 0 \\ i_{dc} \\ 0 \end{vmatrix} \tag{11}$$

4. Simulation

4.1. Simulation Design

In order to assess the proposed control method, a detailed simulation is conducted using a power-system computer aided design (PSCAD). Figure 6 shows a layout of the PV connected distribution network, that utilizes the linked DC section. The PV generation system in the network has a nominal power capacity of 400 kVA. To verify the proposed method, a single PV generation system includes 120 parallel/40 series modules. The PV system is connected to the displayed distribution network by utilizing the central inverter topology. The DC network also includes the ESS model to verify the proposed reference modification method. The distance data about the distribution network is reflected in pi-line which is the base model in PSCAD/EMTDC. The network information, which is used for utility grid construction, is shown in Table 1 including the PI-line distances in Figure 6.

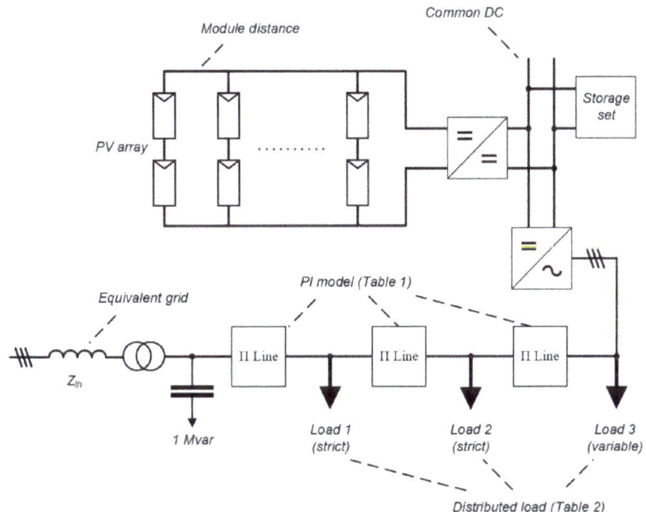

Figure 6. Simulated distribution network including MW-scale PV/ESS.

Table 1. Numerical data for the simulation.

Specific Data	Value	Unit
Rated AC voltage	22.9	kV
Rated DC voltage	500	V
Distance between each module	0.5	Meter
Number of load	3	
Rated energy of ESS	200	kWh
Rated power of ESS	200	kW
Rated power of PV system	1200	kW
PI-line distance	500	Meter
Substation voltage (HV)	154	kV
Short-circuit ratio of utility grid	15	
X/R ratio of utility grid	15	

Figure 7 describes the simulations of the designed PV production. PV system extracts real power based on the radiation data as per the specific coefficient values. A small variation of the extracted power could affect the operation of the connected ESS. To check ESS control effects under certain conditions, unexpected variations need to be checked. Table 2 includes applied load parameters for the case studies. The main objective of the simulation is to confirm that the proposed method shows better reliability compared to rated voltage-based control. The voltage controller is configured based on an existing solar-ESS combined control, which takes into account the sensitivity between voltage and power production [34]. The method generally operates in combination with on-load tap changer (OLTC), yet it depends on the characteristics of the connected power system. Therefore, in this paper, we try to compare the results when only the main controller was applied. To check ESS operation, a number of discrete load variations were designed for each scenario. The simulation mainly considers the charging/discharging processes to confirm profile accuracy in regards to reference signal. The reference signal is designed to change depending on the load condition, while the configured case studies and entire simulation time is set to 10 s including the start-up time. Assuming abrupt load variation, a number of orders changing sections (for charging and discharging) are designed for the simulation. A few of the load changes are executed by stages as displayed in Figure 8. The momentary load changes (75 kW) generates a demand about the order change for reducing the distribution network burden.

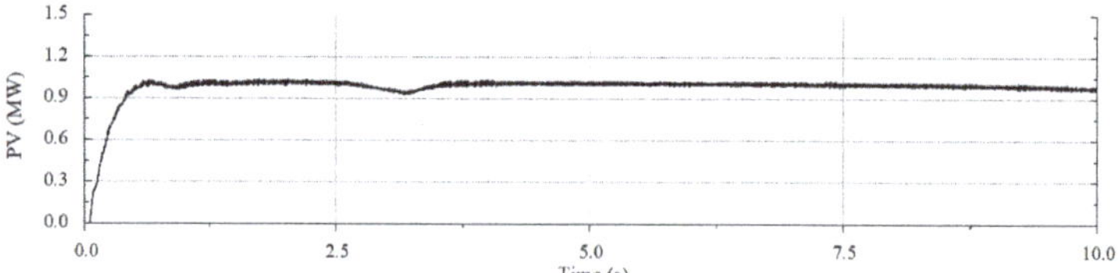

Figure 7. Basic PV production in case study.

Table 2. Simulated case study description.

Imposed Control Method	Strict Voltage Control Method, Proposed Control Method
Simulation time	10 s
Base load condition (Real power)	Load 1: 900 kW Load 2: 1500 kW Load 3: 1380 kW
Load increase sections (Load 3)	2.5–3, 3.5–4, 8–8.5 s (75 kW) 3–3.5, 4–4.5 s (150 kW)
Load decrease sections (Load 3)	5–5.5, 6–6.5, 8.5–9 s (75 kW) 5.5–6, 6.5–7 s (150 kW)

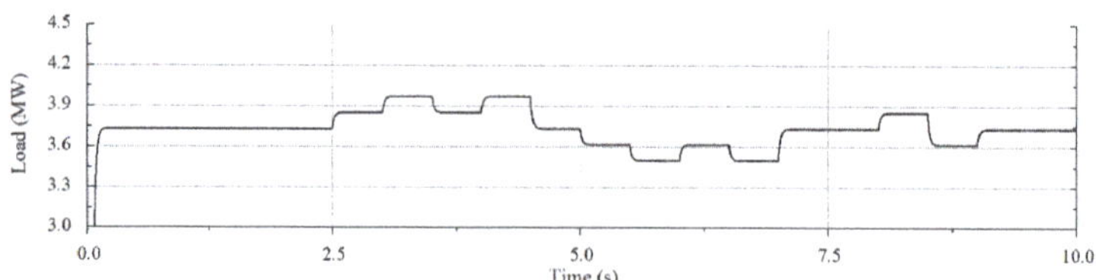

Figure 8. Simulated load variation in case study.

Without ESS, the connected substation should supply the required power based on the change of the distribution network. However, with ESS, a distribution system operator would attempt to cover the unexpected variation through the support of ESS due to operation responsibility (i.e., rate of power change). This forms the basis of the operational objective of the ESS in this scenario. The total power supplied from the grid with non-ESS condition is shown in Figure 9. Based on this condition, ESS charging/discharging would be handled through applied methods. The basic algorithm using the power command and the rated voltage is entered to obtain an effect firstly, and a comparison analysis with the proposed use is advanced.

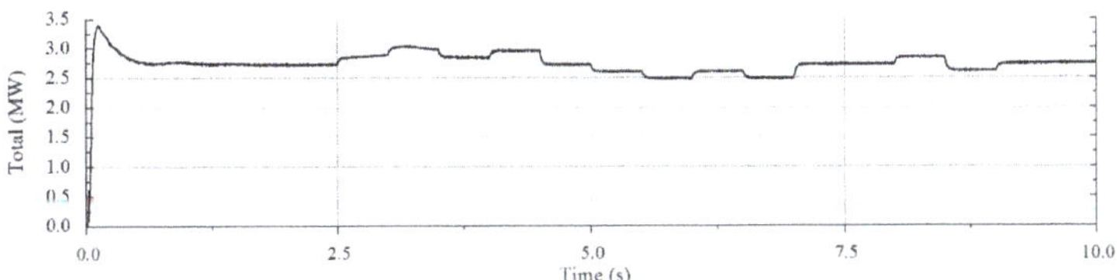

Figure 9. Power supplied quantity from main grid (non-ESS).

4.2. Simulation Results

Figure 10 shows a power production curve of the common DC system including ESS output with rated voltage level consideration. The ESS charges and discharges depending on the load variation on Figure 8. It can be confirmed that the power supplied from DC, which depends mainly on the amount of PV generation, has been modified according to the complementary power of ESS corresponding to the increase in load of each designed section. Depending on the total output power including extracted power from the ESS, unexpected voltage fluctuations can occur in the DC system, and the main objective of the simulation is to compare the response ability. Since the control was established to reduce the variations of the power supplied, the quantity from the grid can be attenuated as illustrated in Figure 11. Compared with the general conditions in Figure 9, the supplied power from the grid can avoid discrete changes through ESS compensation. However, a minor variation is still exhibited on the power curve, hence more precise compensations are still required.

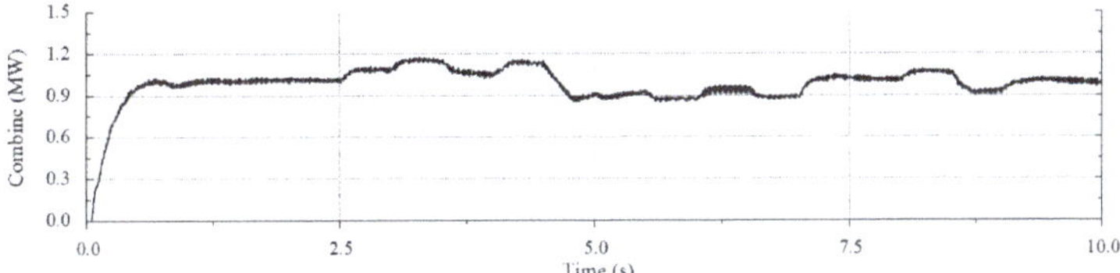

Figure 10. Combined power production of DC network (ESS-strict voltage control).

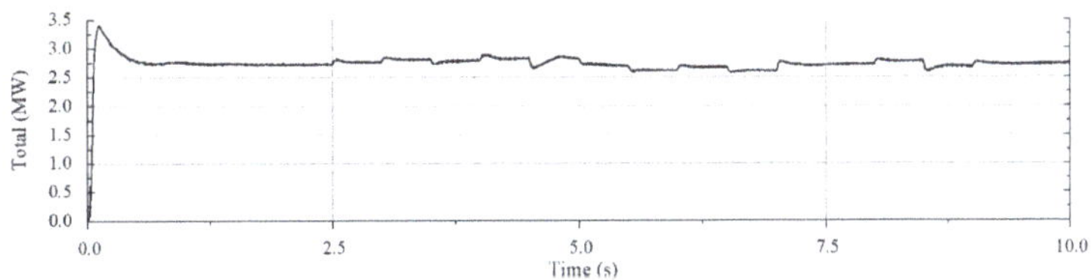

Figure 11. Quantity of power supplied from main grid (ESS-strict voltage control).

The proposed flow expectation method is applied to the same simulating condition. In regards to precise voltage condition, the proposed control attempts to reduce the previous errors and smoothen the power supplied quantity from the grid.

The power production is handled within the rated capacity and the differences between control methods are generated depending on the precision of controllers in terms of voltage. The highlighted charging and discharging sections (2.5–4.5 and 5–7 s) are described in Figure 12. In both cases Figure 12a,b, a strict voltage control method finds it difficult to detect DC voltage variations, and generate mismatches for both sections. The method triggers ESS operation with signal and supports the grid according to the voltage mismatch through comparison of the rated voltage (500 V). Since storage sets perform according to the power capacity, there could be double error generated which could affect the grid power balance. The ESS output curves of the proposed method are displayed together with strict voltage control. The mismatches are reduced in both control modes as described in the figures, and the reliability of the compensation significantly improved in terms of solution. Without the initialization section, the power supplied by ESS almost coincides with the increased variable load value. The power supplied quantity from the main substation is illustrated in Figure 13. Compared with the previous state, as shown in the figure, the fluctuation generated is significantly reduced. The control error via the general method is modified in the case of both charging and discharging processes.

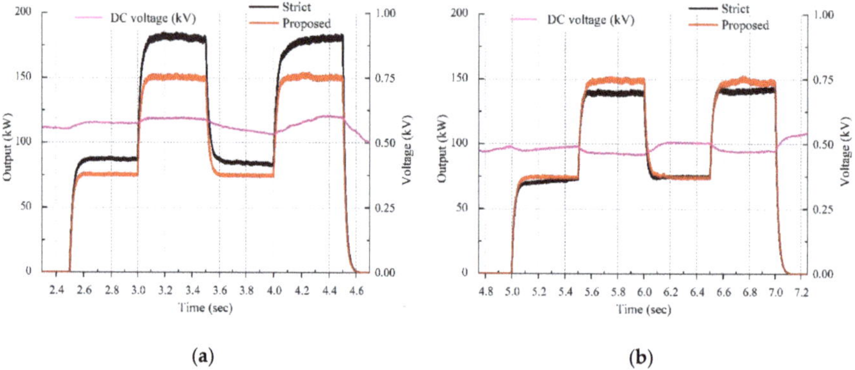

Figure 12. Detailed power compensation quantities for ESS with DC voltage variations (**a**) extracted power quantity in the discharging section (2.5 to 4.5 s); (**b**) absorbed power quantity in the charging section (5 to 7 s).

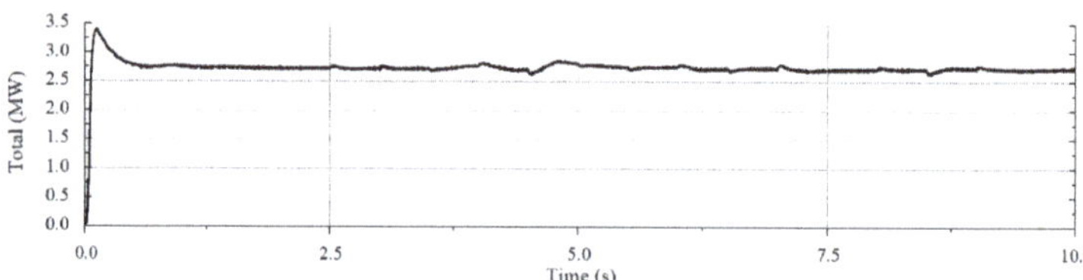

Figure 13. Quantity of power supplied from main grid (ESS-proposed control).

The total mismatch quantity for both methods in terms of power support is presented in Table 3. The difference between the demanded control quantity (calculated by strict power references) and the profile is reduced for both control modes, and the reliability of the profiled quantity is significantly improved in regard to energy compensation. In particular, error generated in the discharging state is significantly reduced with the proposed controller.

Table 3. Numerical results for imposed order and profiled quantity.

Control	ESS Mode	Demanded Quantity (Wh)	Accomplished Quantity (Wh)	Gap(Wh)	Error (%)
Strict voltage	Charging	145.83	136.304	9.526	6.53
	Discharging	145.83	167.877	22.047	15.12
Proposed	Charging	145.83	140.912	4.918	3.37
	Discharging	145.83	143.834	1.996	1.37

5. Conclusions

This paper proposes a modified compensation method for the MW-scaled AC–DC hybrid systems such as storage devices. To perform the ESS operation with an effective reference order, a voltage-estimation method was formulated and adopted in the designed generation system. Verification of the performance improvement in regard to the real power management that responds to the order is verified through case studies. According to the changed state in terms of load, the proposed controller generated a precise order considering the real power extraction from the DC network. The MW-scaled DC networks with different sources experience voltage fluctuation when interworking with the grid. In case of the proposed method, however, it is required to establish the power production of the module as an input signal since credibility is required not only on the PV module but also on the applied sensors. When pursuing detailed management of large-scale DC systems, the application could be considered by the ESS operator. In general, the large-scale DC networks with diverse sources have voltage fluctuation when interworking with the grid. The designed method is based on a multiple PV system, the active support plan using ESS is able to further expand large DC system. Even if the error mitigation on the basis of the order is a minor issue in certain system, the improvement can be useful for the ESS management process with respect to state of charge. The controller will improve estimation accuracy of the remaining amount for the ESS, and this can help support operational accuracy.

Author Contributions: Conceptualization, Y.Y. and S.J.; Methodology, S.J.; Software, Y.Y.; Validation, Y.Y., and S.J.; Formal Analysis, J.-H.K., I.N. and M.Y.; Investigation, J.-H.K., I.N. and M.Y.; Data Curation, J.-H.K., I.N. and M.Y.; Writing-Original Draft Preparation, S.J.; Writing-Review and Editing, S.J. and G.J.; Supervision, S.J.; Project Administration, S.J. and G.J.; Funding Acquisition, S.J.

Funding: This work was supported by the National Research Foundation Grant (No. 2018R1C1B5030524) and the Korea Institute of Energy Technology Evaluation and Planning/MOTIE (No. 20184030201900) funded by the Korean government.

Conflicts of Interest: The authors declare no conflict of interest.

Nomenclature

g_n	Equivalent admittance of nth module
g_{uno}	Admittance of positive side DC cable between nth module and collector
g_{dno}	Admittance of negative side DC cable between nth module and ground
$g_{dc\text{-}dc}$	Equivalent admittance about DC/DC convertor for ESS
g_{ESS}	Equivalent admittance of ESS module
G	Solar irradiation (W/m^2)
I_{PV}	PV current in single diode model
I_r	Reverse saturation current of PV
I_{sc}	Short circuit current of PV
I_o	Output current in single diode model
R_P	Shunt resistance of PV
R_S	Series resistance of PV
V_{un}	Upper-side voltage of nth module
V_{dn}	Lower-side voltage of nth module
V_{dc}	Collector voltage
V_{ESS}	Induced ESS voltage

References

1. Rakhshani, E.; Rouzbehi, K.; Sánchez, A.J.; Tobar, A.C.; Pouresmaeil, E. Integration of Large Scale PV-Based Generation into Power Systems: A Survey. *Energies* **2019**, *12*, 1425. [CrossRef]
2. Biggest Solar Power Plants. Mechanical, Electrical and Electronics Engineers. Available online: https://meee-services.com/biggest-solar-power-plants (accessed on 30 August 2018).
3. Su, M.; Luo, C.; Hou, X.; Yuan, W.; Liu, Z.; Han, H.; Guerrero, J.M. A Communication-Free Decentralized Control for Grid-Connected Cascaded PV Inverters. *Energies* **2018**, *11*, 1375. [CrossRef]
4. Osório, G.J.; Shafie-khah, M.; Lujano-Rojas, J.M.; Catalão, J.P. Scheduling Model for Renewable Energy Sources Integration in an Insular Power System. *Energies* **2018**, *11*, 144. [CrossRef]
5. Subramani, G.; Ramachandaramurthy, V.K.; Padmanaban, S.; Mihet-Popa, L.; Blaabjerg, F.; Guerrero, J.M. Grid-tied photovoltaic and battery storage systems with Malaysian electricity tariff—A review on maximum demand shaving. *Energies* **2017**, *10*, 1884. [CrossRef]

6. Zeng, Z.; Yang, H.; Zhao, R.; Cheng, C. Topologies and control strategies of multi-functional grid-connected inverters for power quality enhancement: A comprehensive review. *Renew. Sustain. Energy Rev.* **2013**, *24*, 223–270. [CrossRef]
7. Miao, L.; Wen, J.; Xie, H.; Yue, C.; Lee, W. Coordinated control strategy of wind turbine generator and energy storage equipment for frequency support. *IEEE Trans. Ind. Appl.* **2015**, *51*, 2732–2742. [CrossRef]
8. Bird, L.; Cochran, J.; Wang, X. *Wind and Solar Energy Curtailment: Experience and Practices in the United States*; NREL: Golden, CO, USA, 2014.
9. Chen, P.; Thiringer, T. Analysis of Energy Curtailment and Capacity Over installation to Maximize Wind Turbine Profit Considering Electricity Price—Wind Correlation. *IEEE Trans. Sustain. Energy* **2017**, *8*, 1406–1414. [CrossRef]
10. Saez-de-Ibarra, A.; Milo, A.; Gaztañaga, H.; Debusschere, V.; Bacha, S. Co-Optimization of Storage System Sizing and Control Strategy for Intelligent Photovoltaic Power Plants Market Integration. *IEEE Trans. Sustain. Energy* **2016**, *7*, 1749–1761. [CrossRef]
11. Stimoniaris, D.; Tsiamitros, D.; Dialynas, E. Improved Energy Storage Management and PV-Active Power Control Infrastructure and Strategies for Microgrids. *IEEE Trans. Power Syst.* **2016**, *31*, 813–820. [CrossRef]
12. Shi, G.; Zhang, J.; Cai, X.; Zhu, M. Decoupling control of series-connected DC wind turbines with energy storage system for offshore DC wind farm. In Proceedings of the IEEE 7th International Symposium on Power Electronics for Distributed Generation Systems (PEDG), Vancouver, BC, Canada, 27–30 June 2016.
13. Vargas, L.S.; Bustos-Turu, G.; Larraín, F. Wind Power Curtailment and Energy Storage in Transmission Congestion Management Considering Power Plants Ramp Rates. *IEEE Trans. Power Syst.* **2015**, *30*, 2498–2506. [CrossRef]
14. Knap, V.; Chaudhary, S.K.; Stroe, D.; Swierczynski, M.; Craciun, B.; Teodorescu, R. Sizing of an Energy Storage System for Grid Inertial Response and Primary Frequency Reserve. *IEEE Trans. Power Syst.* **2016**, *31*, 3447–3456. [CrossRef]
15. Zhang, H.; Yue, D.; Xie, X. Robust Optimization for Dynamic Economic Dispatch under Wind Power Uncertainty with Different Levels of Uncertainty Budget. *IEEE Access* **2016**, *4*, 7633–7644. [CrossRef]
16. Byungdoo, J.; Hyun, K.; Heechan, K.; Hansang, L. Development of a Novel Charging Algorithm for On-board ESS in DC Train through Weight Modification. *J. Electr. Eng. Technol.* **2014**, *9*, 1795–1804.
17. Yang, J.; Fletcher, J.E.; O'Reilly, J. Multiterminal DC wind farm collection grid internal fault analysis and protection design. *IEEE Trans. Power Del.* **2010**, *25*, 2903–2912. [CrossRef]
18. Kaipia, T.; Salonen, P.; Lassila, J.; Partanen, J. Possibilities of the low voltage DC distribution systems. In Proceedings of the NORDAC Conference, Stockholm, Sweden, 30 August–2 September 2006.
19. Dzamarija, M.; Keane, A. Autonomous curtailment control in distributed generation planning. *IEEE Trans. Smart Grid* **2016**, *7*, 1337–1345. [CrossRef]
20. Jung, S.; Yoon, Y.-T.; Jang, G. Adaptive Curtailment Plan with Energy Storage for AC/DC Combined Distribution Systems. *Sustainability* **2016**, *8*, 818. [CrossRef]
21. Yazdani, A.; Dash, P.P. A Control Methodology and Characterization of Dynamics for a Photovoltaic (PV) System Interfaced with a Distribution Network. *IEEE Trans. Power Deliv.* **2009**, *24*, 1538–1551. [CrossRef]
22. Cabrera-Tobar, A.; Bullich-Massagué, E.; Aragüés-Peñalba, M.; Gomis-Bellmunt, O. Topologies for large scale photovoltaic power plants. *Renew. Sustain. Energy Rev.* **2016**, *59*, 309–319. [CrossRef]
23. Mansur, A.A.; Amin, M.R.; Islam, K.K. Performance Comparison of Mismatch Power Loss Minimization Techniques in Series-Parallel PV Array Configurations. *Energies* **2019**, *12*, 874. [CrossRef]
24. Widén, J.; Wäckelgård, E.; Paatero, J.; Lund, P. Impacts of distributed photovoltaics on network voltages: Stochastic simulations of three Swedish low-voltage distribution grids. *Electr. Power Syst. Res.* **2010**, *80*, 1562–1571. [CrossRef]
25. Paatero, J.V.; Lund, P.D. Effects of large-scale photovoltaic power integration on electricity distribution networks. *Renew. Energy* **2007**, *32*, 216–234. [CrossRef]
26. Koutroulis, E.; Blaabjerg, F. A New Technique for Tracking the Global Maximum Power Point of PV Arrays Operating Under Partial-Shading Conditions. *IEEE J. Photovolt.* **2012**, *2*, 184–190. [CrossRef]
27. Karanayil, B.; Ceballos, S.; Pou, J. Maximum Power Point Controller for Large Scale Photovoltaic Power Plants Using Central Inverters under Partial Shading Conditions. *IEEE Trans. Power Electron.* **2018**, *34*, 3098–3109. [CrossRef]

28. Garces, A. A Linear Three-Phase Load Flow for Power Distribution Systems. *IEEE Trans. Power Syst.* **2016**, *31*, 827–828. [CrossRef]
29. Karthikeyan, V.; Gupta, R. Multiple-Input Configuration of Isolated Bidirectional DC–DC Converter for Power Flow Control in Combinational Battery Storage. *IEEE Trans. Ind. Inform.* **2018**, *14*, 2–11. [CrossRef]
30. Chew, B.S.H.; Xu, Y.; Wu, Q. Voltage Balancing for Bipolar DC Distribution Grids: A Power Flow Based Binary Integer Multi-Objective Optimization Approach. *IEEE Trans. Power Syst.* **2019**, *34*, 28–39. [CrossRef]
31. Jabr, R.A.; Džafić, I. Solution of DC Railway Traction Power Flow Systems Including Limited Network Receptivity. *IEEE Trans. Power Syst.* **2018**, *33*, 962–969. [CrossRef]
32. Jayalakshmi, N.S.; Gaonkar, D.N.; Adarsh, S.; Sunil, S. A control strategy for power management in a PV-battery hybrid system with MPPT. In Proceedings of the 2016 IEEE 1st International Conference on Power Electronics, Intelligent Control and Energy Systems (ICPEICES), Delhi, India, 4–6 July 2016; pp. 1–6.
33. Ardashir, J.F.; Sabahi, M.; Hosseini, S.H.; Blaabjerg, F.; Babaei, E.; Gharehpetian, G.B. A Single-Phase Transformerless Inverter with Charge Pump Circuit Concept for Grid-Tied PV Applications. *IEEE Trans. Ind. Electron.* **2017**, *64*, 5403–5415. [CrossRef]
34. Zhang, D.; Li, J.; Hui, D. Coordinated control for voltage regulation of distribution network voltage regulation by distributed energy storage systems. *Prot. Control Mod. Power Syst.* **2018**, *3*, 3. [CrossRef]

© 2019 by the authors. Licensee MDPI, Basel, Switzerland. This article is an open access article distributed under the terms and conditions of the Creative Commons Attribution (CC BY) license (http://creativecommons.org/licenses/by/4.0/).

Article

Practical Application Study for Precision Improvement Plan for Energy Storage Devices Based on Iterative Methods

Jaewan Suh [1], Minhan Yoon [2,*] and Seungmin Jung [3,*]

[1] Department of Electrical Engineering, Dongyang Mirae University, Seoul 08221, Korea; jwsuh@dongyang.ac.kr
[2] Department of Electrical Engineering, Kwangwoon University, Seoul 01897, Korea
[3] Department of Electrical Engineering, Hanbat National University, Daejeon 305-719, Korea
* Correspondence: minhan.yoon@gmail.com (M.Y.); seungminj@hanbat.ac.kr (S.J.); Tel.: +82-42-821-1096 (S.J.)

Received: 29 December 2019; Accepted: 1 February 2020; Published: 4 February 2020

Abstract: In the aspect of power grid, attention is being given to conditions of environmental variation along with the need for precise prediction strategies based on control elements in recently designed large-scale distributed generation systems. With respect to distributed generators, an operational prediction system is used to respond to the negative impacts that could be generated. As an active response plan, efforts are being made by system operators to cover fluctuations with utilization of battery-based storage devices. Solar or ocean energy that shares electrical structure with an energy storage system has recently being seen as a combined solution. Although this structure is supported by a state analysis plan, such methods must be performed within the range where the response is possible under consideration of the power requirements of the electronic devices. This paper focuses on an iterative based solution for enhancing response of storage that included in DC generation system, to check its availability in terms of possible calculation load. A previous storage management plan was utilized and tested using a commercially available transient electromagnetic simulation tool that focused on possible delays. Case studies were performed sequentially on the time delays based on utilizable inverter topologies.

Keywords: PV diagnosis; ESS application; DC power flow; calculation load; iterative methods

---

1. Introduction

Recent advances in the field of energy has shown that the penetration rate of photovoltaic (PV) power generation is expected to increase steadily over the next decade [1]. The existing small-scale PV systems dominating local demand has been replaced with bulk farm systems to take advantage of environmentally friendly policies as described in [2]. Until now, network operators have focused on the characteristics of large-scale power generation systems and have focused on increasing its capacity. To add to this, the demand for properly managed generated power has been derived due to grid expansion. The recently revised IEEE-1547, particular momentary cessation, illustrates these requirements of related industrial sectors [3]. Currently, PV systems with a capacity of one gigawatt or more are implemented worldwide, including an operating system to enable compliance with directives of system operators. In a PV farm, several sets of megawatt (MW) arrays are built, which should interact with centralized topologies. The need for a control strategy that focuses on the detailed specification of PV is increasing and may be a major challenge in the power system industry that focuses on stochastic uncertainty.

Although renewable energies supply smoothed energy by composing a farm network, the operators still demand advanced solutions for flexible power management with additional compensators [4,5].

However, most major storage applications are still geared to play a role in responding to demand rather than providing real-time compensation [6]. As described in [7], since the requirements in terms of response between the farm and operators have expanded, the power system industry has considered advanced real-time solutions. The study in [8], depending on the various practical conditions, emphasizes the importance of real/reactive compensation response. Reactive power compensation generally focuses on voltage fluctuations in the local energy system, but conversely, real power compensation requires control that focuses on supply and demand of the entire system. These real power reactions are achieved with the aid of detailed backup control of the energy storage system (ESS) based on power conversion systems (PCSs) which are performed with optimized signals generated according to the main system criteria as in [9–11].

Newly developed renewable energy has been generalized to connect to a grid through full-converter based PCS. A process for combining renewable sources has been commercialized with a study on how to treat an independent DC section as a cluster as in the concept of multi-terminal DC [12]. The main objective of this approach is to improve its control capability by sharing the imposed order from a transmission system operator (TSO) with each distributed energy source (DES) in the cluster. Since PV requires a relatively small geographical area, the concept has been preferentially applied to existing farms along with possible storage system [11]. In a PV based hybrid DC system, which adopts a storage application, to perform precise control in a region where there is a number of DESs, a compensatory calculation process for each signal may be necessary. With regards to this, a proposal about a voltage calculation method based on a power flow analysis that considers the current flow between the PV modules is presented in [13], with detailed description of a practical ESS application. The possible voltage fluctuations generated between cables which were reported in [14,15] was reflected and analyzed in [13].

In this paper, in the signal dispatch process of a hybrid system, a feasibility analysis about proceeding with order management while calculating an appropriate voltage was derived. The main objective was to check the feasibility of an ESS compensation scheme which applies a voltage signal calibration in order to implement a power management plan for a hybrid system according to the applied load. Previous formula descriptions were expanded to enable it to deal with practical voltage variations. To carry out a practical approach in which an included PCS controller manages not only the voltage of each DC section but also the calculation delays, a simulation analysis was studied and confirmed with the consideration of real power that changes continuously. Focusing on a distribution network, the voltage impact was established and analyzed to check the availability of additional signal compensation plan. The ESS model configurations, including control topologies, are composed as in [13] to implement a reasonable case design.

2. Electrical Components

2.1. General Objective

The power flow calculation is a conventional method used in the power system analysis to confirm not only the power supply condition but also the status of each bus [16]. In general, it is based on the Newton–Raphson method of AC power system analysis without the consideration of grid scale as investigated in some literatures [17–21]. The optimal power flow based on an iterative method have been modified to consider recently composed DC components as revealed in [22]. The study in [22] considers a voltage-sourced converter (VSC) based on high voltage DC in the power flow equation to improve accuracy and stability in power system analysis. Meanwhile, the pure DC network also utilized an iterative method in a nodal analysis [23]. The methods are normally formed to find the DC voltage level at each node in state analysis. These processes are influenced by the scale of the system matrix in terms of calculation time, which is a major problem in the order management plan of a real-time controller [24].

The aim of this paper is to present a feasibility study on the power flow analysis approach to improve accuracy of managing a hybrid system which consist of several DESs, incorporating the characteristics of main controller. Figure 1 illustrates the typical figure of the controller structure highlighting the target processes. A storage system is included in the structure connected to the main PCS with a boost chopper device. Together with the storage device, the PV modules are expected to connect to the main PCS as a hybrid form. Several current flows from DESs will form DC current (i_{dcn}) and be imposed on the main PCS and the measured signals (i_n) used in the highlighted section for voltage calculation.

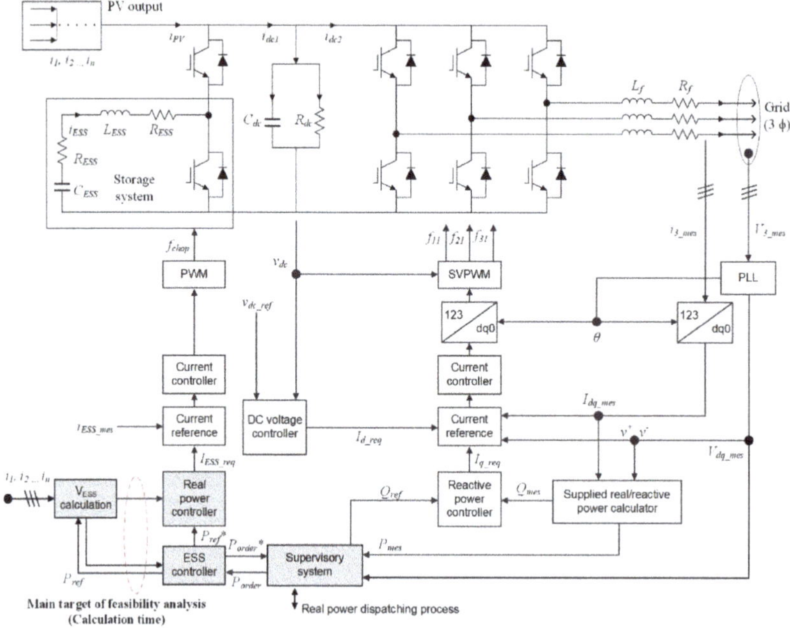

Figure 1. Typical structure of a photovoltaic (PV)-energy storage system (ESS) power conversion system. PLL, phasor locked loop.

A typical control flow for VSC is applied. The measured DC voltage (v_{dc}) is used in the PCS controller for grid-side, and the controller is independent of the real power control for connected DESs in the DC section. The determined reference for DC system operation (v_{dc_ref}) handles PCS signals based on measured values at grid side phasor locked loop (PLL). These control flows are for composing a PV generation system that could follow maximum power point tracking process. When a monitoring system receives a real power order from a TSO, a real power control for the entire DC system is operated independently. Without a signal correction scheme, the real power command for the ESS ignores a voltage variation in the DC section. However, based on a PV current flow signal, a modified value could be generated in this concept to decide the final reference signal for the power controller. The likely issues that may be encountered when the calculation process is applied to existing controls are illustrated in Figure 2. After constructing the line matrix for power flow analysis, a delay can be generated according to the size of the matrix for its complementation process. As the size of PV increases, the possibility of time delay is expected to increase since the calculation load for signal correction depends on the number of modules.

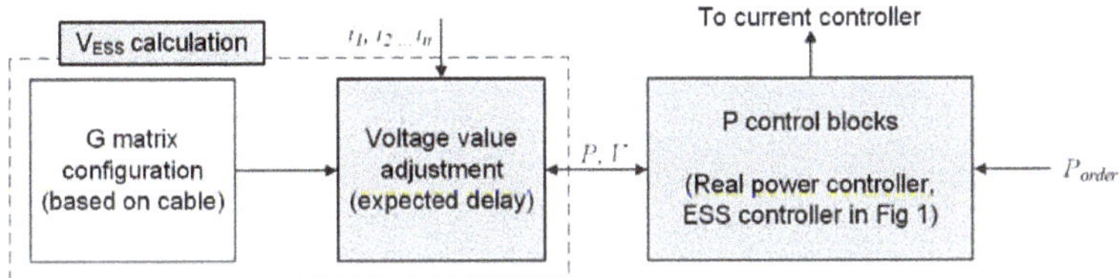

Figure 2. Description of added block for accuracy improvement method.

2.2. Hybrid System

The hybrid model used in this paper focuses on the structure of combined energy resources. It is expected that all the power generated by the DES passes through the main PCS. If an ESS is included, it must be able to comply with the TSO's order according to the current flow variation in the main PCS. Sectional losses and errors are frequent in coupled DC networks as described in [25] and approaches to these issues were being studied [26]. However, if a combined DC system expands its size, the low DC voltage may generate unstable conditions and require precise techniques to comply with imposed orders. In this study, the hybrid model was employed and a compensation technique that can reflect the voltage fluctuations of each DC section applied.

2.3. PV Plant Configuration

Typical information of a PV generation system is shown in Table 1. A PV system contains inverters by based on own topology: central, string, multistring, and module integrated. The general form of the hybrid system takes a large-scale PV farm that is suitable for the ESS compensation plan. Thus, central and multistring topologies (more than 30 kilowatts (kW)) would be appropriate in this analysis. The PV connection that includes the inverter and transformer based on both topologies is described in Figure 3. As shown in the figure, in the case of a multistring topology, a common DC area is formed at connection point of several single strings. Since the DC-DC converters could measure each connection point of the string, a power flow analysis for generating a modified ESS signal could be simplified. In other words, the feasibility study for the multistring topology can be progressed based on one single array that makes up a string (the matrix for the modules is larger than that of the converters). The generalized string structure in [27] is used in this analysis. To secure the robustness in terms of power extraction, a MW-scaled PV is connected to the grid through a central inverter topology [28]. With the topology, a mega-voltage-ampere (MVA) class DC/AC inverter can accommodate several thousands of PV panels. However, unlike other topologies, the central topology exhibits low levels of flexibility and high mismatching losses due to its huge configuration characteristics.

As mentioned earlier, the main focus of this paper is to check control constraint of the accuracy improvement plan. A more detailed description of the structure can be found in [13]. In order to explain the ESS operation plan, a DC power flow method applied on two topologies is described in Section 3. In addition, simulation studies to find availability according to expected delays are discussed in Section 4.

Table 1. Numerical information of PV inverter topologies.

Topology	P (kW)	V_{dc} Range (V)	V_{ac} Range (V)	f (Hz)
Central	100–1500	400–1000	270–400	50, 60
String	0.5–5	200–500	110–230	50, 60
Multistring	2–30	200–800	270–400	50, 60
Module integrated	0.06–0.5	20–100	110–230	50, 60

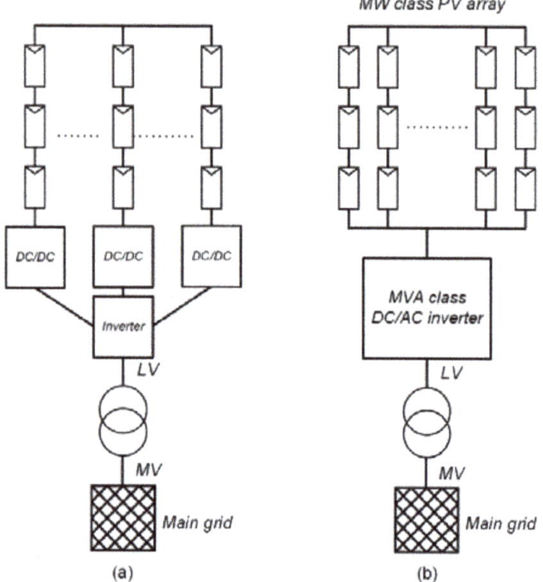

Figure 3. Configuration of PV inverter topologies (**a**) multistring (**b**) central.

3. DC Flow Analysis

To perform a current flow analysis of the defined DC network, an electricity-based circuit model is required. The nodal analysis method based on the presented PV circuit was used. In this method, an admittance matrix which includes cable components is adjusted in power flow analysis according to the PV power extraction.

For the general analysis of PV circuits for current estimation, the admittance of the components was divided into two categories (consumption and supply) for simplicity. With these simplifications, the PV current can then be reflected in the matrix representation. In a PV system, a current flows through the cathode for grounding. According to [13], a method was used to eliminate the current leakage in consideration of ground impedance. Since this study examined the impact of computational load with a focus on a previously used method of deriving current expectation based on real-time power output, the detailed resistance component between each PV module is applied to the admittance matrix. A consideration is made of the resistance component of the DC pole of the cathode which is generally considered as the ground section.

In this study, the nodal analysis was applied to formulate the detailed current flow in DC section. With this, the PV system can convert the energy extraction state of each module into a current flow. The output power of the PV module can be changed as a negative admittance component to be included in the matrix. In the case of an ESS, the charging and discharging power can be implemented as equivalent components according to the amount of profile. Based on a single module, each component can be represented by an admittance matrix as follows:

$$g = \begin{vmatrix} G_P & -G_{PN} & -G_{P0} & 0 \\ -G_{PN} & G_N & 0 & 0 \\ -G_{P0} & 0 & G_{PCS} & -G_{DC} \\ 0 & 0 & -G_{DC} & G_{ESS} \end{vmatrix} \tag{1}$$

The diagonals of the admittance matrix for positive/negative node of module and connection points of PCS, ESS (G_P, G_N, G_{PCS}, G_{ESS}) includes power extractions along with connected cable component as follows:

$$G_P = g_n + g_{Pnn-1} \tag{2}$$

$$G_N = g_n + g_{Nnn-1} \tag{3}$$

$$G_{PCS} = g_{Nnn-1} + g_{PCS} + g_{dc-dc} \tag{4}$$

$$G_{DC} = g_{dc-dc} + g_{ESS} \tag{5}$$

The non-diagonals of the admittance matrix for positive/negative section are solely composed with regarded cable (g_{Pnn-1}, g_{Nnn-1}) or equivalent admittance of own conversion system ($g_{dc\text{-}dc}$, g_{ESS}). When adding n modules, the size of the admittance matrix adds $n \times 2$ columns and rows on the basis of generalized matrix (1). Based on this, it is possible to analyze a mean conversion time for a large PV system and organize a feasibility study whether the expected delay is within a constraint.

In the iterative process, it is necessary to define an input variable with previous state (k) that can reflect the amount of output power to estimate the voltage level of the DC section which is considered as an unknown value. The corresponding input parameter must be converted to an admittance value, and the power extracted from each PV module connected in series can be updated repeatedly as follows:

$$g_{n\cdot k} = -\frac{P_{n\cdot k}}{\left(V_{Pn\cdot k} - V_{Nn\cdot k}\right)^2} \tag{6}$$

The admittance component for ESS is able to be expressed as shown in Equation (7) considering grounding.

$$g_{ESS\cdot k} = -\frac{P_{ESS\cdot k}}{V_{ESS\cdot k}^2} \tag{7}$$

The voltage in each section for next state (k + 1) is affected by the admittance factor due to the modified value until it converges within the available range. The main PCS current is fixed for convergence as a slack element. To perform the iterations considered as the main calculation, the inverse matrix is used to advance the iteration method as described in (8). The generalized control diagram is illustrated in Figure 4.

$$[V_{k+1}] = [g_k]^{-1} \times [I_k] \tag{8}$$

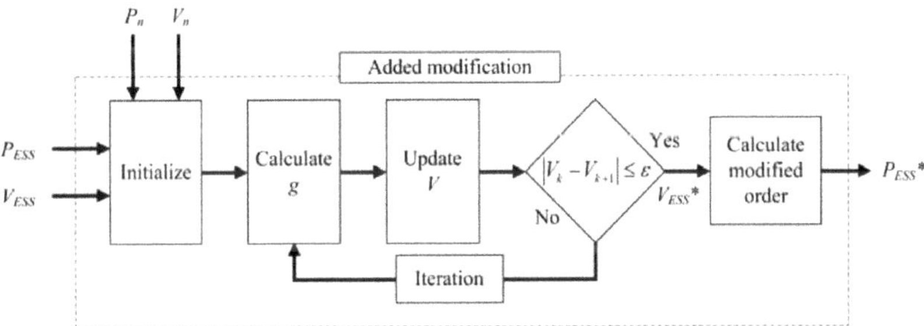

Figure 4. Modified control diagram of iterative method.

The contents of (1) depends on the number of PV modules, hence, a large-scale PV generation system requires further dimensional matrix. If n modules are added, the circuit will be expanded as mentioned earlier which results in a change of the basic equation as given in (9). Based on the configured sparse matrix, an analysis of the possible computational load and delay effect on main controller was performed using case studies.

$$\begin{vmatrix} g_1 + g_{P1n} & -g_1 & \cdots & -g_{P1n} & 0 & \cdots & 0 & 0 \\ -g_1 & g_1 + g_{N1n} & \cdots & 0 & -g_{N1n} & \cdots & 0 & 0 \\ \vdots & \vdots & \ddots & \vdots & \vdots & \iddots & \vdots & \vdots \\ -g_{P1n} & 0 & \cdots & g_{P1n} + g_n + g_{Pn0} & -g_n & \cdots & -g_{Pn0} & 0 \\ 0 & -g_{N1n} & \cdots & -g_n & g_{N1n} + g_n + g_{Nn0} & \cdots & 0 & 0 \\ \vdots & \vdots & \iddots & \vdots & \vdots & \ddots & \vdots & \vdots \\ 0 & 0 & \cdots & -g_{Pn0} & 0 & \cdots & g_{Pn0} + g_{PCS} + g_{dc-dc} & -g_{dc-dc} \\ 0 & 0 & \cdots & 0 & 0 & \cdots & -g_{dc-dc} & g_{dc-dc} + g_{ESS} \end{vmatrix} \begin{vmatrix} V_{u1} \\ V_{d1} \\ \vdots \\ V_{dn} \\ V_{dn} \\ \vdots \\ V_{dc} \\ V_{ESS} \end{vmatrix} = \begin{vmatrix} 0 \\ 0 \\ \vdots \\ 0 \\ 0 \\ \vdots \\ i_{dc} \\ 0 \end{vmatrix} \quad (9)$$

4. Simulation

4.1. Simulation Design

To verify the effectiveness of the control method, a simulation was performed using PSCAD (power system computer-aided design). Figure 5 shows a design of the network with PV sources configuring a common DC section that includes ESS. The basic rated power capacity of a single PV is 400 kVA. The designed PV system was connected to a displayed distribution network through an inverter topology (multistring and central). Preferentially, the individual unit of a PV system is composed of a single string to check calculation load of the multistring scheme (three strings are used for PV). Next, an expanded feasibility test was then carried out with an integrated 1.2 MW PV system with consideration of the maximum computational load as well. The distance of the pi line was reflected using generic model in PSCAD (library with *R*, *Xl*, *Xc* elements), and the residual information is given in Table 2.

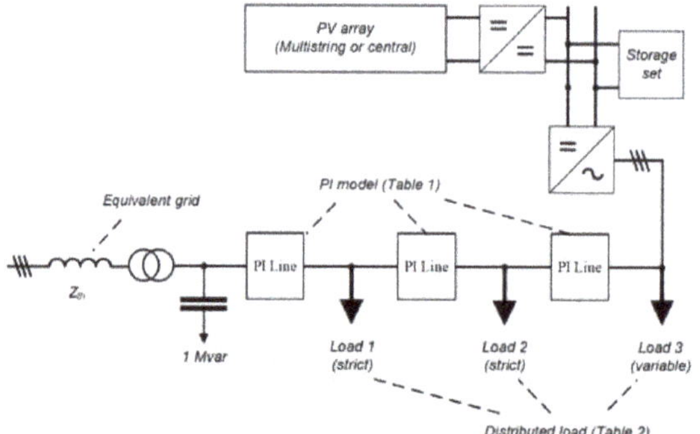

Figure 5. Simulated distribution network including MW scale PV/ESS.

The configured PV system extracts real power based on environmental variables, as in the previous study. DC voltage fluctuations due to the extracted power may affect the operation of the connected ESS. To continue the feasibility studies based on the verified charging/discharging pattern of ESS, load fluctuations (Figure 6) and ESS control effects (Figure 7) were used. The AC grid voltages at each load connection site are described in Figure 8 to confirm if it causes an impact on the DC network in the simulation. The RMS magnitudes are classified according to the distance from the substation and each section is maintained with stable conditions. The main objective of the simulation is to confirm the reliability of the proposed method. For this reason, individual load fluctuations were designed and ESS operating signal configured to change according to load conditions. An analysis focusing the ESS control according to each load change proceeded.

Table 2. Numerical data for the simulation.

Specific Data	Value	Unit
Rated AC voltage	22.9	kV
Rated DC voltage	500	V
Distance between each module	0.5	Meter
Amount of load	3	
Rated energy of ESS	200	kWh
Rated power of ESS	200	kW
Rated power of PV system (one unit)	400	kW
PI (π) line distance	500	Meter
Substation voltage (HV)	154	kV
Substation MVA	60	MVA
Short-circuit ratio of utility grid	15	
X/R ratio of utility grid	15	

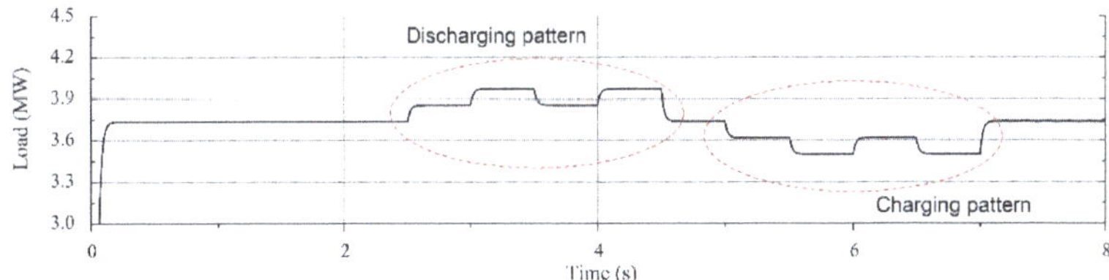

Figure 6. Imposed load variating condition for case study.

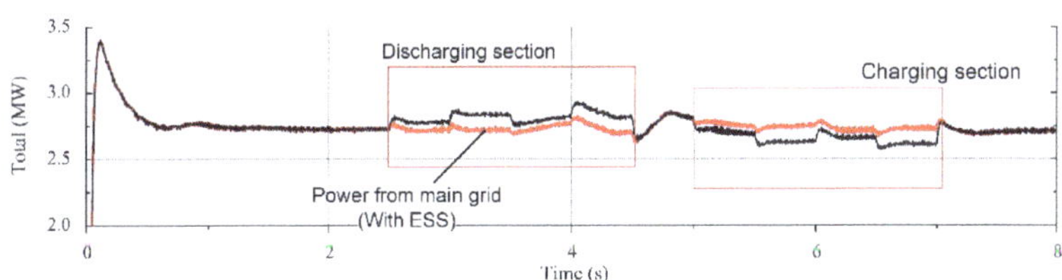

Figure 7. Power supply quantity from main grid in designed scenario (with and without ESS).

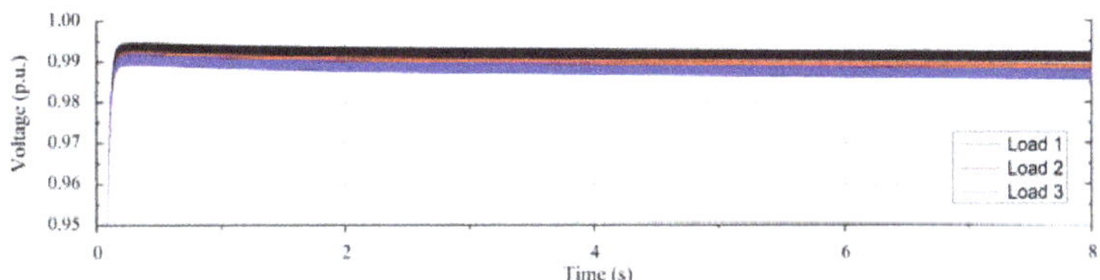

Figure 8. AC side root mean square (RMS) voltage at connection point for loads.

Table 3 details the load parameters applied to the case study. The total simulation time was defined in 8 seconds including initialization. Load changes were made in stages as shown in Figure 6. The ESS was configured to charging/discharging in response to an instantaneous load change (75 kW). The derived average iteration number (adjusting the input values) and expected average solution time, for each topology were represented together. Taking into account the configuration that a single string has 40 modules, the multistring topology configured a system matrix. In the case of the central topology, it was designed with three parallel units that were used for a single string. The number of iterations were not change in each topology, however, the expected solution time increased and the regarded delay increased as well. In this paper, the plan was to apply the 100 ms (10 Hz calculation frequency) interval to the multistring topology and apply not only 200 ms (5 Hz) to the central topology, but 400 ms (2.5 Hz) to consider the additional computational load. If the calculation is not terminated within the imposed constraint, the previous value must be used until the calculation is completed.

Table 3. Simulated case study description.

Simulation Time	8 Seconds (s)
Base load condition (Real power)	Load 1: 900 kW, Load 2: 1,500 kW, Load 3: 1,380 kW
Load increase sections (Load 3)	2.5–3, 3.5–4 s (75 kW) 3–3.5, 4–4.5 s (150 kW)
Load decrease sections (Load 3)	5–5.5, 6–6.5 s (75 kW) 5.5–6, 6.5–7 s (150 kW)
Iterations	3 (multistring), 3 (central)
Required solution time	95 millisecond (ms), (multistring) 185 ms (central)
Imposed maximum interval of calculation	100 ms (multistring) 200, 400 ms (central)

Within the improvement effect shown in Figure 7, the simulation was configured to deduce availability by showing how computational delay affects control. One of the objectives of this study was to confirm if the utilized control can reduce the possible error compared to the strict voltage control as well perform an adequate power supply with expected delay.

4.2. Simulation Results

An availability study should ensure that voltage corrections can be managed with limited calculation capability. The analysis of the charging and discharging sections (2.5–4.5 and 5–7 seconds), which were the main focus of this analysis are explained sequentially. Figure 9 shows the delay effect that can be derived from the multistring topology compared to the pre-analyzed method (strict-voltage, voltage-estimation). It can be confirmed that both charging and discharging are not significantly affected by the delay. The output control was observed to be performed substantially in the same manner as the accuracy improvement method without any delay. In the multistring topology, it was confirmed that the control signal can be derived within the maximum calculation load. Figure 10; Figure 11 show the delay effects that can be derived from the central topology. It can be confirmed that the influence of the delay occurs in both places, however, the control proceeds within a range that did not reach the strict-voltage method. In the case of Figure 11, including the maximum calculation load, an error occurs with a high probability, but it was confirmed that the occurrence of the delay was not frequent.

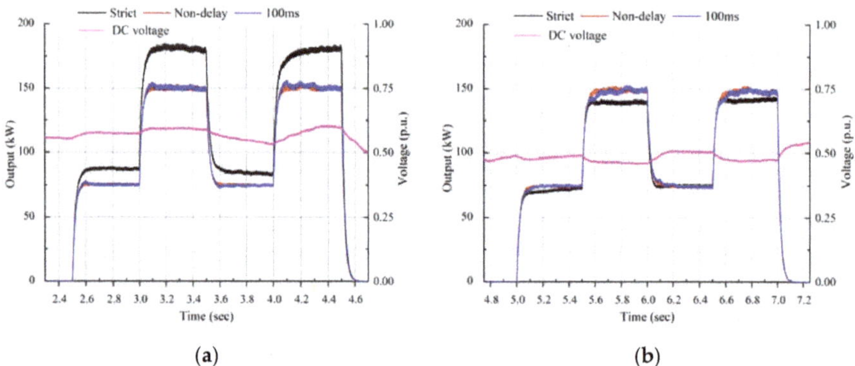

Figure 9. ESS states for multistring topology with 100 ms interval (**a**) extracted power quantity in the discharging section (2.5 to 4.5 s); (**b**) absorbed power quantity in the charging section (5 to 7 s).

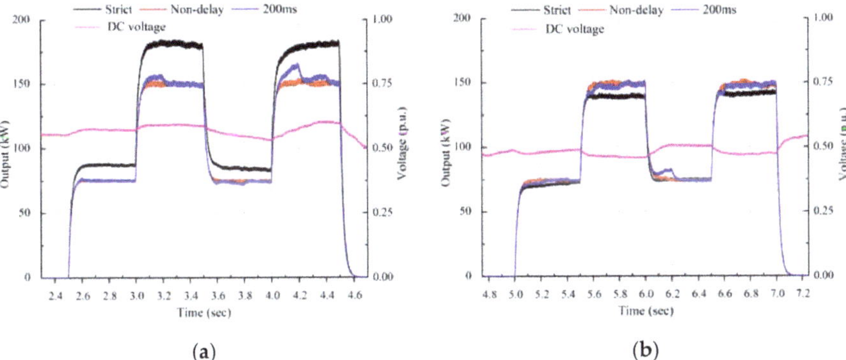

(a) (b)

Figure 10. ESS states for central topology with 200 ms interval (**a**) extracted power quantity in the discharging section (2.5 to 4.5 s); (**b**) absorbed power quantity in the charging section (5 to 7 s).

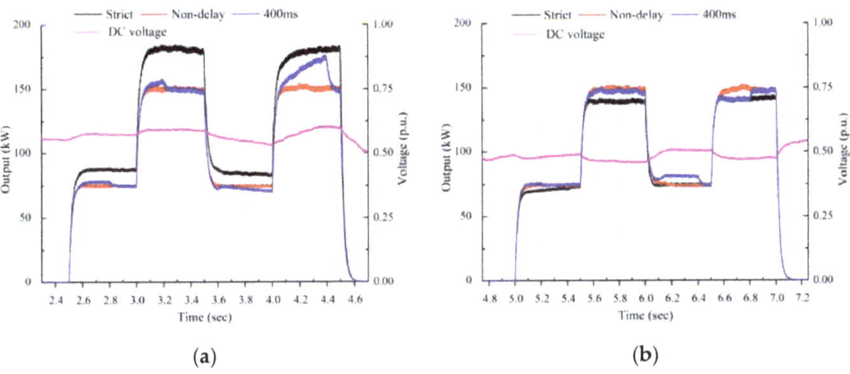

(a) (b)

Figure 11. ESS states for central topology with 400 ms interval (**a**) extracted power quantity in the discharging section (2.5 to 4.5 s); (**b**) absorbed power quantity in the charging section (5 to 7 s).

Table 4 shows the errors in terms of power support for designed scenarios. Compared to the ideal case (voltage-estimation without delay), the multistring topology could lose minor accuracy in terms of power support, and the central topology with 5 Hz calculation frequency shows less than 5 percent mismatch. It seems that a large mismatch can be induced with the central topology when considers maximum calculation load, however, it is expected to be generated on the oversized PV scale. The utilized signal correction scheme seems to be available with a normalized PV system in consideration of the possible delay ranges.

Table 4. Comparison data for designed simulation.

Imposed Constraint	ESS Mode	Demanded Quantity (Wh)	Accomplished Quantity (Wh)	Supply Accuracy (%)
Non-delay	Charging	125	121.212	96.97
	Discharging	125	123.731	98.98
10 Hz	Charging	125	121.101	96.88
	Discharging	125	123.911	98.98
5 Hz	Charging	125	119.303	95.34
	Discharging	125	126.282	98.97 (exceed)
2.5 Hz	Charging	125	117.92	94.33
	Discharging	125	134.9	92.08 (exceed)

5. Conclusions

This paper confirms the feasibility of the voltage-estimation method for a MW-scaled hybrid system. Existing signal correction methods to assist ESS operation were analyzed and utilized. The possible delays in calculations that occur with real power management were used in case studies. In designed simulation, the voltage-estimation method showed robustness in target topologies (multistring, central).

In response to the changed load condition, the voltage-estimation method acts to generate a modified order considering the actual required power extraction and calculation load. Focusing on the computational load caused by the expansion of a single PV system, we have analyzed the convergence of the PCS signal with the iteration method. With respect to this, a calculation delay could be generated not only by the PV module, but also by the applied sensor. Although this can be improved with detailed gain adjustments or by using a correction method according to the response characteristics, these circumstances need to be considered in the ESS management plan in advance. Therefore, a possible situation about signal calculation was derived, tested, and listed based on previous simulation design. In the aspect of the inverter topology, the possible delays need to be considered in advance when configuring the system. This operation can improve the accuracy both of the state of charge estimation and response operation.

Author Contributions: Conceptualization, J.S. and S.J.; Methodology, S.J.; Software, J.S.; Validation, J.S., and S.J.; Formal analysis, J.S., S.J.; Investigation, M.Y.; Data curation, M.Y.; Writing—original draft preparation, S.J.; Writing—review and editing, J.S. and S.J.; Supervision, S.J.; Project administration, S.J.; Funding acquisition, S.J. All authors have read and agreed to the published version of the manuscript.

Funding: This work was supported by the National Research Foundation Grant (No. 2018R1C1B5030524) and the Korea Electric Power Corporation Grant (R18XA06-40) funded by the Korean government.

Conflicts of Interest: The authors declare no conflict of interest.

Nomenclature

g_n	Equivalent admittance of nth module
g_{Pnn-1}	Admittance of positive side DC cable between modules
g_{Nnn-1}	Admittance of negative side DC cable between module
$g_{dc\text{-}dc}$	Equivalent admittance about DC/DC convertor for ESS
g_{ESS}	Equivalent admittance of ESS module
P_n	Injected power wt nth node
V_{Pn}	Upper-side voltage of nth module
V_{Nn}	Lower-side voltage of nth module
V_{dc}	Collector voltage
V_{ESS}	Induced ESS voltage
k	Iteration number

References

1. Rakhshani, E.; Rouzbehi, K.; J. Sánchez, A.; Tobar, A.C.; Pouresmaeil, E. Integration of Large Scale Pv-Based Generation into Power Systems: A Survey. *Energies* **2019**, *12*, 1425. [CrossRef]
2. Jo, B.-K.; Jung, S.; Jang, G. Feasibility Analysis of Behind-The-Meter Energy Storage System According to Public Policy on an Electricity Charge Discount Program. *Sustainability* **2019**, *11*, 186. [CrossRef]
3. IEEE Standard for Interconnection and Interoperability of Distributed Energy Resources with Associated Electric Power Systems Interfaces. In *IEEE Std 1547-2018 (Revision of IEEE Std 1547-2003)*; IEEE: Piscataway, NJ, USA, 2018.
4. Osório, G.J.; Shafie-khah, M.; Lujano-Rojas, J.M.; Catalão, J.P. Scheduling Model for Renewable Energy Sources Integration in an Insular Power System. *Energies* **2018**, *11*, 144. [CrossRef]
5. Subramani, G.; Ramachandaramurthy, V.K.; Padmanaban, S.; Mihet-Popa, L.; Blaabjerg, F.; Guerrero, J.M. Grid-tied Photovoltaic and Battery Storage Systems with Malaysian electricity tariff—A review on maximum demand shaving. *Energies* **2017**, *10*, 1884. [CrossRef]

6. Zeng, Z.; Yang, H.; Zhao, R.; Cheng, C. Topologies and Control Strategies of Multi-Functional Grid-Connected Inverters for Power Quality Enhancement: A Comprehensive Review. *Renew. Sustain. Energy Rev.* **2013**, *24*, 223–270. [CrossRef]
7. Miao, L.; Wen, J.; Xie, H.; Yue, C.; Lee, W. Coordinated Control Strategy of Wind Turbine Generator and Energy Storage Equipment for Frequency Support. *IEEE Trans. Ind. Appl.* **2015**, *51*, 2732–2742. [CrossRef]
8. Bird, L.; Cochran, J.; Wang, X. *Wind and Solar Energy Curtailment: Experience and Practices in the United States*; NREL: Golden, CO, USA, 2014.
9. Chen, P.; Thiringer, T. Analysis of Energy Curtailment and Capacity Over installation to Maximize Wind Turbine Profit Considering Electricity Price - Wind Correlation. *IEEE Trans. Sustain. Energy* **2017**, *8*, 1406–1414. [CrossRef]
10. Saez-de-Ibarra, A.; Milo, A.; Gaztañaga, H.; Debusschere, V.; Bacha, S. Co-Optimization of Storage System Sizing and Control Strategy for Intelligent Photovoltaic Power Plants Market Integration. *IEEE Trans. Sustain. Energy* **2016**, *7*, 1749–1761. [CrossRef]
11. Stimoniaris, D.; Tsiamitros, D.; Dialynas, E. Improved Energy Storage Management and PV-Active Power Control Infrastructure and Strategies for Microgrids. *IEEE Trans. Power Syst.* **2016**, *31*, 813–820. [CrossRef]
12. Vargas, L.S.; Bustos-Turu, G.; Larraín, F. Wind Power Curtailment and Energy Storage in Transmission Congestion Management Considering Power Plants Ramp Rates. *IEEE Trans. Power Syst.* **2015**, *30*, 2498–2506. [CrossRef]
13. Yoo, Y.; Jang, G.; Kim, J.-H.; Nam, I.; Yoon, M.; Jung, S. Accuracy Improvement Method of Energy Storage Utilization with DC Voltage Estimation in Large-Scale Photovoltaic Power Plants. *Energies* **2019**, *12*, 3907. [CrossRef]
14. Zhang, H.; Yue, D.; Xie, X. Robust Optimization for Dynamic Economic Dispatch Under Wind Power Uncertainty with Different Levels of Uncertainty Budget. *IEEE Access* **2016**, *4*, 7633–7644. [CrossRef]
15. Byungdoo, J.; Hyun, K.; Heechan, K.; Hansang, L. Development of a Novel Charging Algorithm for On-board ESS in DC Train through Weight Modification. *J. Electr. Eng. Technol.* **2014**, *9*, 1795–1804.
16. Padrlőn, J.F.M.; Lorenzo, A.E.F. Calculating Steady-State Operating Conditions for Doubly-Fed Induction Generator Wind Turbines. *IEEE Trans. Power Syst.* **2010**, *25*, 922–928.
17. Li, S. Power Flow Modeling to Doubly-fed Induction Generators (DFIGs) Under Power Regulation. *IEEE Trans. Power Syst.* **2013**, *28*, 3292–3301. [CrossRef]
18. Yu, H.; Rosehart, W.D. An Optimal Power Flow Algorithm to Achieve Robust Operation Considering Load and Renewable Generation Uncertainties. *IEEE Trans. Power Syst.* **2012**, *27*, 1808–1817. [CrossRef]
19. Liu, S.; Xu, Z.; Hua, W.; Tang, G.; Xue, Y. Electromechanical Transient Modeling of Modular Multilevel Converter Based Multi-Terminal HVDC Systems. *IEEE Trans. Power Syst.* **2014**, *29*, 72–83. [CrossRef]
20. Sun, H.; Guo, Q.; Zhang, B.; Guo, Y.; Li, Z.; Wang, J. Master-slave-splitting Based Distributed Global Power Flow Method for Integrated Transmission and Distribution Analysis. *IEEE Trans. Smart Grid.* **2015**, *6*, 1484–1492. [CrossRef]
21. Mumtaz, F.; Syed, M.H.; Hosani, M.A.; Zeineldin, H.H. A Novel Approach to Solve Power Flow for Islanded Microgrids Using Modified Newton Raphson with Droop Control of DG. *IEEE Trans. Sustain. Energy* **2016**, *7*, 493–503. [CrossRef]
22. Yang, Z.; Zhong, H.; Bose, A.; Xia, Q.; Kang, C. Optimal Power Flow in AC–DC Grids with Discrete Control Devices. *IEEE Trans. Power Syst.* **2018**, *33*, 1461–1472. [CrossRef]
23. Cai, Y.; Irving, M.R.; Case, S.H. Iterative Techniques for the Solution of Complex DC-rail-traction Systems including Regenerative Braking. *IEE Proc. Gener. Transm. Distrib.* **1995**, *142*, 445–452. [CrossRef]
24. Pyzara, A.; Bylina, B.; Bylina, J. The Influence of a Matrix Condition Number on Iterative Methods' Convergence. In Proceedings of the 2011 Federated Conference on Computer Science and Information Systems (FedCSIS), Szczecin, Poland, 18–21 September 2011; pp. 459–464.
25. Karthikeyan, V.; Gupta, R. Multiple-Input Configuration of Isolated Bidirectional DC–DC Converter for Power Flow Control in Combinational Battery Storage. *IEEE Trans. Ind. Informat.* **2018**, *14*, 2–11. [CrossRef]
26. Chew, B.S.H.; Xu, Y.; Wu, Q. Voltage Balancing for Bipolar DC Distribution Grids: A Power Flow Based Binary Integer Multi-Objective Optimization Approach. *IEEE Trans. Power Syst.* **2019**, *34*, 28–39. [CrossRef]

27. Planning of a PV Generator 2013, SMA Solar Technology AG. Available online: https://files.sma.de/dl/1354/DC-PL-en-11.pdf (accessed on 1 December 2019).
28. Cabrera-Tobar, A.; Bullich-Massagué, E.; Aragüés-Peñalba, M.; Gomis-Bellmunt, O. Topologies for Large Scale Photovoltaic Power Plants. *Renew. Sustain. Energy Rev.* **2016**, *59*, 309–319. [CrossRef]

© 2020 by the authors. Licensee MDPI, Basel, Switzerland. This article is an open access article distributed under the terms and conditions of the Creative Commons Attribution (CC BY) license (http://creativecommons.org/licenses/by/4.0/).

Article

Development of Floquet Multiplier Estimator to Determine Nonlinear Oscillatory Behavior in Power System Data Measurement

Namki Choi, Hwanhee Cho * and Byongjun Lee *

School of Electrical Engineering, Anam Campus, Korea University, 145 Anam-ro, Seongbuk-gu, Seoul 02841, Korea; fleminglhr@korea.ac.kr

* Correspondence: whee88@korea.ac.kr (H.C.); leeb@korea.ac.kr (B.L.); Tel.: +82-2-3290-3697 (H.C.); +82-2-3290-3242 (B.L.)

Received: 4 April 2019; Accepted: 7 May 2019; Published: 14 May 2019

Abstract: Measurement-based technology has been developed in the area of power transmission systems with phasor measurement units (PMU). Using high-resolution PMU data, the oscillatory behavior of power systems from general electromagnetic oscillations to sub-synchronous resonances can be observed. Studying oscillations in power systems is important to obtain information about the orbital stability of the system. Floquet multipliers calculation is based on a mathematical model to determine the orbital stability of a system with the existence of stable or unstable periodic solutions. In this paper, we have developed a model-free method to estimate Floquet multipliers using time series data. A comparative study between calculated and estimated Floquet multipliers has been performed to validate the proposed method. The results are provided for a sample three-bus power system network and the system integrated with a doubly fed induction generator.

Keywords: power systems; floquet multiplier; poincaré map; time series data; DFIG

---

1. Introduction

Over the past few decades, power systems have experienced significant changes regarding the amount of power consumed as well as the complexity of the network. Voltage instabilities tend to occur in power systems that are heavily loaded or in faulty condition. There are various indicators of voltage instability, which are determined by the generator, load dynamics, and network structure [1]. Nonlinear oscillatory behaviors are usually noticeable even before the voltage collapse occurs. These oscillatory behaviors can be observed with high-resolution devices. After detecting nonlinear oscillatory behavior, the type of oscillation needs to be determined to check whether the condition will be harmful to the power system.

Nonlinear oscillatory behaviors are an intrinsic characteristic of power systems caused by the structure of the system under a specific condition. To analyze nonlinear behavior, system topology and dynamics are necessary to describe the system mathematically. Then, mathematical expressions of the system, such as state equations or system Jacobian matrix, can provide detailed information on the current status of the system. In power systems, practical issues such as special nonlinear oscillatory behavior can be revealed by time series data measurements. However, without the input of other measurement data or conditions, it is hard to assess the system state since the information about the system is limited to local measurements. Fortunately, studies related to limited time series data applications have been conducted using mathematical models.

Based on the mathematical modeling of nonlinear system local stability, calculation of the maximal Lyapunov exponent in time series data was proposed by Wolf et al. [2]. Other researchers improved the calculation efficiency of the maximal Lyapunov exponent in time series data [3,4]. The maximal

Lyapunov exponent was also modified for applications in time series data on power systems [5,6]. The maximal Lyapunov exponent is a stability index that is strongly connected to the fluctuation of data. If the maximum Lyapunov exponent of the system is negative (positive), then the nearby system trajectories will converge (diverge) toward each other [7]. In the results of [5], the maximal Lyapunov exponent in time series data was reasonable compared to the original signal of the bus voltage after the fault was cleared. In other words, all the simulations in [5] involved increased oscillation or oscillation with large amplitudes of approximately 0.4 p.u. However, in reality, the maximal Lyapunov exponent in time series data is difficult to apply since it strongly depends on the initial values and data size. For example, when the bus voltage oscillates for a long time interval, the maximal Lyapunov exponent will be different for the initial values and data size selected. When oscillation occurs, the maximal Lyapunov exponent also fluctuates, regardless of whether it is positive or negative, which makes it hard to determine system stability. Therefore, the characteristics or stability of the oscillatory behavior of nonlinear systems differ from local stability, especially for time series data. There are some certain approaches (real-time or near-to-real-time) in large amounts of literature to detect local stability such as positively or negatively damped oscillations (for example, Voltage stability indices or maximum Lyapunov exponent [5,6]). However, there are few existing solutions or approaches in power systems or other applicable engineering field to detect uncertain response as marginal stability (mathematically defined but not practically). However, still marginal stability issues such as sustained oscillation or forced oscillation could lead some damages or instability to power system [8]. So, this paper focuses on the feature of periodic stability of the power system dynamics.

Knowing the type of oscillatory behavior is important to predict how the oscillatory behavior of the system will change. In practical applications, filter-based approaches are often used to detect and classify the oscillation [9]. For mathematically based applications, the stability of nonlinear oscillatory behaviors is determined with a Floquet multiplier. A Floquet multiplier is the eigenvalue of a matrix that gives orbital stability for the periodic solution of the system. A solution is stable when all the calculated moduli of the eigenvalues in the monodromy matrix are below the unity. A monodromy matrix is a matrix that influences whether the periodic solution decays or grows for the initial perturbation [7]. In classical texts, a geometric concept called a Poincaré map has been introduced to discuss some behaviors of periodic solutions in terms of a Floquet multiplier [7]. One study [10] introduced a method for nonlinear time series analysis that provided some guidelines on constructing a Poincaré map from data-based signals. However, the method reduced the phase space dimensionality one at a time to turn the continuous time flow into a discrete time map. It notes that the method is the intersection count and not simply proportional to the original time t of the flow. The number of intersections that are counted might be very small, since the parameter is over specific value or interval of unstable region (the chaos), the system may collapse. Therefore, surfaces should be carefully selected or else they will not contain enough information on the signal.

From the perspective of biomechanical engineering, orbital stability is defined using estimated Floquet multipliers from measured data of physical rotation and orbital movement [11–13]. The study by [11] provided a method that generated accurate estimates with noisy experimental data. However, there is no generalized method to choose the proper Poincaré section. Moreover, the verification of the estimated Floquet multiplier against the calculated Floquet multiplier in an actual system model is required.

In this paper, the selection of Poincaré sections for time series dimensions is proposed. Then, the Floquet multiplier is estimated for a one-dimensional Poincaré section. The estimation technique is based on linear regression applied not only to a simple three-bus system but a complex mathematical model of a power system with a wind generator modeled as a doubly fed induction generator (DFIG).

The paper starts with an introduction of the stability of a periodic solution. Then, mathematical concepts of the Poincaré section and Floquet multiplier will be expanded to the proposed method. Comparison of the proposed method and calculated Floquet multipliers will be performed in a test power grid that is integrated with DFIG. Three noteworthy summaries are provided followed by the conclusion.

2. Stability of Periodic Solutions

Power grid dynamics with initial values t_0 are shown as follows:

$$\dot{x} = f(t, x), x(t_0) = x_0 \tag{1}$$

System stability can be determined by the real value of eigenvalue as system is linear. However, for a nonlinear system, there are plenty of ways to determine system stability. The maximal Lyapunov exponent is an indicator that gives information on stability of time-dependent solution from Equation (1). The fluctuation of solution can be decreased as maximal Lyapunov exponent is negative and vice versa. However, different approaches are required to analyze behavior as the solution is oscillating. A monodromy matrix gives the characteristics of the oscillatory solution of the nonlinear system (1).

2.1. Monodromy Matrix

Stability of periodic solutions in dynamic systems can be represented by using a mathematical tool. Suppose the solution x^* is a periodic feature of constant frequency f and its period T as time evolves. The problem of instability of periodic solutions have been studied within the framework of the methods developed by [7]. The trajectories of Equation (1) can be defined as $x := \phi(t, z)$. Equation (1) has periodic solution with z as an initial value, i.e., $\phi(t + T, z) = \phi(t, z)$. The trajectory progresses to the regular orbit $x^* = \phi(t, z^*)$ as Equation (1) is perturbed with $z^* + d_0$. The distance between trajectory progression and the periodic orbit is

$$d(t) = \phi(t, z^* + d_0) - \phi(t, z^*) \tag{2}$$

where z^* is the specific initial value for a specific solution. The distance with each period T is calculated by $d(T)$ and the linear representation with Taylor expansion becomes

$$d(T) = \frac{\partial \phi(T, z^*)}{\partial z} d_0 \tag{3}$$

This linear approximation is not applicable to large disturbances such as d_1 in Figure 1. In Equation (3), the matrix

$$M = \frac{\partial \phi(T, z^*)}{\partial z} \tag{4}$$

governs the growth or decay as the initial disturbance d_0 is applied to the periodic solution. The matrix form of Equation (4) is the monodromy matrix. The characteristics of monodromy matrix are directly related to the behavior of the periodic solution and determined by its eigenvalues [7].

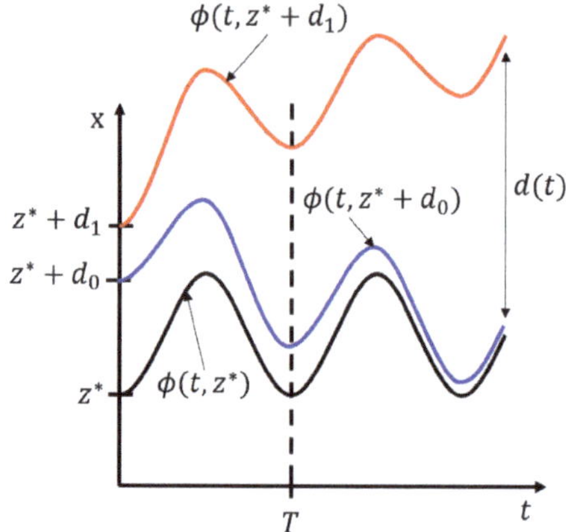

Figure 1. A periodic response of system and stable and unstable solution for initial disturbance d_0 and d_1.

2.2. Poincar é Map

The Poincaré map construction is powerful geometric approach for studying the dynamics of various periodic phenomena. Specifically, some approaches start with phase portrait and its Poincaré section can be related to periodic stability results. The Poincaré map can be applied to n-dimensional differential equations. The set Ω must be an $(n-1)$-dimensional hypersurface, satisfying a specific condition based on n-dimensional vector space. All orbits crossing Ω in a $q^* \in \Omega$ should meet two requirements:

(a) The Ω is intersected by orbits transversally
(b) Orbits cross Ω in the same direction

The Ω can be characterized by the requirements as a local set of the trajectory. For instance, another Ω for each period T can be driven by choosing another q point. The hypersurface Ω is the Poincaré section. The subsets of planes are important class of Ω.The Ω in $z \in \mathbb{R}_n$ is intersected by periodic trajectory y with period T. The z^* can be represented as q^* in a coordinate system on Ω, where q is $(n-1)$-dimensional. As ϕ is restricted to Ω, this can be summarized as

$$q^* = \phi(T; q^*) \tag{5}$$

The time taken for an orbit $\phi(t; q)$ to first return to Ω is defined as $T_\Omega(q)$ with $q \in \Omega$.

$$\phi(T_\Omega(q); q) \in \Omega, \phi(t; q) \notin \Omega, \quad \text{for} \quad 0 < t < T_\Omega(q) \tag{6}$$

Poincaré map or return map $P(q)$ can be defined by

$$P(q) := P_\Omega(q) = \phi(T_\Omega(q); q), \quad \text{for} \quad q \in \Omega \tag{7}$$

Figure 2a,b illustrate the geometric representation of $P(q)$.

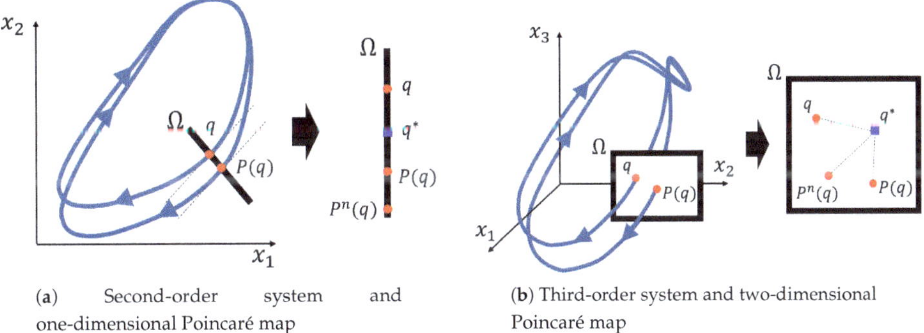

(**a**) Second-order system and one-dimensional Poincaré map

(**b**) Third-order system and two-dimensional Poincaré map

Figure 2. A conceptual sketch of Poincaré map.

2.3. Stability on the Periodic Orbits

In this section, the stability of one particular solution x^* or $\phi(t, z^*)$ with period T along the periodic branch is investigated. The monodromy matrix $M(\lambda)$ has n eigenvalues $\mu_1(\lambda), \mu_2(\lambda), \cdots, \mu_n(\lambda)$ with a value of λ in (1), t; these eigenvalues are called Floquet multipliers. The magnitude in one of them can be always equal to unity. The other $(n-1)$ Floquet multipliers can determine local stability by applying following rule [7]:

$x(t)$ is stable if $|\mu_j| < 1$ for $j = 1, \cdots, n-1$.
$x(t)$ is unstable if $|\mu_j| > 1$ for some j.

The $(n-1)$ multipliers should be always inside the unit circle on the stable periodic trajectory. The multipliers are functions of the variables under deliberation. Crossing points between some of multipliers and the unit circle may exist as the parameter is varied. The critical multiplier can be defined as the multiplier crossing the unit circle. The multiplier crossing the unit circle is called the critical multiplier.

For a geometric interpretation, these Floquet multipliers can be expressed for the Poincaré section Ω. As defined by Equation (6), $P(q)$ not only q takes values in Ω, which is $(n-1)$-dimensional. The interpretation of both q and $P(q)$ is to have $(n-1)$ elements with respect to an appropriately chosen basis. The Poincaré map should satisfy

$$P(q^*) = q^* \tag{8}$$

where q^* is a secured point of P. The q is closer to q^* as $T_\Omega(q)$ is closer to the period T. The behavior of the Poincaré map can be near its secured point q^* as a reduction of stability in the periodic trajectory x^* occurs. Hence, the fixed point q^* provides data on stability and can be an indicator to distinguish whether it is attracting or repelling. Similar to Equation (2), the unknown $P(q)$ can be described in Taylor series expansion.

$$P(q) = P(q^*) + \frac{\partial P(q^*)}{\partial q}(q - q^*) + \text{higher-order terms} \tag{9}$$

As q and $P(q)$ are in the hypersurface Ω, the number of elements in the linear approximated matrix $\partial P(q^*)/\partial q$ is $(n-1) \times (n-1)$. The $\mu_1, \cdots, \mu_{n-1}$ is defined as the eigenvalues in linearization of P as it is near the fixed point q^*,

$$\text{eigenvalue of}(\mu_j) \quad \frac{\partial P(q^*)}{\partial q}, j = 1, \cdots, n-1 \tag{10}$$

Monodromy matrix and Poincaré surface are defined for the periodic solution. Thus, the selection of Ω is not dependent on eigenvalues of the matrix. i.e., the eigenvalues will be the same regardless of the choice of point q. Hence, corresponding one set of eigenvalues μ_j exists for each periodic trajectory. Then, the eigenvalues determined from Ω are also Floquet multipliers or (characteristic) multipliers, and a property of stability can be treated equally.

The n^2-monodromy matrix (8) has $+1$ as an eigenvalue with eigenvector $\dot{x}^*(0)$ tangent to the intersecting curve $x^*(t)$. The eigenvector $\dot{x}^*(0)$ is not in hypersurface Ω since the property of that intersection point should cross transversally. The eigenvalue $+1$ in the monodromy matrix matches up with an agitation along $y^*(t)$ leading out of Ω. while the other $n-1$ eigenvalues in the monodromy matrix can decide what occurs to small agitation within Ω. In summary, choosing the proper basis for the n-dimensional space shows that the remainder of $n-1$ eigenvalues of M match the eigenvalues of $\partial P(q^*)/\partial q$.

3. Estimated Floquet Multipliers in Time Series Data

In the previous section, a system is dealt with a specified mathematical model. Here, we address the case of measurement-based or time series data when the equation or model is unknown. First, deciding on the proper form of the Poincaré map for time series data that corresponds to the original Poincaré map is required. Then, we can estimate the Floquet multiplier by the linearized form of the equation.

3.1. *Poincar é Map Construction for Time Series Data*

As discussed in Section 2, Reference [7] gives some guidelines for constructing a Poincaré map for time series data. These focus on the data quality of the constructed Poincaré map. It is a fact that the Poincaré surface should include information on the cycle. A specific procedure to construct a data-based Poincaré surface is presented in the following subsection.

3.1.1. Poincar é Surface Decision for Time Series Data

Two requirements in choosing hyperplane Ω have been provided in the previous section. At Ω, transversally intersecting points should have the same direction. For continuous values, these conditions are conceptually reasonable. However, it is hard to choose a Poincaré surface when discrete values are given. The direction corresponds to differential values of the trajectory and the transversality condition matches the local intersection. Thus, the Poincaré map in time series data should fulfill these two conditions, where ϕ is a periodic trajectory in discrete values, x is all the points in the trajectory ϕ, and α is a differential value that has to be determined.

(a) Transversality $x_{\Omega 1} = \{x | x_a \le x \le x_b, x \in \phi\}$
(b) Direction $x_{\Omega 2} = \{x | \dot{\phi} \approx \alpha, x \in \phi\}$

Therefore, the points included in the Poincaré surface are the intersection of two sets $x_{\Omega 1}$ and $x_{\Omega 2}$. The problem of constructing the Poincaré surface for time series data has been converted into a problem of choosing x_a, x_b, and α. First, using engineering-based judgment, it is safer for discrete data to set α as zero. Reference [8] supports the notion that the (numerical) time derivative of the signal is a legal coordinate in a reconstructed state space for scalar data that contains some information on the original state space. Hence, the time derivative for scalar s might provide information on the direction. $\dot{s} = 0$ is precisely given by the local minima (or maxima) of the time series. In addition, the local minima (or maxima) are interpreted as the special measurement function which projects onto the first component of a vector applied to the state vectors inside surface Ω. It is experimentally acceptable that the local peak values when $\dot{\phi} \approx 0$ have less errors than other slopes of the trajectory.

Once the trajectory crosses Ω, the next intersecting point in the same direction is acquired after the 1 cycle. In Figure 2a, the period leading point $P(q)$ is on the same line as the period lagging point q. Similarly, third-order systems can be generalized to the two-dimensional hyperplane Ω. Since there is neither a mathematical model nor cycle T given for the time series data, specific points that can be regarded as intersection points of Ω need to be selected. For the second-order system in Figure 2a, the line might be clearly chosen at two peak points in $\dot{x}_2 = 0$ from a point of the requirements for choosing the Poincaré surface. Similar behaviors in Figure 2b, however, show that there are four peak points $\dot{x}_3 = 0$ to be selected. In this case, the lower local minima are the best candidates to choose the Poincaré surface.

For measured data for an arbitrary system, the values of x_a and x_b for condition (a) transversality are naturally decided, such that x_a are values near zero and x_b is the median value of measured data from the conceptual sketch in Figure 3. Then, two conditions for Poincaré map decision-making can be expressed as

All local minima sets less than the median value are included in the Poincaré map Ω.

Figure 3 explains the concept of Poincaré map construction for time series data in an arbitrary system. When the measured data for x_1 are given, the peak points marked in red stars and blue rectangular points are found by applying the peak searching algorithm. Then, the sorted peak points under the computed median of x_1 are ready to estimate the Floquet multiplier.

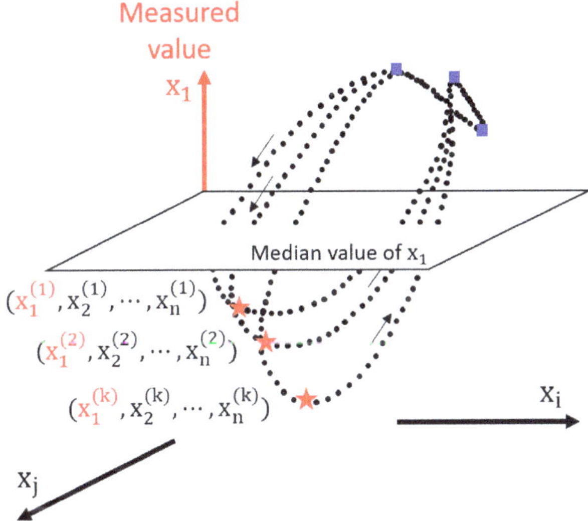

Figure 3. Selecting Poincaré section for time series data.

For example, for the three-bus system given in [14,15], when setting the load bus reactive power of 10.946 MVar as a parameter, a pure four-dimensional periodic return map is acquired. Then, for 100 samples of local minima set p_l, the standard deviation of the state variables δ, ω, δ_L, and V are 5.43×10^{-6}, 9.80×10^{-5}, 1.79×10^{-6}, and 7.16×10^{-8}, respectively. These small standard deviations show that the 100 sample points of local minima are a fixed point. As the fixed point is defined in the Poincaré surface Ω, the local minima set is in the Poincaré surface Ω.

3.1.2. Effect of Dimension Reduction by Projection

For a higher-order system which is hard to visualize, such as a fourth-order system, it is still possible to imagine the three-dimensional Poincaré section. Assume that all local minima are the values when one of the state variables satisfies $x_4 = c$, where c is a constant value. Then, the three-dimensional axis of the Poincaré section will be x_1, x_2, and x_3. If the measured oscillatory behavior is purely periodic, not only x_4 but also the three variables x_1, x_2, and x_3 would be fixed. However, when the three variables change for every cycle, checking the three variables at once and monitoring the measured value of local minima will not be the same.

Using the local minima searching concept stressed above, a set of local minima or intersection points are identified inside a three-dimensional box. At the first projection in Figure 4, the average distance between each point decreases. Then, another projection makes the points stick to the axis of the measured value. Hence, there apparently exists an inevitable gap between the three-dimensional Poincaré map and the double projected local minima of the time series data. As mentioned in Section 3.1.1, the standard deviation is small when the trajectory is stable, and projection will have no effect. The gap will only affect the results when the trajectory is unstable, especially for higher-order systems.

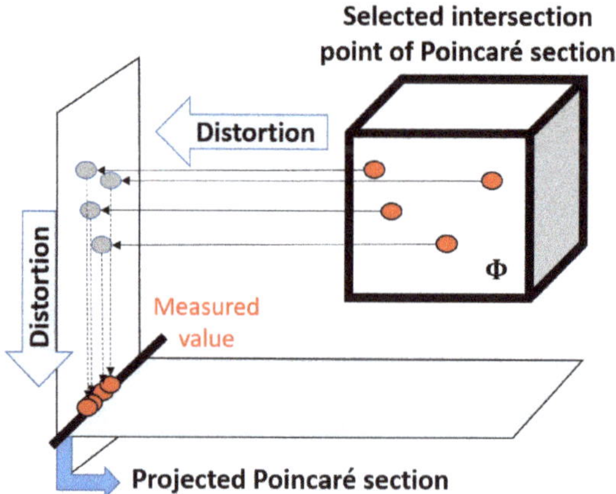

Figure 4. Effect of projection of three-dimensional intersection points in Poincaré section of fourth-order system.

3.2. Floquet Multiplier Estimation

In Section 2.3, the matrix $\partial P(q^*)/\partial q$ is provided by the monodromy matrix with restriction to the $n-1$-dimensional hyperspace Ω. The $P(q)$ can be expressed by Taylor series in the vicinity of the fixed point q^*. Assuming that periodic solution is stable, higher-order term can be negligible. Recalling that $P(q^*) = q^*$ at the fixed point, Equation (10) is linearized as

$$P(q) - q^* = [\frac{\partial P(q^*)}{\partial q}](q - q^*) \tag{11}$$

For the specific vector q_1, the corresponding critical eigenvalue μ_1 is calculated using the linear approximated equation

$$P(q_1) - q^* = \mu_1(q_1 - q^*) \tag{12}$$

If we have series data x, the relationship between the mapping P of point q and point q can be followed by sequence $x[k]$ and $x[k+1]$. Hence, the computational form gives Equation (11) by

$$x[k+1] - x^* = \hat{\mu}(x[k] - x^*) \tag{13}$$

Equation (12) is a typical form of linear regression, where $\hat{\mu}$ is a constant to be determined by series data x.

For application to time series data, the equation should be modified to a scalar form. In that case, the calculated Floquet multiplier might not be the same as the actual value since the scalar value cannot reflect the direction. However, there will be negligible differences when selecting the acceptable component of the state variable. Using engineering judgment, it will be natural to select the component with the largest variance among the state variables x.

$$s[k+1] - s^* = \hat{\mu}_F(s[k] - s^*) \tag{14}$$

The estimated Floquet multiplier $\hat{\mu}_F$ can be easily computed with Equation (10). The estimated value, however, is not the exact value of the calculated Floquet multiplier because it is estimated and projected in one dimension, regardless of the number of original system state variables it contains. But as long as the signal is periodic, the estimated Floquet multiplier gives information on stability even if we choose single measurement as a state variable.

For application to time series data, Equation (13) can be expressed after changing scalar s to voltage term V

$$\hat{\mu}_{F,series} = \frac{1}{N}\sum_{k=1}^{N}\frac{V[k+1] - (1-\mu)V^*}{V[k]} \tag{15}$$

Ironically, the linearized form of Equation (13) contradicts the stable orbits when $s[k+1] \approx s[k] \approx s^*$. Hence, the μ term of Equation (14) should be 1 by applying the stable orbit condition.

$$\hat{\mu}_{F,series} = \frac{1}{N}\sum_{k=1}^{N}\frac{V[k+1]}{V[k]} \tag{16}$$

Then, the final estimation Equation (15) further simplifies the problem such that the estimated Floquet multiplier is only the average of the ratio of the previous and current local minima of the voltage at all peak data. In other words, we can set the fixed-point value to zero. It can be guaranteed that $\hat{\mu}_{F,series} = 1$ is a sufficient condition for stable orbit. Using final Equation (15), other unstable cases are demonstrated by two examples of power systems in Section 4.

4. Power System with DFIG

In this section, comparative studies of the calculated and estimated Floquet multipliers for two cases are performed. Starting with the mathematical modeling of a power system with DFIG, the application of the suggested method in test power systems and power system with a complicated DFIG model are studied.

The modeling of a DFIG is not as simple as a general synchronous generator. The state matrix of the system becomes larger, considering not only the characteristics of controllers but also the complexity of the machine. The specified model of the DFIG is provided with network equations in the appendix of [16]. In this paper, Floquet multiplier have been estimated under even more complex system. That is, DFIG model of [16] is added to the three-bus system with induction motor which have been constructed at [15]. The specified model of this system is provided at Appendix A.

Figure 5 shows a three-bus power system network with the DFIG installed at bus 3. Time series voltage data in rms were obtained at the load bus. The Floquet multiplier was estimated for the results of the time domain simulation of V_L. The bifurcation diagram of power grid with DFIG is represented in Figure 6, which was determined by the mathematical model of the system. The mathematical model for the system containing aerodynamics and electric power system dynamics are given as follows:

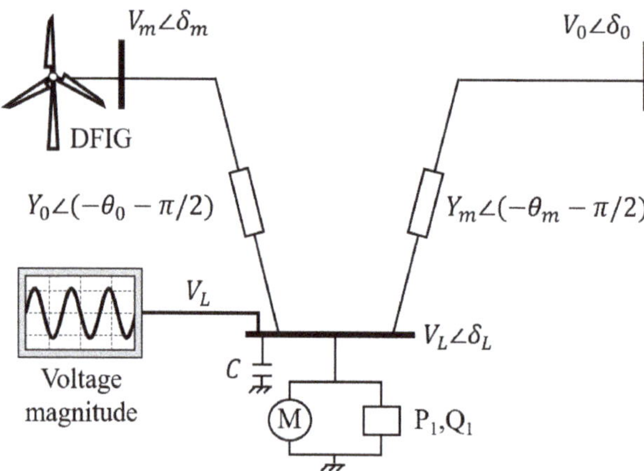

Figure 5. three-bus power system network with doubly fed induction generator (DFIG).

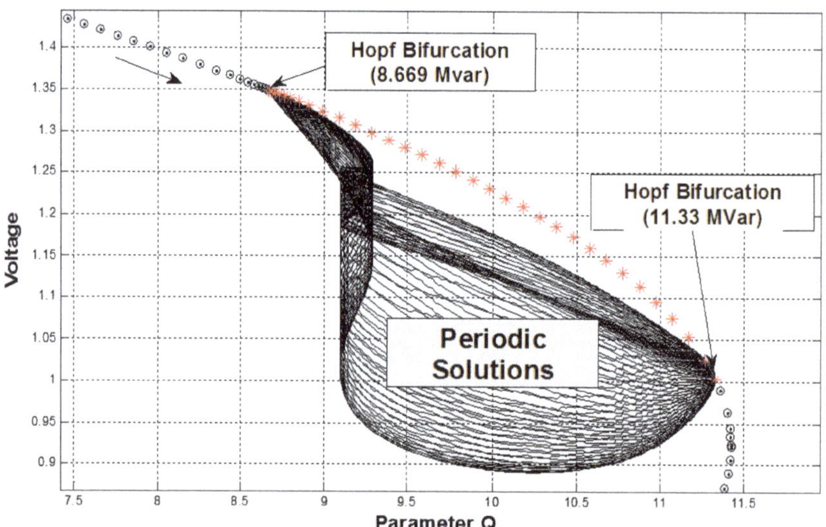

Figure 6. A bifurcation diagram of three-bus power system with DFIG (wind speed 12 m/s).

In Figure 6, from the left Hopf bifurcation point 8.669 MVar to right Hopf bifurcation point 11.33 MVar, periodic solutions exist where the amplitudes of the orbit are large in the middle of the range. Before estimating the Floquet multiplier in the power system with DFIG, the sample three-bus power grid was examined.

Similar to the previous case, the three-bus power grid in [14–16] showed two Hopf bifurcation points. Periodic orbit can be observed in parameter Q_1 from 10.946 MVar to 11.407 MVar in Figure 7. Time domain simulations were performed in between these two bifurcation points to check nonlinear oscillatory behaviors. The Floquet multiplier was calculated by the monodromy matrix of the system and the corresponding Floquet multiplier was estimated for comparison with the critical multiplier.

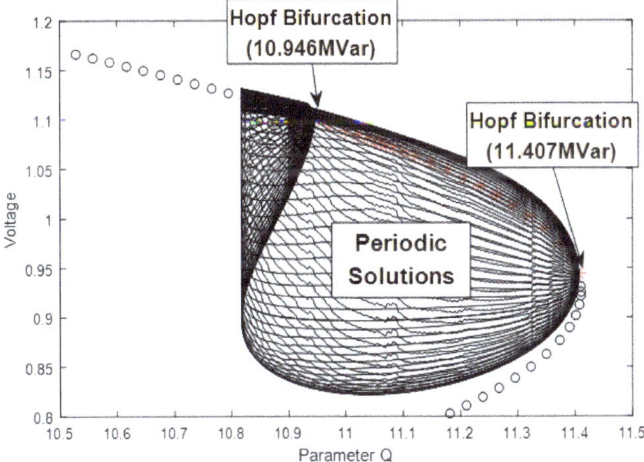

Figure 7. A bifurcation diagram of sample power system.

4.1. Calculated Floquet Multiplier in Sample Power System

In Figure 8, four modes of the Floquet multiplier are calculated and the trajectories of these modes are described. Mode 1 is always zero, while the other modes move inside the unit circle. When the parameter increases at 10.946 MVar, the critical multiplier in mode 4 stays on the right side of the unit circle until 10.8728 MVar; then it moves to the left side of the circle with a slight change in parameter. This is the point where the periodic solution of the system loses stability. Without stopping, the critical multiplier moves to the left along the real axis approximately −30, and then changes direction and heads to the unit circle. The value when the trajectory meets the unit circle is 11.3874 MVar and moves to the right side of the circle with a small change at 11.3887 MVar.

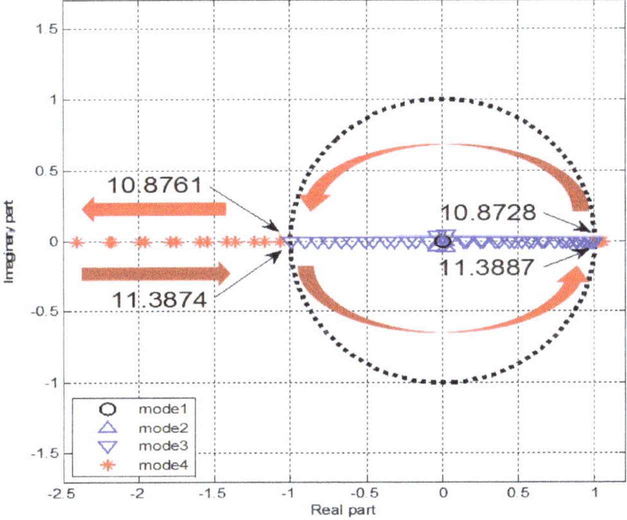

Figure 8. Trajectory of Floquet multiplier for three-bus power system.

4.2. Estimated Floquet Multiplier in Sample Power System

Figure 9a,b shows the rms voltage values at the load bus (left-top), phase portraits of the generator angle and load voltage (left-bottom), and Floquet multiplier estimation (right) for specific parameter values. Stable oscillatory behavior is observed in Figure 9a. The local minima of the rms voltage marked in red circles appear to be steady. The phase portraits on the left-bottom show that the trajectory of the solution is in the attractor. On the right side of the picture, the points marked in red crosses are the values related to $V[k+1]/V[k]$ in Equation (15). The estimated value was determined from the average of these values.

The left-bottom of Figure 9b shows that the periodic solution is still inside the attractor, but the orbit is inconsistent. Figure 9b is obviously an unstable periodic solution such that ratio $V[k+1]/V[k]$, which are marked in red crosses, are scattered from 0.96 to 1.18. The average of these points is 1.002, marked in blue circles. To summarize the sample cases for Figure 9a (Q_1 = 10.931 MVar), the estimated and calculated values are 1 and 1.005, respectively. For Figure 9b (Q_1 = 11.378 MVar), the estimated and calculated values are 1.0023 and -2.317, respectively.

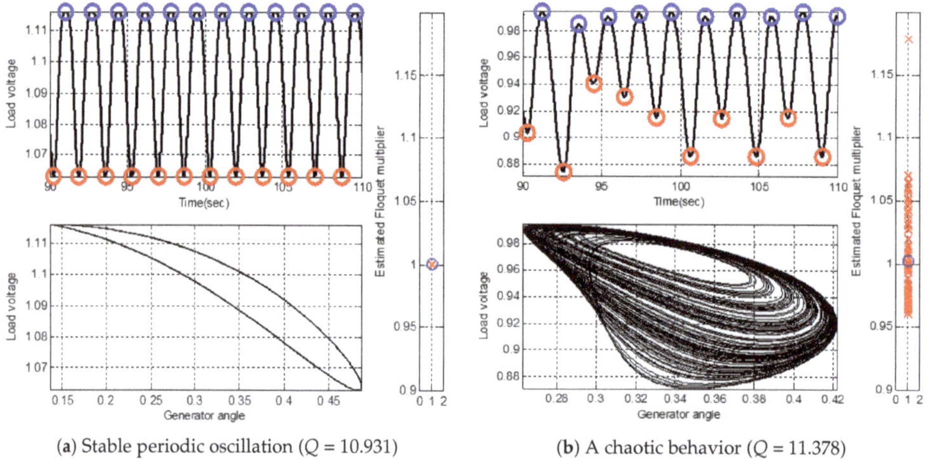

(**a**) Stable periodic oscillation (Q = 10.931) (**b**) A chaotic behavior (Q = 11.378)

Figure 9. Floquet multiplier estimation for rms value of load voltage.

To apply Equation (15), at least two local minima are required. We set a three-cycle time interval for the ordinary differential equation (ODE) tool from MATLAB (R2014a, Mathworks, Netic, MA, USA). For the three-cycle time series data of all parameter ranges, the Floquet multipliers were estimated using Equation (15). In Figure 10a, the overall values are not as large as the moduli of the calculated Floquet multipliers. The points marked with blue dots are the points which are near unity, while the points marked with red dots are the points when the deterministic Floquet multiplier goes outside the unit circle. The estimated Floquet multiplier has unity value when the deterministic Floquet multiplier calculated from monodromy matrix is stable (unity), as shown in Figure 10a. Likewise, the estimated Floquet multiplier is greater than unity when the moduli of the deterministic Floquet multiplier are unstable (greater than one), as shown in Figure 10b. The correlation coefficient between two results Figure 10a,b for partial interval are 0.824.

Comparing the two approaches shows that the estimated Floquet multiplier using Equation (15) gave similar information on a short-term signal. The boundary of stability was less likely to be observed in the estimated Floquet multiplier, while the actual calculated Floquet multiplier changed its sign upon losing stability.

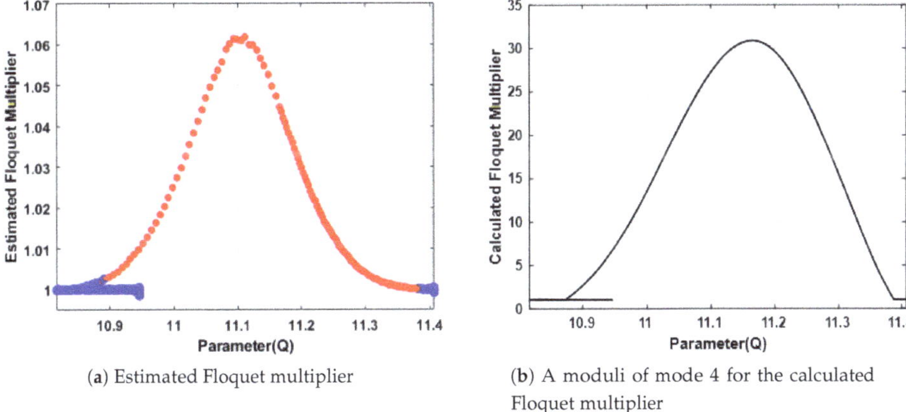

(**a**) Estimated Floquet multiplier

(**b**) A moduli of mode 4 for the calculated Floquet multiplier

Figure 10. Floquet multiplier verification for the sample power system.

4.3. Comparison between Estimated Floquet Multiplier and Calculated Floquet Multiplier

Figure 11a shows an estimated (data-driven) Floquet multiplier and Figure 11b shows the calculated (model-based) Floquet multiplier of test power system with DFIG. For a higher-order system, the estimated Floquet multiplier values remained near unity when these were calculated with the monotomy matrix inside the unit circle. When the calculated Floquet multipliers were about to lose their stability, the estimated Floquet multiplier values stayed at approximately 1 from 10.993 to 11.064 MVar. Then, the estimated Floquet multipliers increased in the positive direction although this was not as large as the calculated values. Likewise, the calculated Floquet multipliers dramatically decreased in the negative direction. Once the estimated Floquet multiplier lost stability, it also lost its increment tendency, i.e., stiff growth was observed from 10.604 to 10.635 MVar, 10.781 to 10.807 MVar, and 10.973 to 10.993 MVar. The correlation coefficient between two results Figure 11a,b for partial interval are 0.758. Specifically, the proposed method accurately estimated the Floquet multiplier as long as the oscillation behavior was purely periodical.

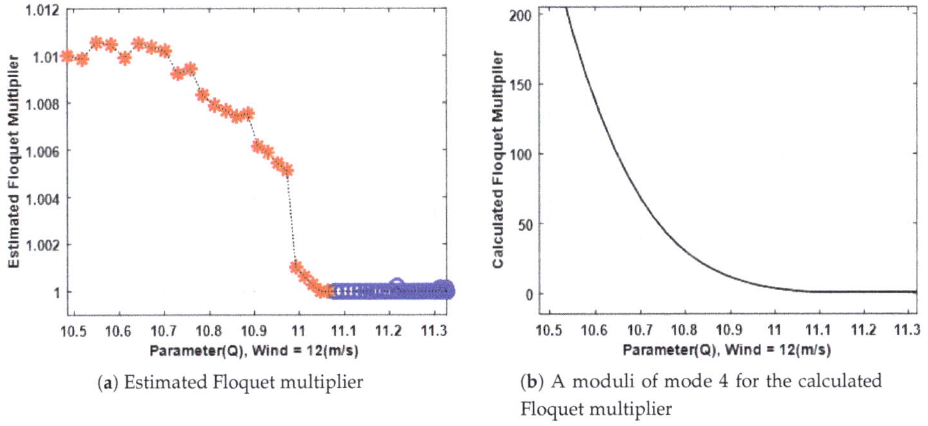

(**a**) Estimated Floquet multiplier

(**b**) A moduli of mode 4 for the calculated Floquet multiplier

Figure 11. Floquet multiplier verification for the power system with wind generation.

5. Actual Oscillation Incident

In this section, estimated Floquet multiplier is calculated to actual oscillation event observed in New England ISO in 17 June 2016. The oscillation was system wide and the frequency of 0.27 Hz, 27 MW peak-to-peak. The location of the source was far away outside of interested area. PMU data was acquired for 3 min, and actual event kept going for 140 s. For data preprocessing, original current data was subtracted to its averaged value. Average values are calculated by 162 samples of moving window considering a period is approximately 8 samples and its quarter at sampling rate of 0.033 s. So the signal oscillates around zero for the current PMU data of 35 locations. And the oscillation is clearly observable at 24 locations. Figure 12a shows example actual current minus average current value with noise at transmission line near substation 6. The local maxima have been well detected but still the peak values are fluctuations of Floquet multiplier of 1.045. Figure 12b shows a simple map of New England ISO with all estimated Floquet multiplier marked on it. It is clear that area marked in yellow has high value than other line currents. Even the line connected substation 9 and external area has the highest value of all. So the oscillation source for this incident can be traced through external area. Overall estimated Floquet multiplier is 1.0396, which can be judged that the system wide oscillation is unstable periodic solution mathematically. However, considering the real power system, Floquet multiplier of the oscillation can be shown slightly higher than 1.0 due to load fluctuation or generator excitor operation.

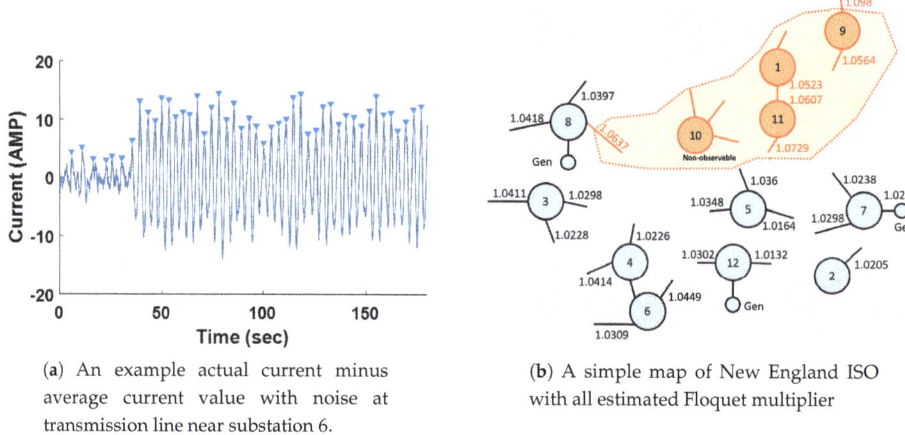

(**a**) An example actual current minus average current value with noise at transmission line near substation 6.

(**b**) A simple map of New England ISO with all estimated Floquet multiplier

Figure 12. An estimation of Floquet multiplier at actual power system oscillation event.

6. Discussion

According to the results of both cases in the previous section, the estimated values followed a similar trend as the calculated values. However, in general, the estimated values were not as large as the calculated values. For the power system with wind generator, the estimated Floquet multiplier lost a pattern of increment once it lost stability. There are several possible reasons that can explain this outcome.

(1) Dimension reduction by projection

The proposed method is targeted to use local phasor measurement unit data from power systems. Thus, estimating the Floquet multiplier is strictly limited to time series data in one-dimensional space. Figure 13a,b shows the three-dimensional intersection in the Poincaré section and their projections onto one-dimensional line and chaotic behavior, respectively. Figure 13a is a typical limit cycle oscillation. Clearly, no distortions occurred. Moreover, when the measured data showed period doubling, the projection of two intersections onto the three-dimensional hyperplane Ω may start to reveal unintended differences.

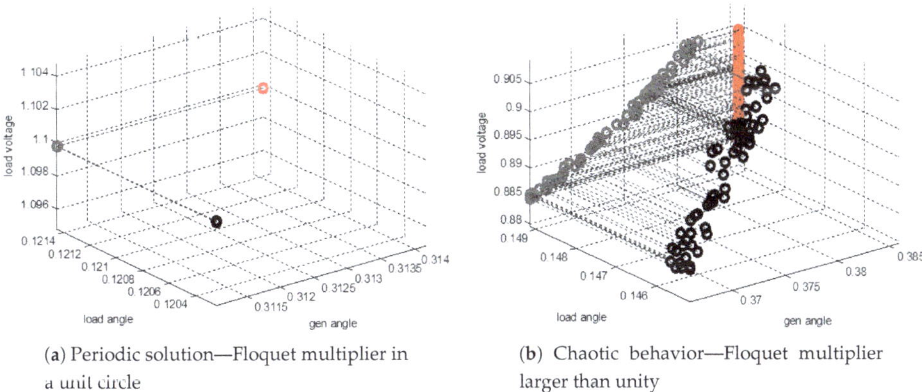

(**a**) Periodic solution—Floquet multiplier in a unit circle

(**b**) Chaotic behavior—Floquet multiplier larger than unity

Figure 13. Projection of three-dimensional Poincaré section into load voltage axis.

Figure 13b shows the results of chaotic behavior. For the sample three-bus power system, distortion occurred for a step of projection. Initially, the three-dimensional intersections were scattered on the space. The projection onto a two-dimensional plane gave different values. Then, the projection onto a line, which was the result for time series data, yielded totally distorted values compared to the original intersection.

The power system with wind generator is a 10-dimensional system. The Poincaré surface would be nine-dimensional, thus projections will be carried out eight times. Then, ignoring the other axes, only the local minima of time series data would give the estimated value of the Floquet multiplier. Therefore, it is natural that the estimated Floquet multiplier is less than the calculated Floquet multiplier.

(2) Assumption of linearization

According to the proposed method introduced in Section 3.2, the estimating Equation (10) is linearized near q^*, which means that the equation assumes a stable periodic solution. However, if the solution loses stability or the Floquet multiplier goes out of the unit circle, then the assumption is void. i.e., the $(1-\mu)$ term of Equation (15) is no longer zero. Therefore, in a stable periodic region, the estimated Floquet multiplier shows the exact value of unity that we intended it to be, but there is a risk of error in the unstable periodic region. Although there are some error risks when the oscillation loses stability, we can easily recognize the unstable region where the estimated Floquet multiplier is greater than 1.

(3) Differential equation calculation limits

The numerical ODE solving tool is required to validate the calculated and estimated Floquet multipliers. For time domain simulation, the ODE45 tool of MATLAB was used. In some parameter ranges, the solution diverged within one or two cycles, which means that not enough local minima were gathered to construct the Poincaré surface. In real power system problems, the proposed method is aimed at sustaining oscillation measured by PMU. The computation of solution of Equation (1) is beyond the scope of the present study.

7. Conclusions

In this work, we compared a proposed method of estimating the Floquet multiplier with time series data and a calculated Floquet multiplier based on a mathematical model. We described how to construct a Poincaré map in time series data, starting with the definition of the Poincaré map.

We stressed the possibility of disagreement between the calculated and estimated Floquet multipliers by projection concept. A linearized form of the Floquet multiplier was provided to estimate the time series data with period.

We found that the estimated Floquet multipliers followed a similar trend as the Floquet multipliers calculated with the monodromy matrix in two power system networks after conducting correlation analysis. A Poincaré map selected by a local peak value searching algorithm provided enough information on an arbitrary system by considering their standard deviation. Thus, the critical multiplier of the estimated value was unity for the stable oscillation. For practical application with noisy measurement, we have conducted Floquet multiplier estimation to actual oscillation incident. Results show that estimated Floquet multiplier is not exactly unity due to load fluctuation or generator excitor response. However, still the estimated Floquet multiplier act as an indicator for feature of sustained oscillation while the value could stand for signification of the oscillation sources. Therefore, the proposed method can successfully function as a period oscillation indicator for time series data acquired from power systems.

Author Contributions: N.C. conceived and build up the research methodology, conducted the system simulations, and wrote this paper. B.L. and H.C. supervised the research, improved the system simulation, and made suggestions regarding this research.

Funding: This research received no external funding.

Acknowledgments: This work was supported by "Human Resources program in Energy Technology" of the Korea Institute of Energy Technology Evaluation and Planning (KETEP)-granted financial resource from the Ministry of Trade, Industry, and Energy, Republic of Korea (no. 20174030201820).

Conflicts of Interest: The authors declare no conflict of interest.

Appendix A. Mathematical Model of System

On the base of three-bus voltage collapse dynamics including induction motor at load bus provided by Chiang et al. [17], DFIG model of [18] is replaced to synchronous generator. Specific model of the system is given as follows:

Appendix A.1. Aerodynamics of DFIG

$$T_m = \frac{1}{2}\rho C_p(\lambda,\theta)\frac{\omega_s}{\omega_r}A_{wt}V_{wind}^3 \quad \text{(A1)}$$

$$C_p(\lambda,\theta) = 0.22(\frac{116}{\lambda} - 0.4\theta - 5)e^{-\frac{12.5}{\lambda}} \quad \text{(A2)}$$

$$\lambda = (\frac{1}{\lambda' + 0.08\theta} - \frac{0.035}{\theta^3+1})^{-1} \quad \text{(A3)}$$

where:
ρ : Air density [kg/m^3]
$C_p(\lambda,\theta)$: Power coefficient
A_{wt} : Wind turbine swept area [m^2]
V_{wind} : Wind speed [m/s]
ω_r : Electrical rotor speed [rad/s]
ω_s : Electrical speed base [rad/s]
λ : Tip speed ratio of a WTG
θ : Pitch angle

Appendix A.2. Differential Equations—DFIG

$$T_0' \frac{dE_{qD}'}{dt} = -(E_{qD}' + (X_s - X_s')I_{ds}) + T_0'(\omega_s \frac{X_m}{X_r} V_{dr} - (\omega_s - \omega_r)E_{dD}') \tag{A4}$$

$$T_0' \frac{dE_{dD}'}{dt} = -(E_{dD}' - (X_s - X_s')I_{qs}) + T_0'(-\omega_s \frac{X_m}{X_r} V_{qr} + (\omega_s - \omega_r)E_{qD}') \tag{A5}$$

$$\frac{2H_D}{\omega_s} \frac{d\omega_r}{dt} = T_m - E_{dD}' I_{ds} - E_{qD}' I_{qs} \tag{A6}$$

Appendix A.3. Differential Equations—Active and Reactive Power Controller

$$\frac{dx_1}{dt} = K_{I1}(P_{ref} - P_{Gen}) \tag{A7}$$

$$\frac{dx_2}{dt} = K_{I2}(K_{P1}(P_{ref} - P_{Gen}) + x_1 - I_{qr}) \tag{A8}$$

$$\frac{dx_3}{dt} = K_{I3}(Q_{ref} - Q_{Gen}) \tag{A9}$$

$$\frac{dx_4}{dt} = K_{I4}(K_{P3}(Q_{ref} - Q_{Gen}) + x_3 - I_{dr}) \tag{A10}$$

Appendix A.4. Differential Equations—Load Dynamics(Induction Motor)

$$\frac{d\delta_L}{dt} = \frac{1}{k_{qw}}(-k_{qv2}V_L^2 - k_{qv}V_L - Q_0 - Q_1 + Q) \tag{A11}$$

$$\frac{dV_L}{dt} = \frac{1}{Tk_{qw}k_{pv}}(k_{pw}k_{qv2}V_L^2 + (k_{pw}k_{qv} - k_{qw}k_{pv})) + k_{pw}(Q_0 + Q_1 - Q) - k_{qw}(P_0 + P_1 - P) \tag{A12}$$

Appendix A.5. Algebraic Equations

$$-V_{qr} + K_{P2}\{K_{P1}(P_{ref} - P) + x_1 - I_{qr}\} + x_2 = 0 \tag{A13}$$

$$-V_{dr} + K_{P4}\{K_{P3}(Q_{ref} - Q) + x_3 - I_{dr}\} + x_4 = 0 \tag{A14}$$

$$-P_{Gen} + E_{dD}' I_{ds} + E_{qD}' I_{qs} - R_s(I_{ds}^2 + I_{qs}^2) = 0 \tag{A15}$$

$$-Q_{Gen} + E_{qD}' I_{ds} - E_{dD}' I_{qs} - X_s'(I_{ds}^2 + I_{qs}^2) = 0 \tag{A16}$$

$$-I_{dr} + \frac{E_{qD}'}{X_m} + \frac{X_m}{X_r} I_{ds} = 0 \tag{A17}$$

$$-I_{qr} - \frac{E_{dD}'}{X_m} + \frac{X_m}{X_r} I_{qs} = 0 \tag{A18}$$

where:

T_0' : Transient open-circuit time constant
X_s' : Transient reactance
X_s : Stator reactance
X_r : Rotor reactance
X_m : Mutual reactance
E_{qD}', E_{dD}' : Transient rotor voltage
x_1, x_2, x_3, x_4 : Active, reactive power controller
δ_L, V_L : Load angle and voltage magnitude

P_0, Q_0 : Constant active and reactive power of the motor
P_1, Q_1 : Constant active and reactive power of the load
$k_{p\omega}, k_{pv}$: Constant impedance parameter related to active power with frequency and voltage
$k_{q\omega}, k_{qv}, k_{qv2}$: Constant impedance parameter related to reactive power with frequency and voltage

References

1. Kundur, P. *Power System Stability and Control*; McGraw-Hill: New York, NY, USA, 1994.
2. Wolf, A.; Swift, J.B.; Swinney, H.L.; Vastano, J.A. Determining Lyapunov exponents from a time series. *Physica D* **1985**, *16*, 285–317. [CrossRef]
3. Sato, S.; Sano, M.; Sawada, Y. Practical methods of measuring the generalized dimension and the largest Lyapunov exponent in high dimensional chaotic systems. *Prog. Theor. Phys.* **1987**, *77*, 1–5. [CrossRef]
4. Rosenstein, M.T.; Collins, J.J.; De Luca, C.J. A practical method for calculating largest Lyapunov exponents from small data sets. *Phys. D Nonlinear Phenom.* **1993**, *65*, 117–134. [CrossRef]
5. Dasgupta, S.; Paramasivam, M.; Vaidya, U.; Venkataramana, A. PMU-based model-free approach for short term voltage stability monitoring. In Proceedings of the 2012 IEEE Power and Energy Society General Meeting, San Diego, CA, USA, 22–26 July 2012.
6. Cho, H.; Oh, S.; Nam, S.; Lee, B. Non-linear dynamics based sub-synchronous resonance index by using power system measurement data. *IET Gener. Transm. Distrib.* **2018**, *12*, 4026–4033. [CrossRef]
7. Seydel, R. *Practical Bifurcation and Stability Analysis*; Springer: New York, NY, USA, 2010.
8. Guideline, R. *Forced oscillation Monitoring & Mitigation*; North American Electric Reliabilty Corporation: Atlanta, GA, USA, 2017.
9. Baudette, M. *Fast Real-Time Detection of Sub-Synchronous Oscillations in Power Systems Using Synchrophasors*; KTH Electric Engineering: Stockholm, Sweden, 2013.
10. Kantz, H.; Schreiber, T. *Nonlinear Time Series Analysis*; Cambridge University Press: Cambridge, UK, 2004.
11. Ahn, J.; Hogan, N. Improved Assessment of Orbital Stability of Rhythmic Motion with Noise. *PLoS ONE* **2015**, *10*, e0119596. [CrossRef] [PubMed]
12. Dingwell, J.B.; Kang, H.G. Differences between local and orbital dynamic stability during human walking. *ASME Trans. Biol. Mech.* **2007**, *129*, 586–593. [CrossRef] [PubMed]
13. Hurmuzlu, Y.; Basdogan, C. On the Measurement of Dynamic Stability of Human Locomotion. *ASME Trans. Biol. Mech.* **1994**, *116*, 30–36. [CrossRef]
14. Ajjarapu, V.; Lee, B. Bifurcation theory and it's application to nonlinear dynamical phenomena in an electrical power system. *IEEE Trans. Power Syst.* **1991**, *7*, 424–431. [CrossRef]
15. Ajjarapu, V.; Lee, B. Period-doubling route to chaos in an electrical power system. *IEE Proc. C Gener. Transm. Distrib.* **1993**, *140*, 490–496.
16. Ajjarapu, V.; Lee, B. Nonlinear Oscillations and voltage collapsse phenomenon in electrical power system. In Proceedings of the Twenty-Second Annual North American Power Symposium, Auburn, AL, USA, 15–16 October 1990.
17. Chiang, H.-D.; Dobson, I.; Thomas, R.J.; Thorp, J.S.; Fekih-Ahmed, L. On voltage collapse in electric power systems. *IEEE Trans. Power Syst.* **1990**, *5*, 601–611. [CrossRef]
18. Pulgar-Painemal, H.A.; Sauer, P.W. Dynamic modeling of wind power generation. In Proceedings of the 41st North American Power Symposium, Starkville, MS, USA, 4–6 October 2009.

© 2019 by the authors. Licensee MDPI, Basel, Switzerland. This article is an open access article distributed under the terms and conditions of the Creative Commons Attribution (CC BY) license (http://creativecommons.org/licenses/by/4.0/).

Article

Preliminary Design of Multistage Radial Turbines Based on Rotor Loss Characteristics under Variable Operating Conditions

Zheming Tong [1,2,3,*,†], Zhewu Cheng [1,2,†] and Shuiguang Tong [1,2,*]

1 State Key Laboratory of Fluid Power and Mechatronic Systems, Zhejiang University, Hangzhou 310027, China
2 School of Mechanical Engineering, Zhejiang University, Hangzhou 310027, China
3 Department of Mechanical Engineering, Massachusetts Institute of Technology, Cambridge, MA 02139, USA
* Correspondence: tzm@zju.edu.cn (Z.T.); cetongsg@zju.edu.cn (S.T.)
† These authors contributed equally to this work.

Received: 9 June 2019; Accepted: 28 June 2019; Published: 2 July 2019

Abstract: The loading-to-flow diagram is a widely used classical method for the preliminary design of radial turbines. This study improves this method to optimize the design of radial turbines in the early design phase under variable operating conditions. The guide vane outlet flow angle is a key factor affecting the off-design performance of the radial turbine. To optimize the off-design performance of radial turbines in the early design phase, we propose a hypothesis that uses the ratio of the mean velocity of the fluid relative to the rotor passage with respect to the circumferential velocity of the rotor as an indicator to indirectly and qualitatively estimate the rotor loss, as it plays a key role in the off-design efficiency. Theoretical analysis of rotor loss characteristics under different types of variable operating conditions shows that a smaller design value of guide vane outlet flow angle results in a better off-design performance in the case of a reduced mass flow. In contrast, radial turbines with a larger design value of guide vane outlet flow angle can obtain a better off-design performance with increased mass flow. The above findings were validated with a mean-line model method. Furthermore, this study discusses the optimization of the design value of guide vane outlet flow angle based on the matching of rotor loss characteristics with specified variable operating conditions. It provides important guidance for the design optimization of multistage radial turbines with variable operating conditions in compressed air energy storage (CAES) systems.

Keywords: preliminary design; optimization; rotor loss; guide vane outlet flow angle; radial turbine; CAES

1. Introduction

Multistage radial turbines—usually referring to the multistage radial turbo expanders—are the key component of power generation systems in compressed air energy storage (CAES) and are also employed in the recycling of waste heat, residual pressure, and gas in the petrochemical industry. A schematic diagram of the conventional structure of a multistage radial turbine in CAES systems is shown in Figure 1. It consists of several single-stage radial turbines in separate casings, connected in series. In addition, heat exchangers are connected between stages to preheat the compressed air. Depending on the inlet pressure of the multistage radial turbine, which is generally from 3 to 10 MPa, the number of turbine stages is typically between three and five to ensure that the pressure ratio of each turbine stage falls in the range of two to five. The specific value of the pressure ratio of each turbine stage is related to the turbine inlet temperature. When cold energy production is not required, it is often necessary to meet the requirements of the turbine outlet temperature close to atmospheric temperature. Multistage radial turbines can realize the efficient energy release of a large expansion ratio from the high pressure of compressed air storage to the atmospheric pressure, and have the advantages of high efficiency, compact structure, and large power capacity. Thus, they are widely used in CAES systems [1,2].

In recent years, CAES technology has received increasing attention as one of the most promising solutions to the problems of intermittency and lack of control in renewable energy generation [3,4]. When multistage radial turbines in CAES systems are integrated with renewable energy generation, the fluctuating power output and inlet pressure make them able to operate under variable working conditions. Guide vane control is a mature and efficient mass flow regulation method that is widely used in turbomachinery. simulation studies have demonstrated the superiority of guide vane control for multistage radial turbines in CAES systems [5–7]. The challenge of multistage radial turbine design thus becomes to maintain high efficiency over a broader range of variable operating conditions involving guide vane opening changes.

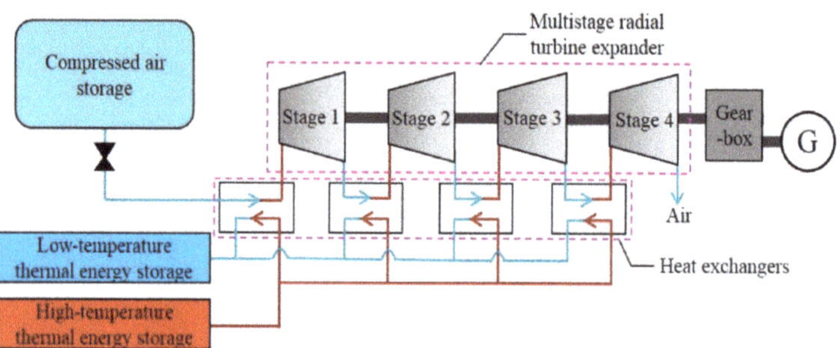

Figure 1. Schematic diagram of a multistage radial turbine in a compressed air energy storage (CAES) system.

As shown in Figure 1, the intermediate turbine duct of the multistage radial turbine is usually a circular pipe in an inter-stage heat exchanger. The pressure change of the internal flow is small, and the flow characteristic is simple. Thus, understanding the internal flow pattern at each turbine stage is more important than that in the intermediate turbine duct between turbine stages. It is the primary factor affecting the performance of multistage radial turbines. Therefore, the design optimization of a single radial turbine stage is still the focus for the design of multistage radial turbines. Computational fluid dynamics (CFD) is the currently preferred method of turbine design optimization, as it enables accurate internal flow analysis to guide the detailed turbine design [8–11]. However, prior to its application, a reasonable one-dimensional preliminary design is the necessary first step in radial turbine design [12–14]. Having such a preliminary design is particularly important for the overall performance analysis of multistage radial turbines.

The loading-to-flow diagram first proposed by Chen and Baines is a classical preliminary design method which has several advantages [15,16]. It relates to the operating conditions of the radial turbine. More importantly, it allows for the creation of a contour map of the expected turbine efficiency, with the loading coefficient and flow coefficient as variables, based on a large amount of data from different radial turbine tests (Figure 2). Thus, it has been widely used for the preliminary design of radial turbines [17–19]. However, the loading-to-flow diagram was developed to achieve an efficient preliminary design of a radial turbine under a single operating condition with fixed geometry. It cannot guarantee an optimal radial turbine design for variable operating conditions, especially with guide vane opening changes. Thus, it is necessary to update the current preliminary design method to meet the needs of variable operating conditions. So far, there has been little public research on this issue apart from the work of Lauriau et al. [20]. They provide some theoretical bases for the preliminary design of variable guide vane geometry-based radial turbines, taking into consideration the need for multi-point specifications. They also show how the loading-to-flow map can be modified for different optimal target regions.

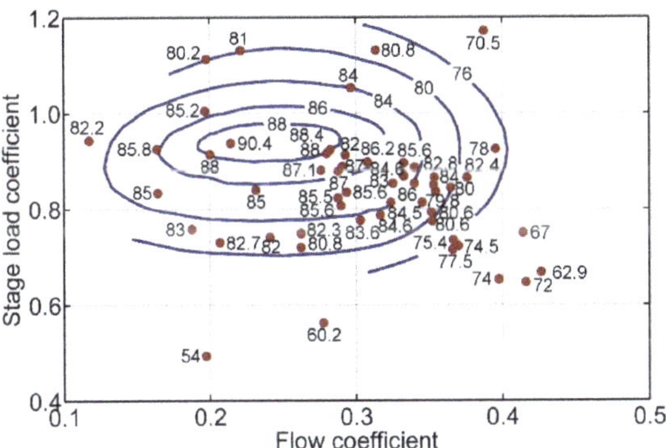

Figure 2. The loading-to-flow diagram [13].

By considering the off-design performance optimization, this paper updates the loading-to-flow diagram method for the preliminary design of radial turbines to accommodate variable operating conditions. The influence of the design value of guide vane outlet flow angle on the rotor loss characteristics was investigated in the continuity of the work presented by Lauriau et al. [20]. Subsequently, aiming at the preliminary design of multistage radial turbines in CAES systems, the optimal design of the guide vane outlet flow angle is discussed from the perspective of the matching of variable operating conditions with rotor loss characteristics. As far as the authors are aware, no similar studies have been performed.

2. Preliminary Design Method Based on the Loading-to-Flow Diagram

According to the definition of the loading-to-flow diagram [17], the loading and flow coefficient can be explained by using the velocity triangles as shown in Figure 3, where the absolute velocity c is a vector addition of the circumferential velocity u and the relative velocity w in the direction of the blade (as a formula: $\vec{c} = \vec{u} + \vec{w}$). The absolute velocity c can be split into a circumferential component c_u and a meridian component c_m. According to Chen and Baines [16], the meridional component of flow velocity at the rotor inlet and outlet can be considered approximately equal. Thus, from Equations (1) and (2), the loading coefficient (ψ) and flow coefficient (ϕ) can be expressed as a function of the guide vane outlet flow angle α_4 and the relative flow angle β_4 at rotor inlet, respectively:

$$\psi = \frac{c_{u4}}{u_4} = \frac{\tan(\alpha_4)}{\tan(\alpha_4) - \tan(\beta_4)}, \tag{1}$$

$$\phi = \frac{c_{m6}}{u_4} \approx \frac{c_{m4}}{u_4} = \frac{1}{\tan(\alpha_4) - \tan(\beta_4)}, \tag{2}$$

where subscripts 4 and 6 denote the rotor inlet and outlet, respectively.

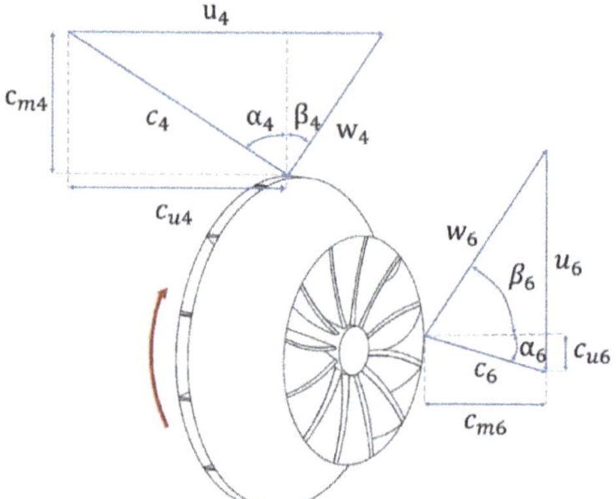

Figure 3. The velocity triangles of the radial turbine rotor inlet and outlet.

The existing literature shows that the optimal values for β_4 are in the range of −40° to −20° [21,22], and the recommended values for α_4 are in the range of 60° to 80° [14]. Thus, by substituting the recommended values of α_4 and β_4 into Equations (1) and (2), the optimum loading-to-flow coefficient range with high expected turbine efficiency could be obtained for the radial turbine preliminary design.

On the other hand, rotor loss is a major factor affecting turbine efficiency, and is directly related to the flow velocity of the fluid in the rotor [23,24]. According to Lauriau et al. [20], the rotor loss can be minimized by reducing the mean velocity of the fluid relative to the rotor passage, which can be expressed as a function of α_4, ϕ, and the mean blade angle at rotor outlet (β_{6m}), as shown in Equation (3). Furthermore, the optimal value of the flow coefficient corresponding to the minimum rotor loss can be expressed by Equation (4).

$$Minimize : \overline{W}_r^2 = \frac{w_4^2 + w_6^2}{2u_4^2} = \frac{1}{2}\phi^2\left(\frac{1}{\cos^2(\alpha_4)} + \frac{1}{\cos^2(\beta_{6m})}\right) + \frac{1}{2} - \phi\tan(\alpha_4), \tag{3}$$

$$\phi_{opt,r} = \frac{\tan(\alpha_4)}{\left(\frac{1}{\cos^2(\alpha_4)} + \frac{1}{\cos^2(\beta_{6m,opt})}\right)}, \tag{4}$$

where $\overline{W}_r$ denotes the ratio of the mean velocity of the fluid relative to the rotor passage with respect to the circumferential velocity of the rotor, $\beta_{6m,opt}$ denotes the recommended values for the mean blade angle at rotor outlet in the range of −60° to −45° [16,25], and w_4 and w_6 denote the relative velocity of air flow at the rotor inlet and outlet, respectively.

Figure 4 depicts the variation of the flow coefficient corresponding to the minimum total loss (corresponding to the optimal values for β_4) and minimum rotor loss (corresponding to the optimal values for $\beta_{6m,opt}$) of the radial turbine as a function of the guide vane outlet flow angle. The former is generally used as the design flow coefficient. It can be seen that the two tended to coincide above a relatively large guide vane outlet angle range ($\alpha_4 \geq 80°$). However, the design flow coefficient significantly increased with the decrease of the guide vane outlet flow angle, but the change of the flow coefficient corresponding to minimum rotor loss was relatively small. The difference between the two gradually increased with the decrease of the guide vane outlet flow angle. Experimental studies have indicated that changes in rotor losses are the most critical factors affecting turbine efficiency under variable operating conditions including pressure ratio change and guide vane opening change [26,27]. It can be inferred that the deviation between the flow coefficient corresponding to minimum rotor loss and the design flow coefficient caused by the different guide vane outlet flow angles will result in different turbine loss characteristics. Therefore, the value of the guide vane outlet flow angle becomes the key to the preliminary design that aims at optimizing the off-design performance of the radial turbine.

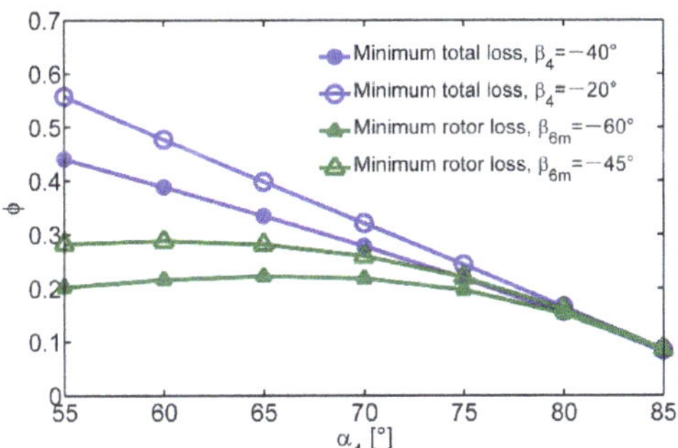

Figure 4. Effect of guide vane outlet flow angle on the optimal flow coefficient value.

3. Analysis of Rotor Loss Characteristics

In this section, the relationship between the design value of guide vane outlet flow angle ($\alpha_{4,d}$) and the rotor loss of the radial turbine is investigated for two typical operating conditions (i.e., variation in pressure ratio and variation in guide vane opening change from the designed value). The ratio of the mean velocity of the fluid relative to the rotor passage with respect to the circumferential velocity of the rotor ($\overline{W}_r$) was determined to infer rotor loss in the early phase of preliminary design without the need of detailed turbine parameters. Since the rotor loss plays a key role in the off-design efficiency of radial turbines, we assumed that $\overline{W}_r$ is a reasonable indictor to qualitatively estimate the rotor loss and turbine efficiency in the preliminary design phase.

3.1. Pressure Ratio Change

For radial turbines with fixed geometry guide vane, the rotor loss under off-design operating conditions is dominated by viscous loss with a nearly constant loss coefficient [24]. Therefore, the value of $\overline{W}_r$ is used to predict the rotor loss characteristics of radial turbines with different $\alpha_{4,d}$ indirectly in the case of pressure ratio change.

In the case where only the change of pressure ratio is considered, the off-design flow coefficient can be estimated using the improved Flügel formula [28] and the definition of flow coefficient (Equation (5)). It can be seen that the off-design flow coefficient is approximately proportional to the pressure ratio change.

$$\left\{\begin{array}{l} \frac{\dot{m}_t}{\dot{m}_{t,d}} = \frac{P_{in}}{P_{in,d}}\sqrt{\frac{1-1/\beta_t^2}{1-1/\beta_{t,d}^2}} \\ \frac{\dot{m}_t}{\dot{m}_{t,d}} = \frac{c_{m6}A_6\rho_6}{c_{m6,d}A_6\rho_{6,d}} \approx \frac{(\frac{c_{m6}}{u_4})\cdot P_6}{(\frac{c_{m6,d}}{u_4})\cdot P_{6,d}} = \frac{\phi\cdot P_6}{\phi_d\cdot P_{6,d}} \end{array}\right. \rightarrow \frac{\phi}{\phi_d} \approx \sqrt{\frac{\beta_t^2-1}{\beta_{t,d}^2-1}}, \quad (5)$$

where $\dot{m}_t$ denotes the mass flow, subscript d denotes the design value, P_{in} denotes the inlet pressure of turbine, β_t denotes the pressure ratio, A_6 denotes the rotor outlet area, ρ_6 denotes the density of gas at the rotor outlet, and P_6 denotes the rotor outlet pressure.

Substituting the off-design flow coefficient into Equation (3), the change characteristics of $\overline{W}_r$ in the case of pressure ratio change could be obtained as shown in Figure 5. The figure shows that for a radial turbine with a small design value of guide vane outlet flow angle (e.g., $\alpha_{4,d} = 60°$), the corresponding $\overline{W}_r$ had a characteristic of decreasing first and then increasing when the pressure ratio was reduced relative to the design point. However, for a radial turbine with a large design value of guide vane outlet flow angle (e.g., $\alpha_{4,d} = 80°$), an increase or decrease in the pressure ratio relative to the design point resulted in an increase in $\overline{W}_r$. The rotor losses also exhibited the characteristics described above due to the change in $\overline{W}_r$. The above results can be explained as follows. For turbines with a small

$\alpha_{4,d}$, the flow coefficient corresponding to minimum rotor loss is less than the design flow coefficient (Figure 4). Therefore, in the process of decreasing the pressure ratio, the off-design flow coefficient (which is approximately proportional to the pressure ratio) is close to the minimum rotor-loss-based flow coefficient first and then gradually deviates from it, resulting in rotor loss exhibiting similar change characteristics. Similarly, for turbines with a large $\alpha_{4,d}$, the flow coefficient corresponding to minimum rotor loss is coincident with the design flow coefficient. An increase or decrease in the pressure ratio will cause the off-design flow coefficient to deviate from the minimum rotor-loss-based flow coefficient, resulting in an increase in rotor loss.

Moreover, within a wide range of pressure ratio change ($0.5 < \beta_{t,i}/\beta_{t,d} < 1.5$), the larger the guide vane outlet flow angle, the smaller $\overline{W_r}$ is. It can be concluded that the turbine efficiency under the same range of pressure ratio change increases proportionally to the design value of guide vane outlet flow angle.

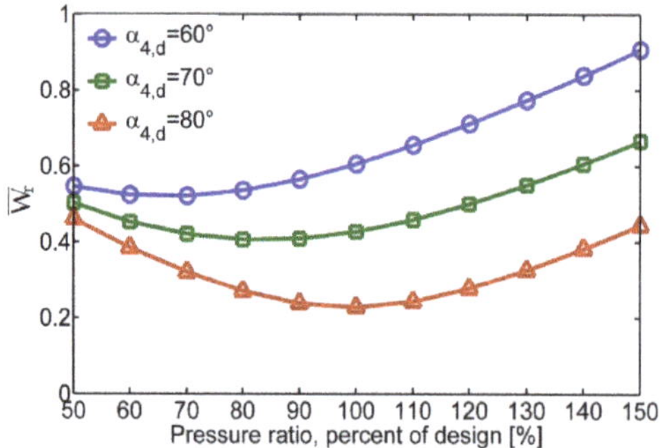

Figure 5. Effect of design guide vane outlet flow angle on the change characteristics of $\overline{W_r}$ under pressure ratio change ($\beta_{6m} = -52.5°$, $\beta_4 = -30°$, $\beta_{t,d} = 3$).

3.2. Guide Vane Opening Change

For radial turbines with variable geometry guide vane, the rotor loss under off-design operating conditions becomes more complex due to the change in the internal flow of the rotor with guide vane opening. Meitner et al. [29] show a relationship between the loss coefficient of rotor kinetic energy and guide vane opening where the loss coefficient increased significantly at small guide vane opening. This occurs because the rotor loss mechanism is not only viscous loss but also increased secondary flow loss at small guide vane opening, according to Otsuka et al. [30]. Therefore, in the case of decreasing guide vane opening (<100% design value), the relative velocity $\overline{W_r}/\overline{W_{r,d}}$ is proposed to indirectly and qualitatively estimate rotor loss for comparison between radial turbines with different design values of guide vane outlet flow angle ($\alpha_{4,d}$). On the other hand, the value of $\overline{W_r}$ is used for increasing guide vane opening (>100% design value).

According to Spence et al. [26,27], the mass flow rate of the radial turbine is approximately proportional to the guide vane opening. As a result, the existing Flügel formula [28] can be updated by multiplying a guide vane opening ratio (O_r) to estimate the mass flow rate under guide vane opening change. From the updated Flügel formula and the definition of flow coefficient, the off-design flow coefficient can be obtained as shown in Equation (6).

$$\left\{\begin{array}{l} \frac{\dot{m}_t}{\dot{m}_{t,d}} \approx O_r \frac{P_{in}}{P_{in,d}} \sqrt{\frac{1-1/\beta_t^2}{1-1/\beta_{t,d}^2}} \\ \frac{\dot{m}_t}{\dot{m}_{t,d}} = \frac{c_{m6} A_6 \rho_6}{c_{m6,d} A_6 \rho_{6,d}} \approx \frac{(\frac{c_{m6}}{u_4}) \cdot P_6}{(\frac{c_{m6,d}}{u_4}) \cdot P_{6,d}} = \frac{\phi \cdot P_6}{\phi_d \cdot P_{6,d}} \end{array}\right. \rightarrow \frac{\phi}{\phi_d} \approx O_r \sqrt{\frac{\beta_t^2 - 1}{\beta_{t,d}^2 - 1}}, \tag{6}$$

where O_r is defined as the ratio of the off-design guide vane opening to the design value.

Moreover, from the guide vane geometric (Figure 6), a sinus rule can be employed to estimate the updated α_4 caused by the guide vane opening change as shown in Equation (7) [23]. Finally, by substituting the off-design flow coefficient and the updated α_4 into Equation (3), the off-design $\overline{W}_r$ the corresponding $\overline{W}_r/\overline{W}_{r,d}$ can be obtained.

$$\alpha_4 = \cos^{-1}(\frac{O}{S}) = \cos^{-1}(\frac{O}{O_d} \times \frac{O_d}{S}) = \cos^{-1}(O_r \cdot \cos(\alpha_{4,d})), \tag{7}$$

where O denotes the guide vane opening and S denotes the span of the guide vane at the outlet.

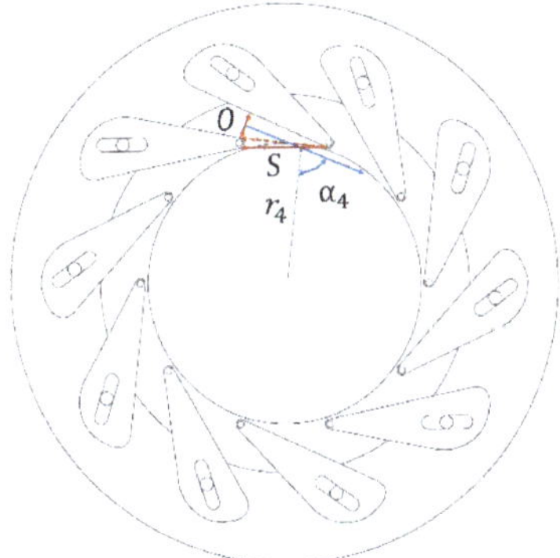

Figure 6. Typical variable-geometry guide vanes for radial turbines.

From Equation (6), the variation in flow coefficient was closely related to the pressure ratio in the case of guide vane opening change. Thus, the rotor loss analysis of two representative cases of guide vane opening change was carried out respectively; these were the constant pressure ratio case and a more complicated case where the pressure ratio changed inversely to the guide vane opening.

Figure 7a shows that in the case where the pressure ratio remained constant, $\overline{W}_r/\overline{W}_{r,d}$ decreased significantly as the guide vane opening decreased. Although the difference between different $\alpha_{4,d}$ was not notable, it can be seen that the $\overline{W}_r/\overline{W}_{r,d}$ of the radial turbine with a smaller $\alpha_{4,d}$ was smaller, which means a smaller increase in the rotor loss with decreased guide vane opening. In the case where the guide vane opening was increased, the larger the $\alpha_{4,d}$, the smaller the $\overline{W}_r$. This means a smaller increase in the rotor loss with the increased guide vane opening. The above results can be explained as follows. In the case of a constant pressure ratio, the flow coefficient is proportional to the guide vane opening ratio (Equation (6)). Since smaller $\alpha_{4,d}$ leads to a larger corresponding design flow coefficient (Figure 4), it can be inferred that the relative velocity of air flow at the rotor inlet (w_4) is also larger (Figure 3). Thus, within the same regulation range of the guide vane opening, the changes in w_4 caused by the flow coefficient change are more pronounced in both the increasing and decreasing directions. This directly leads to a more significant change in the rotor loss.

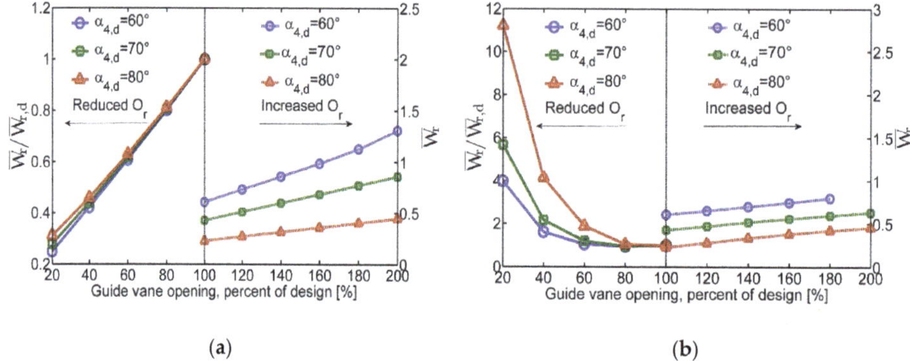

Figure 7. Effect of design guide vane outlet flow angle on the change characteristics of $\overline{W}_r$ with variable guide vane opening: (**a**) constant pressure ratio; (**b**) constant flow coefficient ($\beta_{6m} = -52.5°, \beta_4 = -30°$).

Figure 7b shows that in the case where the pressure ratio changed inversely with the guide vane opening and maintained a constant flow coefficient, $\overline{W}_r/\overline{W}_{r,d}$ increased significantly as the guide vane opening decreased. In contrast, the radial turbine with smaller $\alpha_{4,d}$ had a smaller increase in $\overline{W}_r/\overline{W}_{r,d}$, which means a relatively smaller increase in the rotor loss with decreased guide vane opening. In the case where the guide vane opening was increased, the larger the $\alpha_{4,d}$, the smaller the $\overline{W}_r$. This means a smaller increase in the rotor loss with the increased guide vane opening. The above results can be explained as follows. According to Figure 4, the larger α_4, the more obvious the change (decreased) of the flow coefficient corresponding to the minimum rotor loss. Thus, for the radial turbine with a smaller $\alpha_{4,d}$, the deviation of the off-design flow coefficient (approximately equal to the design value) from the flow coefficient (decreased) corresponding to the minimum rotor loss caused by the decrease of the guide vane opening is smaller. It can also be inferred that the increase in the rotor loss would be smaller. In contrast, in the case where the guide vane opening was increased, the increased flow coefficient corresponding to the minimum rotor loss tended to be gentle with the increase of α_4. The deviation of the off-design flow coefficient (approximately equal to the design value) from the flow coefficient (increased) corresponding to the minimum rotor loss caused by the increase of the guide vane opening can be ignored for radial turbines with a different $\alpha_{4,d}$. Therefore, due to the smaller $\overline{W}_r$ at the design point, the larger the $\alpha_{4,d}$, the smaller the off-design rotor loss.

Furthermore, by comparing Figure 7a,b, we can find that the inverse change of pressure ratio with the change in the guide vane opening will increase the rotor loss in the case of reduced guide vane opening. On the contrary, it can mitigate the increase of the rotor loss in the case of increased guide vane opening.

In conclusion, the difference in the rotor loss characteristics under guide vane opening change for radial turbines with different $\alpha_{4,d}$ is determined by the directionality of the guide vane regulation. It can be concluded that the optimum design of the guide vane outlet flow angle is closely related to the directionality of the guide vane regulation.

4. Optimal Design of the Guide Vane Outlet Flow Angle for Multistage Radial Turbines

This section presents a discussion of the optimal design of the guide vane outlet flow angle ($\alpha_{4,d}$) for a multistage radial turbine from the perspective of matching the rotor loss characteristics with variable operating conditions. In addition, the multistage radial turbine of a CAES pilot plant named "TICC-500" [31] was taken as an example for analysis (Table 1).

Table 1. Parameters of the multistage turbine in a "TICC-500" CAES pilot system [31].

Turbine Stage	Input Pressure, MPa	Outlet Pressure, MPa
High-Pressure Turbine (HP)	2.50	1.13
Medium-Pressure Turbine (MP)	1.12	0.40
Low-Pressure Turbine (LP)	0.39	0.10

To obtain the representative variable operating conditions of the multistage radial turbine based on guide vane control, only the case of regulating the guide vane opening of the high-pressure turbine stage was considered here. Since the rotational speed of the multistage radial turbine in a CAES system is generally constant, the mass flow rate of each turbine stage under guide vane control can be estimated by the updated Flügel formula in Equation (6), which includes a guide vane opening ratio (O_r). On this basis, to solve the operating conditions of each turbine stage in the multistage radial turbine at a certain inlet pressure and guide vane opening, the pressure ratios were updated to iteratively calculate the mass flow for each turbine stage until the continuity of mass flow was met. Furthermore, by substituting off-design the operating conditions (guide vane opening and pressure ratio) into Equation (6), the corresponding the flow coefficient could be obtained.

Depending on the directivity of the guide vane opening regulation, the case where the guide vane opening is reduced and increased with respect to the design value is defined as a down-regulation and an up-regulation, respectively. Figure 8a,b presents the variable operating conditions of the multistage radial turbine in the case of down-regulation and up-regulation, respectively. We assumed steady-state operation conditions with different guide vane openings, and the hysteresis during continuous down- and up-regulation was not considered in this study.

(1) Down-Regulation of the Guide Vane Opening

When the inlet pressure of the multistage radial turbine is high while the load demand is low, the down-regulation of the guide vane opening is conducted to reduce the mass flow rate. Figure 8a shows that the pressure ratio of the low-pressure (LP) turbine decreased significantly with the decrease of the guide vane opening of the high-pressure (HP) turbine and relatively small change in the expansion ratio of the medium-pressure (MP) turbine. In contrast, the pressure ratio of the HP turbine changed inversely to the guide vane opening, which significantly increased with the decrease of the guide vane opening.

(2) Up-Regulation of the Guide Vane Opening

When the inlet pressure of the multistage radial turbine is relatively low while the load demand is high, the up-regulation of the guide vane opening is conducted to increase the mass flow rate. Figure 8b shows that the pressure ratio of the LP turbine increased significantly with the increase of the guide vane opening of the HP turbine, while the expansion ratio of the MP turbine remained almost constant. Like the down-regulation of the guide vane opening, the pressure ratio of HP turbine changed inversely to the guide vane opening, which reduced significantly with the increase of the guide vane opening.

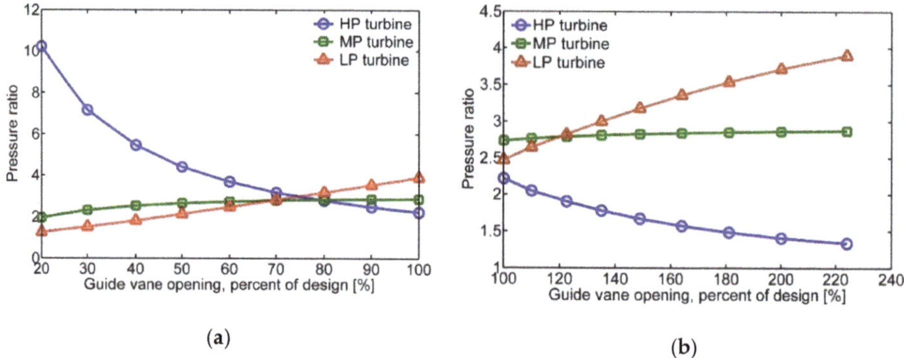

Figure 8. Variable operating conditions for a multistage radial turbine under guide vane opening change: (**a**) down-regulation at rated inlet pressure; (**b**) up-regulation at 60% the rated inlet pressure.

Figure 9a,b presents the flow coefficient variation characteristics of the multistage radial turbine under the variable operating conditions shown in Figure 8a,b, respectively. Combined with the findings regarding the relationship of off-design performance versus the design value of the guide vane outlet flow angle ($\alpha_{4,d}$) for the radial turbine shown in Section 3, the recommendations for the optimum $\alpha_{4,d}$ in the preliminary design of the multistage radial turbine are as follows.

i. The variable operating condition of the LP turbine belongs to the expansion ratio change, and the flow coefficient varies notably within a range smaller than the design value. According to the findings on the relationship of off-design performance versus $\alpha_{4,d}$ for radial turbines under pressure ratio change (Section 3.1), a larger $\alpha_{4,d}$ should be used for the LP turbine to achieve better off-design performance. A value of approximately 80° is recommended.

ii. The variable operating condition of the MP turbine also belongs to the expansion ratio change, but the flow coefficient varies within a small range that is less than the design value. Therefore, turbine efficiency under design conditions dominates the performance of the MP turbine under variable operating conditions. As a result, a larger $\alpha_{4,d}$ should be used for the MP turbine (i.e., about 80°).

iii. The variable operating condition of the HP turbine is a complicated case where the pressure ratio changes inversely with the guide vane opening and maintains a nearly constant flow coefficient. According to the findings on the relationship of off-design performance versus $\alpha_{4,d}$ for radial turbines under guide vane opening change (Section 3.2), the optimum $\alpha_{4,d}$ mainly depends on the direction of the guide vane opening changes relative to the design point. For a multistage radial turbine in a CAES system with constant air storage pressure, the inlet pressure of the HP turbine stage remains at the rated value, so only the down-regulation of the guide vane opening is performed. Accordingly, a smaller $\alpha_{4,d}$ should be used for the HP turbine stage to achieve better off-design performance. Further considering the influence of design efficiency on off-design performance, the recommended design values are in the range of 70–75°. For a multistage radial turbine in a CAES system with varying air storage pressure, the inlet pressure of the HP turbine stage varies with air storage pressure, so there is a simultaneous need for up- and down-regulation of the guide vane opening. In this case, it is necessary to comprehensively consider the influence of the rotor loss characteristics and design efficiency on the off-design performance in determining $\alpha_{4,d}$.

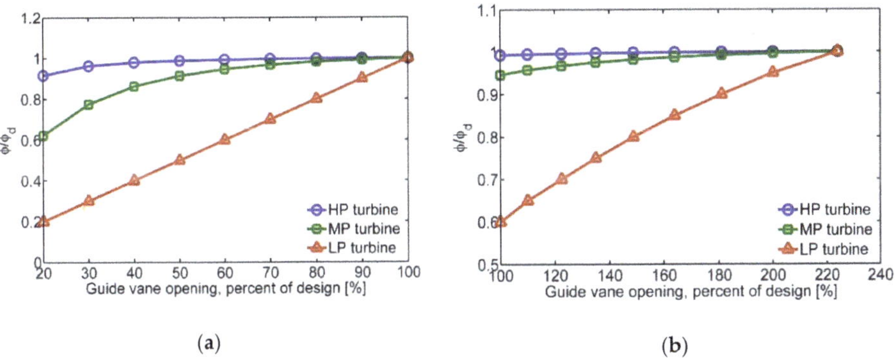

Figure 9. Flow coefficient variation for the multistage turbine under guide vane opening change: (**a**) down-regulation at rated inlet pressure; (**b**) up-regulation at 60% of the rated inlet pressure.

5. Case Study and Discussion

In this section, we compare the off-design performance of a group of radial turbines with different design values of guide vane outlet flow angle. To ensure the comparability of different radial turbine performances, the preliminary design results of radial turbine obtained under the same design conditions were employed for the case study. Moreover, the performance evaluation of the radial turbine was conducted using the mean-line model method.

The mean-line model method previously developed in our group is a proven performance evaluation method for radial turbines. Since the model evaluation of radial turbines is not the focus of this study, please refer to one of our previous publications for more details [7]. The model uses limited turbine geometry information to predict the turbine performance under variable operating conditions, which is especially valuable in the preliminary design process. The total-to-static efficiency of the radial turbine was calculated by the specific work output, isentropic enthalpy drops, and turbine losses, as shown in Equation (8). According to the Euler equation of turbomachines [32], the specific work output can be calculated with Equation (9). Table 2 gives the loss model combination suitable for the turbine loss evaluation under variable guide vane opening, which involves four types of main turbine losses: stator loss, incidence loss, rotor loss, and exit loss.

$$\eta_{t-s} = \frac{W_t}{\Delta h_{1-6s}} = \frac{\Delta h_{1-6s} - \sum h_{loss}}{\Delta h_{1-6s}}, \tag{8}$$

$$W_t = u_4 c_{u4} - u_6 c_{u6}, \tag{9}$$

where W_t denotes the specific work output of the radial turbine, Δh_{1-6s} denotes the isentropic enthalpy drop between the guide vane inlet and rotor outlet, and h_{loss} denotes the turbine loss.

Table 2. The loss model combination of the radial turbine.

Loss	Reference	Model
Stator Loss	[33]	$\Delta h_{stator} = \frac{1}{2} K_n {c_{3s}}^2$
Incidence Loss	[29]	$\Delta h_{incidence} = \frac{1}{2} {w_4}^2 (1 - \cos^n i)$
Rotor Loss	[29]	$\Delta h_{rotor} = \frac{1}{2} K_r ({w_4}^2 \cos^2 i + {w_6}^2)$
Exit Loss	[24]	$\Delta h_{exit} = \frac{1}{2} {c_{m6}}^2$

The design conditions of the radial turbines in this case study are shown in Table 3. Based on the conclusions of the matching analysis in Section 4, three typical design values of guide vane outlet flow angle ($\alpha_{4,d}$= [70°; 75°; 80°]) were employed for the case study. The main parameters of the radial turbines were obtained by a preliminary design algorithm proposed by Ventura et al. [17] (Table 4).

Table 3. The design conditions of the radial turbines for the case study.

Parameter	Value
Inlet Temperature, K	430
Inlet Total Pressure, MPa	10.00
Outlet Pressure, MPa	3.33
Mass Flow Rate, kg/s	24.80

Table 4. The preliminary design results of the radial turbines for the case study.

Parameter	Case A	Case B	Case C
Guide Vane Outlet Flow Angle, degree	70	75	80
Loading Coefficient	0.78	0.82	0.88
Flow Coefficient	0.28	0.23	0.16
Rotor			
Rotational Speed, r/min	50112	39620	32329
Inlet Radius, mm	73.5	90.8	107.0
Inlet Tip Width, mm	8.6	9.3	11.7
Inlet Relative Flow Angle, degree	−38.5	−40.6	−38.2
Blade Number	11	15	18
Outlet Tip Radius, mm	21.1	24.1	29.0
Outlet Hub Radius, mm	52.1	59.6	72.0
Outlet Blade Angle, degree	−62.3	−66.4	−73.2
Stator			
Inlet Radius, mm	99.0	119.4	138.8
Throat Width, mm	11.7	8.4	5.2
Vane height, mm	8.6	9.3	11.7
Outlet Radius, mm	79.2	95.5	111.0
Blade Number	14	18	23
Performance			
Total-to-Static Efficiency	0.881	0.895	0.899

Figure 10 presents the performance curves of radial turbines with different design values of guide vane outlet flow angle operating under pressure ratio change. It can be seen that the larger the design value of guide vane outlet flow angle, the higher the design efficiency of the radial turbine, and the higher the efficiency operating under variable pressure ratio in a certain range ($0.5 < \beta_{t,i}/\beta_{t,d} < 1.5$). These are consistent with the findings of the rotor loss characteristic analysis under pressure ratio change (Section 3.1). It should also be noted that the turbine efficiency deteriorated significantly as the pressure ratio decreased to less than about 60% of the design value. However, simply optimizing the design value of the guide vane outlet angle did not seem to be effective in improving this deterioration.

Figure 11a,b, gives the performance curves of two typical cases of guide vane opening change: the constant pressure ratio case and the constant flow coefficient case. First, it can be seen that the radial turbines with different design values of guide vane outlet flow angle had significant differences in the distribution of the efficient operating range under guide vane opening changes, which verifies the finding of the rotor loss characteristic analysis in Section 3.2. Specifically, the radial turbine with a larger design value of guide vane outlet flow angle (e.g., $\alpha_{4,d} = 80°$) had a broader efficient operating range (e.g., turbine efficiency > 0.8) and higher efficiency for the up-regulation of the guide vane opening, while a smaller design value of guide vane outlet flow angle resulted in broader efficient operating range and higher efficiency for the down-regulation of the guide vane opening. Furthermore, comparing Figure 11a,b, it can be found that the deterioration of the turbine efficiency under the down-regulation of the guide vane opening could be alleviated by reducing the reverse increase in the pressure ratio versus the guide vane opening change. Similarly, for the up-regulation of the guide vane opening, when the design value of the guide vane outlet flow angle was large, diminishing the inverse reduction of the pressure ratio could improve the turbine efficiency. However, for a turbine with a smaller design value of guide vane outlet flow angle (e.g., $\alpha_{4,d} = 70°$), this could worsen efficiency. These are also consistent with the findings in Section 3.2.

Based on the above, for a multistage radial turbine that simultaneously needs up- and down-regulation of the guide vane opening, the design value of guide vane outlet flow angle for the high-pressure turbine stage (or other turbine stages with variable geometry guide vane) is recommended to be about 80°. In this case, higher design efficiency and wide efficient operating range for the up-regulation can be obtained, while the turbine efficiency deterioration under down-regulation can be improved with combined control of the guide vane openings of the multistage radial turbine. The combined control of the guide vane openings of the multistage radial turbine has been proved to be able to alleviate the reverse change in pressure ratio versus the guide vane opening [6]. This could be an effective way to achieve optimum performance for a multistage radial turbine operating under variable working conditions.

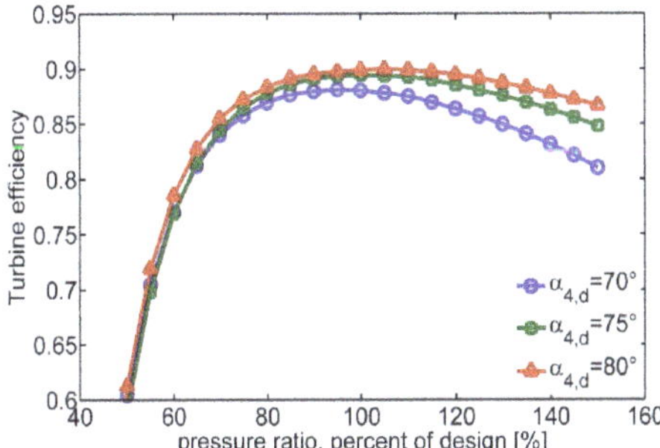

Figure 10. Turbine performance under pressure ratio change.

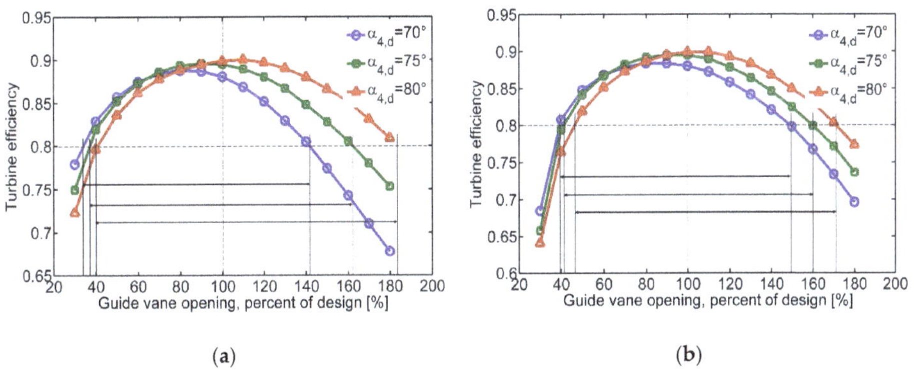

Figure 11. Turbine performances when the guide vane opening changes and: (**a**) the pressure ratio is constant; (**b**) the pressure ratio changes inversely to the guide vane opening.

6. Conclusions

This study improves the classical loading-to-flow diagram method to meet the design needs of radial turbines with variable operating conditions. To optimize the off-design performance of radial turbines in the early design phase, we proposed a hypothesis that uses the ratio of the mean velocity of the fluid relative to the rotor passage with respect to the circumferential velocity of the rotor as an indicator to indirectly and qualitatively estimate the rotor loss, as it plays a key role in the off-design efficiency. This hypothesis is based on the findings from existing studies indicating that rotor loss is a function of the flow velocity within the rotor.

The findings of off-design rotor loss analysis for radial turbines with a different design value of guide vane outlet flow angle are as follows:

(1) For a radial turbine with a smaller design value of guide vane outlet flow angle, the rotor loss first decreased and then increased with the decrease of mass flow. This means better off-design performance in the case of reducing the mass flow. This applies both under conditions of pressure ratio change and of guide vane opening change. However, due to a higher rotor loss at design conditions, the turbine efficiency under pressure ratio changes may be lower than for radial turbines with a larger guide vane outlet flow angle.

(2) A radial turbine with a larger design value of guide vane outlet flow angle not only had higher efficiency at design conditions but also had better off-design performance in the case of increased mass flow. This held for changes in both pressure ratio and guide vane opening.

The above findings were validated with the mean-line model method [7]. Furthermore, based on the findings, this study discusses the optimization of the design value of guide vane outlet flow angle based on the matching of rotor loss characteristics with specified variable operating conditions for a multistage radial turbine in a compressed air energy storage system. It provides important guidance for the design optimization of multistage radial turbines operating under variable working conditions.

Author Contributions: Conceptualization, Z.C., Z.T., and S.T.; Methodology, Z.C.; Formal analysis, Z.C.; Validation, Z.C.; Writing, Z.C. and Z.T.; Supervision, S.T. and Z.T.

Funding: We would like to acknowledge the National Key R&D Program of China (2018YFB0606105), National Natural Science Foundation of China (51708493), Zhejiang Provincial Natural Science Foundation (LR19E050002), Zhejiang Province Key Science and Technology Project (2018C01020, 2018C01060, 2019C01057), and the Youth Funds of State Key Laboratory of Fluid Power & Mechatronic Systems (SKLoFP_QN_1804).

Conflicts of Interest: The authors declare no conflict of interest.

References

1. He, W.; Wang, J.; Ding, Y. New radial turbine dynamic modelling in a low-temperature adiabatic compressed air energy storage system discharging process. *Energy Convers. Manag.* **2017**, *153*, 144–156. [CrossRef]
2. Tong, S.; Cheng, Z.; Cong, F.; Tong, Z.; Zhang, Y. Developing a grid-connected power optimization strategy for the integration of wind power with low-temperature adiabatic compressed air energy storage. *Renew. Energy* **2018**, *125*, 73–86. [CrossRef]
3. Venkataramani, G.; Parankusam, P.; Ramalingam, V.; Wang, J. A review on compressed air energy storage—A pathway for smart grid and polygeneration. *Renew. Sustain. Energy Rev.* **2016**, *62*, 895–907. [CrossRef]
4. Argyrou, M.C.; Christodoulides, P.; Kalogirou, S.A. Energy storage for electricity generation and related processes: Technologies appraisal and grid scale applications. *Renew. Sustain. Energy Rev.* **2018**, *94*, 804–821. [CrossRef]
5. Li, H.; Li, W.; Liu, D.; Zhang, X.; Zhu, Y.; Chen, H. Effect of adjustable guide vanes on the characteristic of multistage reheating radial inflow turbine. *Proc. CSEE* **2016**, *36*, 6180–6186.
6. Liu, D.; Li, W.; Li, H.; Zhang, X.; Zhu, Y.; Chen, H. Characteristic analysis of combined regulation of adjustable guide vanes of multistage radial inflow turbines. *Energy Storage Sci. Technol.* **2017**, *6*, 1286–1294.
7. Cheng, Z.; Tong, S.; Tong, Z. Bi-directional nozzle control of multistage radial-inflow turbine for optimal part-load operation of compressed air energy storage. *Energy Convers. Manag.* **2019**, *181*, 485–500. [CrossRef]
8. Li, P.; Han, Z.; Jia, X.; Mei, Z.; Han, X. Analysis of the effects of blade installation angle and blade number on radial-inflow turbine stator flow performance. *Energies* **2018**, *11*, 2258. [CrossRef]
9. Capata, R.; Hernandez, G. Preliminary design and simulation of a turbo expander for small rated power Organic Rankine Cycle (ORC). *Energies* **2014**, *7*, 7067–7093. [CrossRef]
10. Tong, Z.; Li, Y.; Westerdahl, D.; Adamkiewicz, G.; Spengler, J.D. Exploring the effects of ventilation practices in mitigating in-vehicle exposure to traffic-related air pollutants in China. *Environ. Int.* **2019**, *127*, 773–784. [CrossRef]
11. Tong, Z.; Chen, Y.; Malkawi, A. Defining the Influence Region in neighborhood-scale CFD simulations for natural ventilation design. *Appl. Energy* **2016**, *182*, 625–633. [CrossRef]
12. Daabo, A.M.; Al Jubori, A.; Mahmoud, S.; Al-Dadah, R.K. Development of three-dimensional optimization of a small-scale radial turbine for solar powered Brayton cycle application. *Appl. Therm. Eng.* **2017**, *111*, 718–733. [CrossRef]

13. Fiaschi, D.; Innocenti, G.; Manfrida, G.; Maraschiello, F. Design of micro radial turboexpanders for ORC power cycles: From 0D to 3D. *Appl. Therm. Eng.* **2016**, *99*, 402–410. [CrossRef]
14. Rahbar, K.; Mahmoud, S.; Al-Dadah, R.K. Mean-line modeling and CFD analysis of a miniature radial turbine for distributed power generation systems. *Int. J. Low Carbon Technol.* **2016**, *11*, 157–168. [CrossRef]
15. Chen, H.; Baines, N.C. Analytical Optimization Design of Radial and Mixed Flow Turbines. *Proc. Inst. Mech. Eng. Part A J. Power Energy* **1992**. [CrossRef]
16. Chen, H.; Baines, N.C. The aerodynamic loading of radial and mixed-flow turbines. *Int. J. Mech. Sci.* **1994**, *36*, 63–79. [CrossRef]
17. Ventura, C.A.M.; Jacobs, P.A.; Rowlands, A.S.; Petrie-Repar, P.; Sauret, E. Preliminary design and performance estimation of radial inflow turbines: An automated approach. *J. Fluid Eng. Trans. ASME* **2012**, *134*, 31102. [CrossRef]
18. Shao, S.; Deng, Q.H.; Feng, Z.P. An automated approach for the radial inflow turbine preliminary design and off-design prediction. *J. Eng. Thermophys.* **2016**, *37*, 67–71.
19. Da Silva, E.R.; Kyprianidis, K.G.; Säterskog, M.; Camacho, R.G.R.; Sarmiento, A.L.E. Preliminary design optimization of an organic Rankine cycle radial turbine rotor. In Proceedings of the ASME Turbo Expo, Charlotte, NC, USA, 26–30 June 2017. [CrossRef]
20. Lauriau, P.T.; Binder, N.; Carbonneau, X.; Cros, S.; Roumeas, M. Preliminary design considerations for variable geometry radial turbines with multi-points specifications. In Proceedings of the 12th European Conference on Turbomachinery Fluid Dynamics and Thermodynamics (ETC 2017), Stockholm, Sweden, 3–7 April 2017.
21. Woolley, N.H.; Hatton, A.P. Viscous Flow in Radial Turbomachine Blade Passages. In Proceedings of the Conference on Heat and Fluid Flow in Steam and Gas Turbine Plant, London, UK, 3–5 April 1974; pp. 175–181.
22. Spence, S.W.T.; Artt, D.W. An experimental assessment of incidence losses in a radial inflow turbine rotor. *Proc. Inst. Mech. Eng. Part A J. Power Eng.* **1998**. [CrossRef]
23. Moustapha, H.; Zelesky, M.F.; Japiske, D.; Baines, N.C. *Axial and Radial Turbines*; Concepts NREC: White River Junction, VT, USA, 2003.
24. Suhrmann, J.F.; Peitsch, D.; Gugau, M.; Heuer, T.; Tomm, U. Validation and development of loss models for small size radial turbines. In Proceedings of the ASME Turbo Expo, Glasgow, UK, 14–18 June 2010. [CrossRef]
25. Qiu, X.; Baines, N. Performance prediction for high pressure-ratio radial inflow turbines. In Proceedings of the ASME Turbo Expo, Montreal, QC, Canada, 14–17 May 2007. [CrossRef]
26. Spence, S.W.T.; Artt, D.W. Experimental performance evaluation of a 99.0 mm radial inflow nozzled turbine with different stator throat areas. *Proc. Inst. Mech. Eng. Part A J. Power Eng.* **1997**. [CrossRef]
27. Spence, S.W.T.; Doran, W.J.; Artt, D.W. Experimental performance evaluation of a 99.0 mm radial inflow nozzled turbine at larger stator-rotor throat area ratios. *Proc. Inst. Mech. Eng. Part A J. Power Eng.* **1999**. [CrossRef]
28. Zhang, N.; Cai, R. Analytical solutions and typical characteristics of part-load performances of single shaft gas turbine and its cogeneration. *Energy Convers. Manag.* **2002**, *43*, 1323–1337. [CrossRef]
29. Meitner, P.L.; Glassman, A.J. *Off-Design Performance Loss Model for Radial Turbines with Pivoting, Variable-Area Stators*; NASA TP-1708; NASA: Washington, DC, USA, 1980.
30. Otsuka, K.; Komatsu, T.; Tsujita, H.; Yamaguchi, S.; Yamagata, A. Numerical analysis of flow in radial turbine (Effects of nozzle vane angle on internal flow). *Int. J. Fluid Mach. Syst.* **2016**, *9*, 137–142. [CrossRef]
31. Wang, S.; Zhang, X.; Yang, L.; Zhou, Y.; Wang, J. Experimental study of compressed air energy storage system with thermal energy storage. *Energy* **2016**, *103*, 182–191. [CrossRef]
32. Dixon, S.; Hall, C. *Fluid Mechanics and Thermodynamics of Turbomachinery*; Elsevier Ltd.: Amsterdam, The Netherlands, 2010.
33. Glassman, A.J.; Center, L.R. *Computer Program for Design Analysis of Radial-Inflow Turbines*; NASA TN D-9164; NASA: Washington, DC, USA, 1976.

© 2019 by the authors. Licensee MDPI, Basel, Switzerland. This article is an open access article distributed under the terms and conditions of the Creative Commons Attribution (CC BY) license (http://creativecommons.org/licenses/by/4.0/).

Article

Microgeneration of Electricity Using a Solar Photovoltaic System in Ireland

Vinay Virupaksha [1], Mary Harty [2,*] and Kevin McDonnell [1,2,3]

1 School of Biosystems Engineering, University College Dublin, Belfield, D04 N2E5 Dublin 4, Ireland; vinay.mv2@gmail.com (V.V.); kevin.mcdonnell@ucd.ie (K.M.)
2 School of Agriculture and Food Science, University College Dublin, Belfield, D04 N2E5 Dublin 4, Ireland
3 Biosystems Engineering Ltd, NovaUCD, Belfield, D04 V2P1 Dublin 4, Ireland
* Correspondence: mary.harty@ucd.ie

Received: 30 October 2019; Accepted: 2 December 2019; Published: 3 December 2019

Abstract: Microgeneration of electricity using solar photovoltaic (PV) systems is a sustainable form of renewable energy, however uptake in Ireland remains very low. The aim of this study is to assess the potential of the community-based roof top solar PV microgeneration system to supply electricity to the grid, and to explore a crowd funding mechanism for community ownership of microgeneration projects. A modelled microgeneration project was developed: the electricity load profiles of 68 residential units were estimated; a community-based roof top solar PV system was designed; an electricity network model, based on a real network supplying a town and its surrounding areas, was created; and power flow analysis on the electrical network for system peak and minimum loads was carried out. The embodied energy, energy payback time, GHG payback time, carbon credits and financial cost relating to the proposed solar PV system were calculated. Different crowdfunding models were assessed. Results show the deployment of community solar PV system projects have significant potential to reduce the peak demand, smooth the load profile, assist in the voltage regulation and reduce electrical losses and deliver cost savings to distribution system operator and the consumer.

Keywords: microgeneration; solar energy; photovoltaic; renewable energy; crowd funding

---

1. Introduction

The fifth assessment report of the Intergovernmental Panel on Climate Change [1] has concluded that human influence on the climate system is clear, and anthropogenic emissions of greenhouse gases (GHG) are the highest in history. GHG emissions are driving the increase in global average temperatures by over 1 °C above preindustrial times with this trend projected to continue. In Ireland, the burning of fossil fuels for energy generation is the dominant contributor to total national GHG emissions (60% in 2017) [2]. With limited indigenous fossil energy resources, Ireland is significantly dependent on fossil fuel imports which accounted for over 90% of the primary electricity demand in 2017 [2]. The Irish Government is committed to decrease GHG emissions and advance alternative energy sources to reduce the national dependence on fossil fuels (2009/28/EC Renewable Energy Directive (RED)) [3] and has committed to a target of 40% electricity use from renewables by the year 2020 [4]. In Ireland in 2017 only 10.6% energy supply came from renewable sources and the country was ranked 26th out of the European Union (EU)-28 for progress toward meeting 2020 renewable energy target [5]. Overall renewable energy has displaced 1.8 million tonnes of oil equivalent (Mtoe) of fossil fuel and reduced GHG emissions by 4.2 million tonnes (Mt) CO_2 in 2017 (80% from generation of electricity). The renewable electricity sources include wind, hydro, biomass, renewable wastes, landfill gas, biogas and solar PV, however, the level of electricity generation from solar PV remains very low [5]. Ireland is not on track to meet 2020 renewable energy targets which has cost implications of €100 to

€150 million for each percentage point shortfall [4]. The Irish Government has agreed to the binding renewable energy target for 2030 of 32% in line with EU recast RED 2 [6]. Microgeneration of electricity using solar PV system is expected to contribute to meeting these targets mitigating some of the adverse effects of environmental pollution and climate change. The renewable energy sector technologies are evolving rapidly and ensuring higher levels of renewable energy generation will require substantial investments in new infrastructure which includes wind farms, solar PV systems, grid reinforcement, storage development and interconnection. However, high risks and the high up-front costs associated with developing technologies is a major barrier to securing finance. As a result of rapid growth in the use of social media, crowdfunding is increasingly replacing conventional funding models used as an alternative means of funding renewable energy projects [7].

1.1. Microgeneration

Microgeneration is a form of decentralized or distributed energy supply [8] where: energy generation serves in-situ demand (high degree of self-consumption); installations are deployed at lower-voltage distribution network level; and small-scale technologies are deployed including rooftop solar PV, small wind turbines, small hydro and domestic combined heat and power (CHP) [9]. Benefits include lower electricity bills, hedging against future electricity price rises, lower GHG emissions, reduced reliance on fossil fuels, reduced electrical losses on the electricity network and improved building energy rating (BER) [10].

Photovoltaics is the direct conversion of light into electricity at the atomic level by materials displaying a photoelectric effect causing them to absorb photons of light and release electrons. When these free electrons are captured, it results an electric current [11]. Semi-conductors are treated/doped to form a p-n junction such as in crystalline silicon cells by diffusing phosphorous into the silicon and introducing a small quantity of boron, forming an electric field. When photons are absorbed by a PV cell, electrons under the influence of the field move out towards the surface. This flow or current is 'harnessed' by an external circuit with a load [12]. The electricity generated is direct current (DC), converted to alternating current (AC) using an inverter to synchronise with mains electricity [10]. Solar PV panels do not generate CO_2 emissions during their operation; however, emissions are generated during the production of the solar panels and during their disposal. Solar PV systems can be connected to home for supplemental power, full power and backup supply (off-grid) or as a revenue generating power system [13].

Solar PV panels are installed in residential, commercial and industrial settings or as a stand-alone system for generation of electricity for feeding to the national grid. In 2017 very little renewable electricity in Ireland was produced from solar PV, with installed capacity of around 15.7 MW and around 11 gigawatt hours (GWh) of electricity generated equating to 0.1% of renewable electricity or 0.04% of electricity gross final consumption (GFC) [5]. Households currently account for approximately 1.0 megawatt (MW) of installed residential solar PV systems connected to the grid [5]. The Irish Government, in its climate action plan 2019, has indicated the solar PV system is expected to grow to 1.5 gigawatt (GW) of installed capacity by 2030 [14].

1.2. Solar Energy Potential in Ireland

At the Earth's surface radiation can exist in three forms: direct radiation which comes directly from the sun; diffuse radiation which has undergone scattering during its passage through the atmosphere; or reflected radiation from the ground [15]. Solar radiation distribution and intensity are the key factors in determining the efficiency of solar PV systems and results are highly variable [16]. Ireland typically receives an annual solar radiation of 900 kWh m^{-2} [12] compared to Greece with 1890 kWh m^{-2} and Italy with 1680 kWh m^{-2} [17], where unsurprisingly, solar PV accounted for largest total electricity generation in 2017 (8.7% in Italy and 7.6% in Greece) [18]. As well as solar radiation, module efficiency depends on the type of module and the module temperature [19]. The annual energy output (kWh) of the solar PV systems also depends on the peak rating of the solar PV installation (kWp) [12]. The measured performance of a 1.72 kWp rooftop grid connected PV system in Ireland is 885.1 kWh kWp^{-1} $year^{-1}$ [20]. The location of site and the tilt and orientation of solar PV panels are important for the energy output (kWh) of the solar PV systems [12]. For example, the site should be south facing of have a slight south-east or south-west orientation and should not be overshadowed by obstacles which could prevent sunlight getting to the system [21].

1.3. Barriers to Implementation

Despite the potential of microgeneration technologies to help Ireland meet its energy and emission targets and induce positive shifts in energy consumption, the rate of adoption among homeowners remains low. The reasons include low awareness of microgeneration among homeowners, with intention to install at just over 7% [9] and homeowners' willingness to pay (WTP) falling significantly below market prices. In addition, homeowners purchase, or investment decisions are influenced by factors other than cost–benefit evaluations including the benefits of microgeneration and positive social pressure which can translate into higher uptake [22]. Existing installed microgeneration capacity is very low and a very large increase in installation by 2025 would be required to meet the proposed 5% renewable target. In addition, the network potential to accommodate such an increase in capacity in microgeneration on low voltage network by 2025 not well understood.

Microgeneration policies in other jurisdictions have also encountered issues with growing costs and inadequate incentives i.e., export payment as the only incentive, may not sufficiently stimulate large scale deployment (for example export tariff would need to be 27 cents to have same economic impact as SEAI grant for typical 2 kW system with 20% export) [9]. In 2007, the number of microgeneration installations in the UK was estimated at less than 100,000, but between 2009 to 2014 over 730,000 systems were installed, 88% of which are solar PV [8]. The renewable microgeneration technologies adoption has resulted in significant annual savings in energy running cost [23]. The introduction of feed-in tariff (FiT) support has encouraged greater numbers of installations [24] and the global solar PV market has grown significantly, leading to a reduction in capital costs in the UK between December 2010 and September 2012 of around 50%. Consumer cost reductions are mostly likely to occur through market development with increased number of installations or policies to reduce capital costs such as capital grants and low interest loans which are repaid through FiT payments, potential adopters are also driven by the desire to show others their environmental commitment to reduce GHG emissions and earning or saving money through incentives and reduced fuel bills [24].

The most important barriers to adoption in the UK were the higher capital costs compared to annual energy savings and payback period, the absence of subsidies and the regulatory requirements. Other factors include home ownership, the level of available capital for investment and size of house or the suitability of microgeneration technologies [24]. There is also the loss of utility to households caused by space requirements (e.g., roof top space to install solar PV and/or solar thermal, fuel storage-hot water tanks and gardens dug up to install ground heat pumps etc.). These costs would be reduced by concentrating policy on new houses, where microgeneration technologies could be designed into the house at construction at a lower cost [23].

1.4. Financial Support Mechanism

Many support mechanisms have been employed in France, Germany, Greece, Italy and the UK to help increase the uptake of solar PV systems such as capital subsidies, VAT reduction, tax credits, renewable portfolio standards, net-metering, FiT etc. In 2012 the most popular support mechanism in terms of market share were FiTs (60%), capital subsidies and tax rebates (20%), self-consumption (12%), renewable portfolio standards (4%) and net-metering (2%). The electricity compensation schemes (self-consumption and net-metering) have increased their uptake in the last decade, rising from a 4% historical value to 14% in 2012. The reduction in PV costs has resulted in the reduction or elimination of the FiT mechanism instead introducing self-consumption rules [25].

In Ireland several initiatives have been taken to improve uptake of solar PV systems with limited success. The primary support mechanisms for installation of renewable electricity infrastructure are the Renewable Energy Feed-in Tariff (REFIT) schemes which provides a minimum price for each unit of electricity exported to the grid over a 15-year period giving certainty to renewable electricity generators. Currently solar PV systems are not supported under the REFIT scheme [26] however, the new Renewable Electricity Support Scheme (RESS) will provide opportunities for incorporating solar PV, bioenergy and wind within a cost competitive framework.

To deliver Ireland's renewable electricity ambitions to 2030 including reducing the gap to reach 2020 renewable energy targets and accommodating microgeneration by 2021, the Government of Ireland has indicated the key outcomes in energy sector between the years 2019–2021 will include the increased renewable energy usage in the electricity sector via increased levels of microgeneration. Solar energy has the potential to provide a community dividend while maintaining basic payment schemes, subject to EU commission approval [27].

1.5. Crowdfunding

The European Commission defines crowdfunding as an alternative form of financing that connects those who can give, lend or invest money directly with those who need financing for a specific project and usually refers to public online calls to contribute finance to specific projects [28]. Compared to other major world economies, crowdfunding for the EU market is not well developed as the lack of common rules across member states results in compliance issues and increased operational costs. The European Commission has proposed new regulations to address the barriers to crowdfunding use by small investors and businesses. In Ireland, crowdfunding is not currently a regulated activity constituting only 0.33% to 0.4% of the small to medium sized enterprise (SME) finance market whereas in UK it constitutes 12% [29]. It is planned to regulate crowdfunding in Ireland and enact a domestic regulatory regime which is in parallel with the European Commission regulation, to create an environment for the growth of crowdfunding as one of the alternative source of finance for the Irish SMEs and also to ensure sufficient consumer protection [30].

2. Materials and Methods

2.1. Solar PV System Description

The proposed solar PV microgeneration community-based project consists of 68 residential units located at Belfield, Dublin (herein referred to as "the project"). The solar irradiance data collected from the nearest weather station at Dublin airport is 963 kWh m^{-2}.

The solar PV panels are mounted on the rooftop of each unit with the collector facing south and a tilt angle of 30°. The PV solar panel for use in this system are the Hanwha Q cells (Seoul, South Korea) Q peak G4.1 300 Rev4 monocrystalline modules with dimensions of 1670 × 1000 × 32 mm, with a surface area of 1.67 m^2 and an efficiency of 18%. Annual average solar irradiation received by these modules was 1074 kWh m^{-2}. The roof area available on each unit for mounting of solar PV panels is 10.042 m^2. Solar PV system details are listed in Table 1 [4].

Table 1. Solar PV system details for each unit.

Data for Each Residential Unit	Unit	Quantity
Area of the roof area (A)	m^2	10.042
Panel efficiency (r)	%	18
Installed effect (Wpi)	kWp	1.8
Nominal power of panel (Wp)	kW	0.3
Number of panels (Np)	Piece	6
Annual average irradiation (H)	kWh m^{-2}	1074
Coefficient for losses (C)	Factor	0.8
Annual peak power output (Ep)	kWh $year^{-1}$	1553
Performance ratio (PR)	%	80
Lifetime expectancy	years	30

The total solar PV installed effect of each unit W_{pi} [31] is calculated using Equation (1):

$$\boldsymbol{W_{pi} = A \cdot r} \tag{1}$$

where,

W_{pi} = Total solar PV installed effect of each building in kW
A = Total roof area in m^2;
r = Solar panel efficiency in %.

The number of panels installed in each unit N_p [31] is calculated using Equation (2):

$$N_p = W_{pi}/W_p \tag{2}$$

where,

N_p = Number of panels installed;
W_{pi} = Total solar PV installed effect of each unit in kW;
W_p = Nominal power rating of the panel in kW.

The annual peak output of the solar PV system of each unit E_p [31] is calculated using Equation (3):

$$E_p = A \cdot r \cdot H \cdot C \tag{3}$$

where,

E_p = Annual peak output of the solar PV system in kWh $year^{-1}$
H = Annual average irradiation on tilted panels in kWh $m^{-2.}$
C = Coefficient for losses (range between 0.9 to 0.5).

Coefficient for losses will depend on the site, technology and sizing of the system including inverter losses (6 to 15%), temperature losses (5 to 15%), DC cable losses (1 to 3%), AC cable losses (1 to 3%), shading (0 to 40%), weak irradiation (3 to 7%), losses due to dust, snow (0 to 2%), degradation of modules (0.5 to 1%) and other (1 to 2%). Default value is set to 0.75 [31].

The performance ratio of the solar PV system, PR [31] is calculated using Equation (4):

$$PR = \left(\frac{\frac{E_p}{W_{pi}}}{H} \right) \cdot 100 \tag{4}$$

where,

PR = Performance ratio of the solar PV system in %.

2.2. Solar PV Electricity Generation

There are two options when modelling solar PV electricity for calculating yield [32], either using forecast or measured yields or using a weather profile with irradiance values (Wm^{-2}). The monthly values of solar PV electricity were produced from simulations in PV*SOL software [33] using the weather data from the integrated meteo weather database (meteonorm.com). This system uses World Meteorological Organization (WMO) data including global and regional databases in combination with spatial interpolation methods to generate data for locations between the weather stations [34].

2.3. Electricity Network Model

The electricity network model is based on a typical network supply to a town and its surrounding areas. The model consists of a primary substation, MV feeders, MV to LV distribution substation and other associated equipment. The primary substation is outdoor and air insulated which steps down the voltage from 38 kV to 10.47 kV and is equipped with two units of 5 Mega Volt Amp (MVA) transformers, the on-load tap changers of the 5 MVA transformers automatically regulates and controls the target voltage of 10.7 kV. The stepped down voltage (10.47 kV) is distributed from substation to the 10.47 kV to 0.400 kV substation via feeder circuits [32] Figure 1.

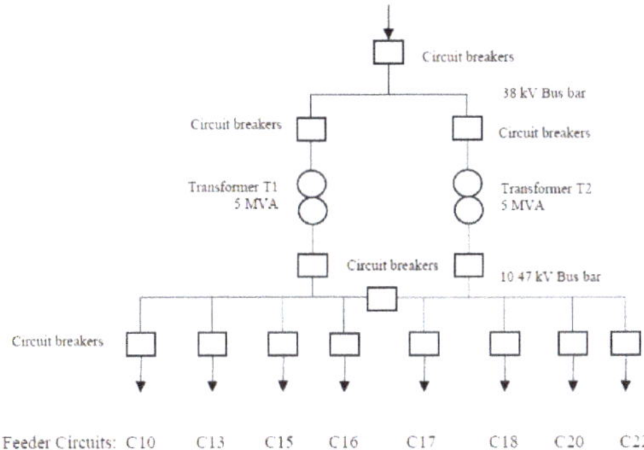

Figure 1. Primary substation single line diagram [32].

The feeder circuit C15 from the primary substation has a total circuit length of 40 km with a mix of overhead lines and underground cables. The project is assumed to be located to the south of the primary substation and main town centre [32] Figure 2.

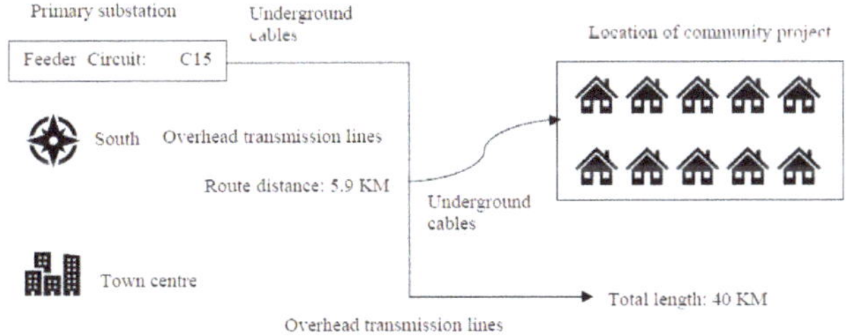

Figure 2. Feeder circuit C15 and location of the project [32].

The transition of the feeder circuit C15 from overhead line to underground cable takes place 5.9 km from the primary substation via spur connection and the underground cable is connected to distribution substation which is located at the edge of the existing project housing estate, where voltage is further stepped down from 10.47 kV to 0.400 kV using a 630 kVA transformer. The LV (0.400 kV) is further distributed using six LV underground cables to 11 street pillars. Over the entire network model, the number of units connected to each phase were balanced with 22 units connected to phase-1, 23 units connected to phase-2 and 23 units connected to phase-3 [32]. At all the 11 load points the solar PV system were connected by each unit.

2.4. Electricity Demand

The electricity demand examines the project demand profile and the demand components including existing demand on the MV circuits without the project; the additional demand on the existing MV including the project and the solar PV electricity output.

2.4.1. Community Project Electricity Demand Profile

To model the demand of the project, the electricity demand data was obtained from the distribution system operator's (Electricity Supply Board (ESB)) standard load profiles which is consolidated samples from residence meters which consists of high, middle and low annual usage from national average values Table 2 [35].

Table 2. Annual household electricity demand [35].

Low Usage	Medium Usage	High Usage
3100 kWh	5300 kWh	8100 kWh

The low usage electricity load profile (3100 kWh) was selected and the load profile of each unit was customized to reflect the monthly electricity consumption based on the approximate number of occupants. Simulations were also performed to determine the annual electricity yields available for self-consumption and for grid feed-in. The total electricity demand profile of the project is composed of variable hourly loads, the timing of usage varies for each unit due to preferences and behaviours of occupants [35].

2.4.2. Demand Components

The electrical network operation downstream of the primary substation was modelled to determine demand at each of the load points on the MV feeder circuit. The demand profile is composed of three components [32]:

(1) The existing demand on the MV circuits excluding the project.
(2) Additional demand on the existing MV circuits including the project based on average peak demand from each unit, (5 kW at 93% power factor in winter and 0.9 kW at 95% power factor in summer).
(3) Solar PV electricity output from community microgenerators.

To model the electrical network system peak and minimum load, the ESB electrical power flow analysis [32] for the following system load conditions are used:

- Peak electricity demand on winter maximum load reading day December 2015, 17:00 pm.
- Midday electricity on summer minimum load reading day in August 2016, 12:00 pm.

Existing Demand on the MV Circuit Excluding the Project

Existing demand on the MV circuits includes the load points (distribution transformer substations) which provide electricity supply to the LV networks consisting of connections to customers. Ideally the actual values of the demands at each of these load points would be metered. In this study known load information was used to calculate the existing demand on the distribution transformers substation and at the large consumer substations which are connected to the existing MV network [32]. The project model includes the LV sections up to the street pillars, (the load points), in this study units were evenly spread across the 3 phases of 11 street pillars, 2 to 4 units per phase [32].

Additional Demand on the MV Circuit Including the Project

The hourly loads of each unit are spikey in nature. In this study along with general loads, a heat pump is assumed to be used in each unit, the heat pump is assumed to consume a power input 2.14 kW to provide a power output 9 kW for an under-floor heating system and a hot water storage system. The timing of the usage or cycling of the heat pumps varies for each residential unit due to personal preferences and behaviours of occupants [32].

Solar PV Electricity Output from Community Microgenerators

The solar PV electricity output from community microgenerators are estimated from weather profiles based on simulations [33] using weather data (with irradiance values) from the integrated weather database.

2.5. Energy Assessment of the Solar PV System

The calculations for embodied energy and the energy payback time of the solar PV system, GHG emissions payback time and carbon credits are outlined.

2.5.1. Embodied Energy of the Solar PV System

Embodied energy of the solar PV system is defined as the energy consumed by the system for materials; manufacturing; transportation and installation [36]. The embodied energy of the solar PV system has been completed by evaluating the total energy required for each process [37].

The embodied energy of each component per m^2 of solar PV module E_{in} [37] for this study was calculated using Equation (5):

$$E_{in} = E_{mfg} + E_{use} + E_{del} \tag{5}$$

where,

E_{in} = Embodied energy of solar PV system (kWh m^{-2});
E_{mfg} = Total manufacturing energy (kWh m^{-2});
E_{use} =Total used energy in installation and operation and maintenance (kWh m^{-2});
E_{del} = Energy requirement to deliver from production to field site (kWh m^{-2}).

The total manufacturing energy E_{mfg} [37] is calculated using Equation (6):

$$E_{mfg} = E_{mpe} + E_{eqp} \tag{6}$$

where,

E_{mpe} = Total material production energy in kWh m^{-2};
E_{eqp} = Total operation and maintenance energy of equipment in kWh m^{-2};

Total material production energy E_{mpe} [37] is calculated using Equation (7):

$$E_{mpe} = \sum_i \left(e_{mpe,i} \cdot m_i\right) \tag{7}$$

where,

$E_{mpe,i}$ = Specific energy to produce ith material;
m_i = Total mass of ith product material.

The total used energy in installation and operation and maintenance E_{use} [37] is calculated using Equation (8):

$$E_{use} = E_{inst} + E_{am} \cdot T_{LS} \tag{8}$$

where,

E_{inst} = Installation energy requirement for the experiment;
E_{am} = Average energy operation and maintenance rate over the life of the PV system;
T_{LS} = Life of the system in years.

The energy requirement to deliver the product materials from production to field site E_{del} is calculated using Equation (9):

$$E_{del} = \sum \left(E_{trans_{i \to i+1}} + E_{pkg\ i \to i+1} \right) \tag{9}$$

where, E_{pkg} and E_{trans} are the packaging and energy requirement for the transfer of the product materials respectively from production to field site.

The balance of system (BOS) components e.g., battery, inverter, electronic components, cables and miscellaneous items should also be included in the calculations [38]. The breakdown of embodied energy of each component per m^2 of solar PV module for this study is in Table A1. (Appendix A).

2.5.2. Energy Payback Time of the Solar PV System (EPBT)

Energy payback time of the solar PV system is defined as the time needed for the system to generate the energy used in its life cycle from the extraction of raw materials to the construction and decommissioning phase. The EPBT [39] is calculated using Equation (10):

$$EPBT = \frac{E_{mat} + E_{manuf} + E_{trans} + E_{inst} + E_{eol}}{\frac{E_{aegen}}{\eta G} - E_{O\&M}} \tag{10}$$

where,

E_{mat} = Primary energy demand to produce materials comprising solar PV system;
E_{manuf} = Primary energy demand to manufacture solar PV system;
E_{trans} = Primary energy demand to transport materials used during the life cycle;
E_{inst} = Primary energy demand to install solar PV system;
E_{eol} = Primary energy demand for end of life management;
E_{aegen} = Annual electricity generation by solar PV system;
$E_{O\&M}$ = Annual primary energy demand for operation and maintenance of solar PV system;
ηG = Grid efficiency, the average primary energy to electricity to electricity conversion efficiency at the demand side.

2.6. GHG Emissions and Carbon Credits

The solar PV power generation is one of the cleanest sources of renewable energy [4]. In 2017, natural gas accounted for 51% of the fuel used for electricity generation in Ireland and the CO_2 intensity of electricity is 437 g CO_2 kWh^{-1} or 0.000437 t CO_2 kWh^{-1} [40].

2.6.1. GHG Payback Time

GHG payback time (GPBT) is defined as the number of years it takes solar PV system to pay back its embodied emissions through solar PV generation. The GPBT [41] is described by Equation (11):

$$GPBT = \frac{CO_2\ equivalents\ (eq) embodied}{CO_2\ eq\ avoided\ (year)} \tag{11}$$

where,

$$CO_2\ eq\ embodied = CO_2\ eq\ modules + CO_2\ eq\ mounting\ structures +$$
$$CO_2\ eq\ electric\ (BOS) + CO_2\ eq\ transport\ etc;$$

$CO_2\ eq\ avoided\ (year)$ = Emissions avoided per year due to the production of electricity from the solar PV system installation.

2.6.2. Carbon Credits

Carbon credits are awarded for the reduction in GHG emissions which can be traded in international market at their current market price. CO_2 has been traded at € 21 per tonne CO_2 eq [42] and each ton reduction in CO_2 is a carbon credit earned [43]. The total carbon credits earned is described by Equation (12):

$$Total\ carbon\ credits\ earned = Net\ CO_2\ mitigation \cdot price \tag{12}$$

where,

price = Current market trading price

The yearly carbon credits earned is described by Equation (13):

$$Yearly\ carbon\ credit\ earned = Total\ carbon\ credits\ earned / T_{LS} \tag{13}$$

2.7. *Financial Assessment of the Solar PV System*

While use of solar PV systems has increased, it is suggested that they need to become more price competitive to sustain further growth [36]. A grid connected solar PV system can reduce capital and maintenance cost by eliminating the need for battery with the grid acting as a storage bank [44]. The financial assessment of the solar PV system is based on current market prices of the project components.

A review of residential solar installers currently active in the Irish market, determined the approximate cost per kWp for a fully installed rooftop solar PV system was €1744 [35]. The Sustainable Energy Authority of Ireland (SEAI) in their payback calculator for domestic solar PV have considered the approximate cost per kWp to be €1900 [45]. Solar PV system cost can vary depending on the quality of solar PV panels and installation and after sales support. The project cost includes [36] the cost of modules; cost of inverters; miscellaneous costs {electrical items such as cables etc., installation cost, packing and freight etc.}; and cost of operation and maintenance.

Net Present Value

The Net present value of the investment per residential unit in the solar PV project can be calculated [35] using Equation (14):

$$NPV = \sum_{n=1}^{30} \cdot \frac{S_n - C_n + NE_n \cdot t}{(1+d)^n} \tag{14}$$

where,

S_n = Savings calculated in year n
C_n = System cost in year n, including capital costs in year 1, operating expenditures and inverter replacement where applicable Table A1 (Appendix A)
NE_n = Net export or amount of excess generation for which the residential building owner is compensated;
t = Rate at which net export is remunerated;

t can be equal to retail rate (r) €0.133 per kWh in case of net metering or

t can be equal to € FiT per kWh amount in case of FiT or
t can be equal to €0 per kWh in case of where no subsidy applies

d = Discount rate (considered 0.55% reflecting mid-range of publicly advertised annual equivalent interest rates on savings account in Ireland).

2.8. Crowdfunding Model for Ireland

A review of crowdfunding alternatives associated with renewable energy projects was conducted, demonstrating the use of different types of crowdfunding model [7].

To forecast the potential of System Dynamics of Solar Crowdfunding (SCF) in Ireland this study incorporates a simulation model which is developed based on system dynamics which uses causal loop diagrams and stock-loop diagrams to express the causal links and relationships between various factors that affect the SCF that represents the SCF market [46] Figure 3.

The SCF potentials refers to the total number of solar projects requiring finance. The SCF adopters are parties looking for private ownership in the project, SCF market saturation can occurs when the SCF adopters decrease as a result of more projects adopting SCF. The SCF adoption is primarily dependent on awareness of the SCF with greater numbers of adopters increasing awareness.

Three risk categories for investors are identified:

- Low risk involving non-material returns (charitable undertakings);
- Low to medium risk involving material returns (rewarding investors);
- High risk involving financial returns (mainly venture capitalists) [7].

The motivation for funders involved in the case studies included: helping people in need of money for energy efficient technology and household appliances; the desire to reap a financial return from their contributions equating to non-trivial investment; desire to reap a financial return for their investment and to be involved as shareholders of the companies [7].

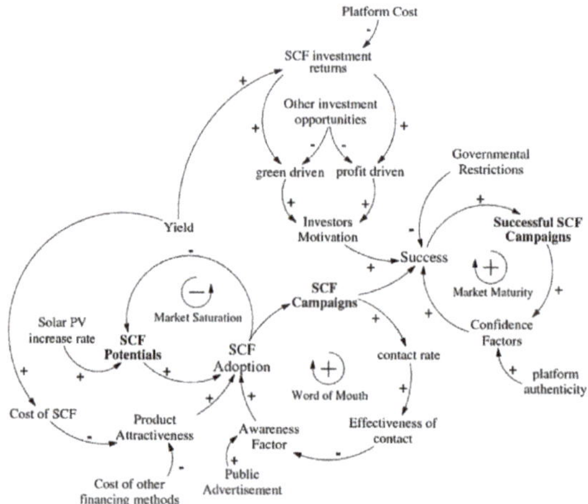

Figure 3. Causal loop diagrams [46].

3. Results

The impacts on electricity demand of the project; on the primary substation, MV feeder circuit, distribution transformer & LV network, network losses and voltage profile; energy pay-back time; GHG pay-back time, carbon credits and the crowdfunding model in Ireland are presented.

3.1. Impact on Electricity Demand of the Project

Using PV*SOL online software [33], (V0.7, Valentin Software GMBH, Berlin, Germany) the annual electricity demand (kWh) for the individual unit (3100 kWh approx.) Figure 4 was scaled to represent the annual electricity load of the project, approximately 210,800 kWh.

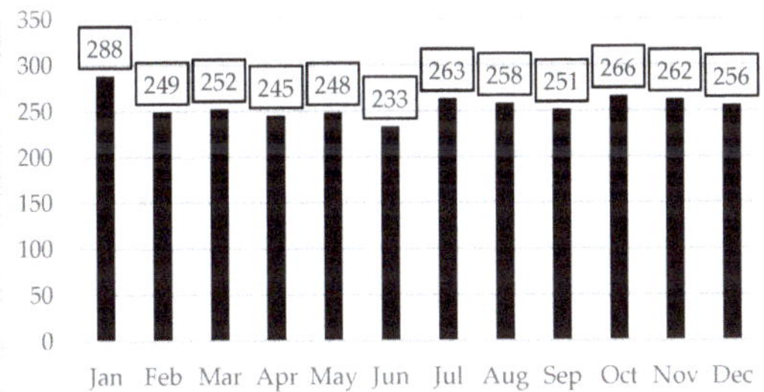

Figure 4. Annual electricity demand (kWh)of a residential unit.

Simulations were performed for a single PV system (the parameters used are listed in Table A2, Appendix C) and the solar PV electricity generation (1553 kWh) (Figure was then scaled to represent the annual solar PV electricity of the project = 105,604 kWh Figure 5.

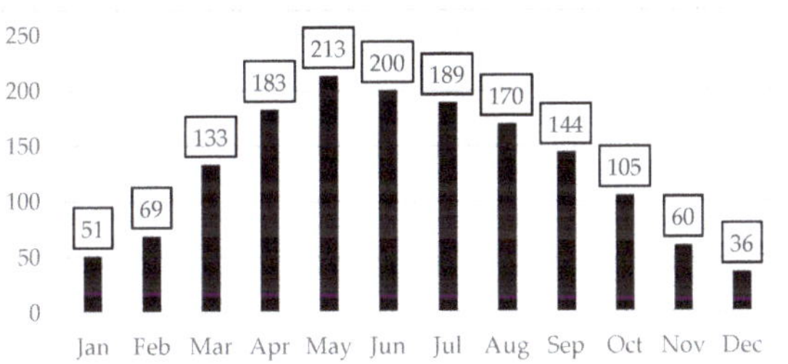

Figure 5. Annual solar PV electricity generation (kWh) of a single residential unit.

There was a low annual match between the annual electricity demand profile of the project and the annual solar electricity generated from the solar PV system. The annual electricity demand of the project without the solar PV system was 210,800 kWh, with the installation of the solar PV system the annual electricity demand of the project was reduced to approximately 166,514 kWh Figure 6.

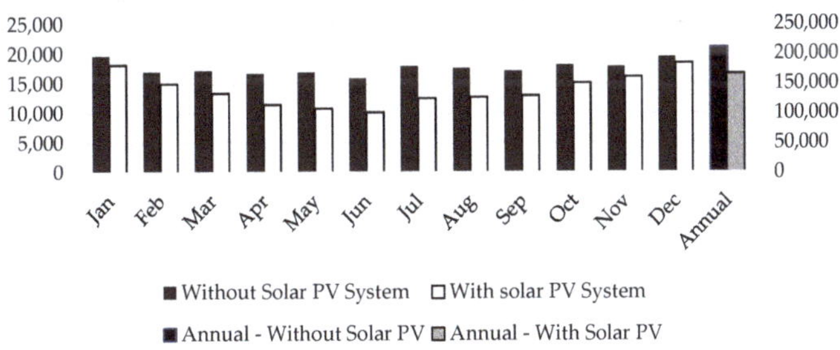

Figure 6. Annual electricity demand on grid of the project without and with solar PV system.

Out of 105,604 kWh annual solar PV electricity, the model shows approximately 44,354 kWh was self-consumed and 61,250 kWh of excess electricity was available to be fed directly into the electricity grid.

3.2. *Impact of Project on Electricity Infrastructure*

To assess the impact of the project electricity demand on the electricity infrastructure, an electricity network model was constructed incorporating, existing demand excluding the project; additional demand including the project; and with solar PV electricity output. Using the peak and minimum load conditions already outlined.

3.2.1. Impact on the Primary Substation

The primary substation (equipped with 2 × 5 MVA, 38/10.47 kV transformers giving a total continuous supply capacity of 10 MVA). Power flow analysis results using feeder circuit demand for winter peak load reading indicated total demand excluding the project was 7.19 MVA, including the project was 7.58 MVA and including the project with solar PV systems was 7.58 MVA. Summer minimum load reading indicated the total demand excluding the project was 4.93 MVA, including the project was 5 MVA and including the project with solar PV systems was 4.93 MVA [32].

3.2.2. Impact on the MV Feeder Circuit

The 10.47 kV feeder circuits consisted of a mixture of underground cables and overhead lines. Each feeder circuit was controlled by a circuit breaker which was installed in a primary substation rated at a capacity at 630 Amps (11.4 MVA). Each feeder circuit left the substation as an underground cable with a rated winter current of 532 Amps (9.65 MVA), the MV feeder circuits (underground cable and overhead lines) conductor size drastically reduced as demand decreased. Results of the power flow analysis demand for the winter peak load reading indicated total demand from all the feeders connected to the primary substation excluding the project was 2.34 MVA, including the project was 2.71 MVA and including the solar PV was 2.71 MVA [32]. Summer minimum load reading indicates total demand excluding the project was 1.91 MVA including the project was 1.98 MVA and including the project with solar PV systems was 1.91 MVA [32].

3.2.3. Impact on the Distribution Transformer and LV Network

The distribution transformer supplied the LV network and had a continuous rating of 630 kVA. Power flow analysis using electricity demand profiles for the winter peak load reading indicated [32]: the total demand on the distribution transformer including the project was 374 kVA; including the project with solar PV systems was 374 kVA. Summer minimum load demand including the project was 68 kVA and including the project with solar PV systems was 121 kVA.

The LV network was equipped 185 mm^2 cross sectional area conductor underground cable which consisted of six feeders to supply the project. The power flow analysis [32] using electricity demand profiles for the winter peak load reading giving total demand on LV feeders with and without solar PV system is shown in Table 3. The summer minimum load reading demand on LV feeders with solar PV is assumed, in this study, to be same as Table 3 above which is the worst-case scenario summer loading of LV feeders.

Table 3. Total demand on LV feeders with and without solar PV system [32].

LV Feeder	Rating (KVA)	Demand (kVA)	Net Demand with Solar PV (kVA)	% Contribution from Solar PV
01	246.64	52.81	29.41	44.3
02	246.64	87.60	64.20	26.7
03	246.64	60.08	37.00	38.4
04	246.64	59.14	46.49	21.4
05	246.64	47.43	24.35	48.7
06	246.64	63.25	51.55	18.5

3.2.4. Impact on the Network Losses

The technical losses occur because of the energy dissipated in feeder circuit conductors and core and windings losses in transformers [32]. Winter peak load reading indicated total technical losses in the existing electrical network excluding the project was 85 kW, including the project was 106 kW and including the project with solar PV systems was 106 kW. For summer minimum load reading the total technical losses excluding the project was 61 kW, including the project was 64 kW and including the project with solar PV systems was 62 kW [32].

3.2.5. Impact on the Voltage Profile

Voltage is the electric potential difference between two points. The voltage drop is the reduction in voltage in an electrical circuit between the source and load [47]. The voltage drop on feeders occurs because of: resistance increase from poor joints and terminations, hot spots, under-sized conductors and non-uniform conductor material or load increases [48]. Results of the power flow analysis carried out by [32] at MV feeder and LV feeder circuits indicated voltage at the MV feeder circuits including the project ranged from a maximum of 100% at primary substation to a minimum of 98.1% 16 km away. The voltage at the distribution substation including the project ranged from a maximum of 97.8% to a minimum of 96.2% at distribution substation. The voltage at the LV network entry point to a point 170 m closer to the project declined by another 2% [32].

3.3. *Energy Payback Time*

To assess the Energy Payback Time (EPBT) of the project with solar PV system, the embodied energy of each component and process of the proposed solar PV system was calculated using experiments conducted by [36] on monocrystalline PV modules as these experiments did not take into account the critical component of BOS i.e., inverter and the embodied energy associated with inverter, these data were extracted from the findings of [38]. This embodied energy value is often used to evaluate the energy balance of the solar PV system [48] i.e., energy metrics e.g., EPBT [36].

$E_{in} = E_{mat} + E_{manuf} + E_{trans} + E_{inst} + E_{eol}$	=	1471.34 kWh m^{-2}
Total area of modules	=	Number of panels (N_p)·(Length of panels)·(Width of panels)
	=	(6)·(1.67 m).(1.0 m)
	=	10.02 m^2 (i.e., A = Total roof area of one unit)
	=	(10.02 m^2)·(68)
	=	681.36 m^2 ((i.e., Total roof area of all units)
Total embodied energy for each unit E_{in}	=	(1471.34 kWh m^{-2})·(10.02 m^2) = 14,743 kWh
Total embodied energy for all units E_{in}	=	(1471.34 kWh m^{-2})·(681.36 m^2) = 1,002,513 kWh
E_{aegen}	=	1553 kWh $year^{-1}$ for one unit or 105,604 kWh $year^{-1}$ for all units.
ηG	=	0.483 Using grid conversion efficiency [4].
$E_{O\&M}$	=	0 (Assumed)

The annual electricity generated by solar PV system (Eagen) was estimated to be 105,604 kWh Figure 5.

The *EPBT* was calculated using Equation (10):

$$EPBT = \frac{E_{mat} + E_{manuf} + E_{trans} + E_{inst} + E_{eol}}{\frac{E_{aegen}}{\eta G} - E_{O\&M}} \quad (15)$$

EPBT =1,000,513

((105,604/0.483) − 0)

EPBT of the proposed project was 4.59 years.

3.4. GHG Payback Time (GBPT) and Carbon Credits

To assess the GBPT of the project, the emissions associated with the embodied energy of the proposed solar PV system were determined [41]. Calculated solar PV system emissions ranged between 50 g to 120 g CO_2 eq. kWh $^{-1}$ which agreed with the findings of [4] (69 g CO_2 eq kWh^{-1}, or 0.000069 tonne CO_2 eq. kWh^{-1}. The CO2 intensity of electricity in Ireland was 0.000437 tonne CO_2 eq. kWh^{-1} [5]. The annual electricity generated by solar PV system was estimated to be 105,604 kWh.

For the project the total embodied emissions, using results calculated from Equation (5):

$$\begin{aligned} CO_2\ eq{\cdot}embodied &= \text{(Total embodied energy for all units } E_{in} \text{ in kWh)}{\cdot}\text{(solar PV system emissions)} \\ &= 1{,}002{,}513 \text{ kWh} * 0.000069 \text{ tonne } CO_2 \text{ eq. kWh}^{-1} \\ &= = 69.2 \text{ t of } CO_2 \text{ eq.} \end{aligned}$$

GHG payback time is defined as the number of years it takes solar PV system to pay back its embodied emissions through solar PV generation. The *GPBT* [41] is described by Equation (11).

For the project, the total GHG emissions avoided in a year:

$$\begin{aligned} CO_2\ eq\ avoided\ (year) &= (E_{aegen}){\cdot}(CO_2 \text{ intensity of electricity}) \\ &= (105{,}604 \text{ kWh})\ 0.000437 \text{ tonne } CO_2 \text{ kWh}^{-1} \\ &= 46.15 \text{ t of } CO_2 \text{ eq} \end{aligned}$$

Using Equation (11) for GHG payback time:

$$\begin{aligned} GPBT &= 69.2 \text{ t of } CO_2 \text{ eq.}/46.15 \text{ t of } CO_2 \text{ eq} \\ &= 1.5 \text{ years.} \end{aligned}$$

The total carbon credits earned from the project was calculated [36] based on the amount of CO_2 mitigated by the project with solar PV systems at its current market trading price (€21 per tonne CO_2 eq.) [43].

$$\begin{aligned} Net\ CO_2\ mitigation &= (CO_2\ eq\ avoided\ (year) \cdot T_{LS}) - CO_2\ eq\ embodied \\ &= (46.15 \text{ t } CO_2 \text{ eq.}){\cdot}(30 \text{ years}) - 69.2 \text{ t } CO_2 \text{ eq.} \\ &= 1315 \text{ t of } CO_2 \text{ eq.} \end{aligned}$$

Carbon credits are awarded for the reduction in GHG emissions which can be traded in international market at their current market price and each ton reduction in CO_2 is a carbon credit earned [43].

The total carbon credits earned is described by Equation (12):

$$Total\ carbon\ credits\ earned = Net\ CO_2\ mitigation{\cdot}price \quad (16)$$

where,

price = Current market trading price (€21 per tonne CO_2 eq.) [41]
= (1315 t of CO_2 eq.)·(€21per t CO_2 eq.)
= €27,615

The yearly carbon credits earned is described by Equation (13):

$$Yearly\ carbon\ credit\ earned = \frac{Total\ carbon\ credits\ earned}{T_{LS}} \quad (17)$$

= €27,615/30 years
= €920

3.5. *Financial Assessment of the Solar PV System*

3.5.1. Cost of Modules

The cost of a Hanwha Q cells Q peak G4.1 300 Rev4 monocrystalline module (with an efficiency of 18% and lifespan of 30 years), is € 0.34 Wp^{-1} in Ireland [49].

Total module cost for each unit = (1800 Wp)·(€ 0.34 Wp^{-1}) = € 612

Total module cost for all units = (1800 Wp)·(68)·(€0.34 Wp^{-1}) = €41,616 (excl Value Added Tax (VAT) @13.5%) [50].

3.5.2. Cost of Inverters

The ABB UNO-2.0-I-OUTD (2 kWp) string inverter can be grid connected and eliminates the need to fit an isolator onto the DC cabling from the solar PV modules to the inverter, has a lifespan of 15 years, the inverter cost is € 920 in Ireland [51].

Total inverter cost for each unit = €920

Total inverter cost for all units = (68)·(€920) = €62,560 (excl VAT @13.5%)

3.5.3. Miscellaneous Cost

The miscellaneous cost includes electrical items (cables etc), installation cost, packing and freight with estimated the installation cost per kWp = €1362 [44]. Assumption of miscellaneous cost per kWp = SEAI approximate cost per kWp—approximate modules cost per kWp—approximate inverters cost per kWp = (€1900)−(€340+€460) = €1100.

Total miscellaneous cost for each unit = (1800 Wp)·(€1.1 Wp^{-1}) = €1980

Total module cost for 68 units = (1800 Wp)·(68)·(€1.1 Wp^{-1}) = €134,640 (excl VAT @13.5%)

3.5.4. Cost of Operation and Maintenance (O&M)

The O&M cost include the inverters replacement cost − €920 (excl VAT @13.5%) [51]. Annual operation, maintenance and insurance costs = €50 [44]. Both costs are subject to annual inflation 0.73% [35].

3.5.5. Other Costs

Annual Standing charge = €132.16 which covers range of electricity supplier and network costs [35]. Annual PSO levy = €41.76 [52]. The standing charge and PSO levy on electricity consumers is used to fund existing support schemes and are subject to changes [35].

3.5.6. Net Present Value (NPV)

The Net present value of the solar project can be calculated [35] using Equation (14):

$$NPV = \sum_{n=1}^{30} \cdot \frac{113.38 - 2302 + (900)\cdot(0.133)}{(1+0.0055)^{n}} \quad (18)$$

where,

S_n = reduction in annual energy costs due to solar energy self-consumption = 652 kWh × unit rate assuming (24-h rate of 17.39 cent per kWh); 652 × 0.174 = €113.38
C_n = year 1 (Table 4 includes cost of PV Modules, Inverters, Miscellaneous Costs, Annual operation, maintenance and insurance = €2302, for subsequent years costs are only annual operation, maintenance and insurance = €50; cost of replacement inverter in year 12 = €920).
NE_n = 900
t = €0.133 (assuming t equal to retail rate (r) €0.133 per kWh in case of net metering)
d = 0.55%

Approximately 44,354 kWh (42%) annual solar PV electricity generated by the project was self-consumed and the remainder (58%) 61,250 kWh available for the grid. This represents 652 kWh usage and 900 kWh for export for each residential unit.

NPV of the investment = −€1918.52 and no of years for financial payback = 12 years.

Table 4. Key cost assumptions of the solar PV system for the project.

Key Assumptions	Each Unit	All 68 Units
Cost of PV Modules (+)	€612	€41,616
Cost of Inverters (+)	€920	€62,560
Miscellaneous Costs (+)	€1980	€134,640
Cost of O&M		
Inverters Replacement Cost (+)	€920	€62,560
Annual operation, maintenance and insurance costs (+)	(€50)·(30 years) = €1500	€102,000
SEAI Grant Level (-)	€1260	€85,680
Total	€4672	€317,696

The SEAI grant level are subject to change and the above total amount excludes the annual rate of inflation (0.73%) and VAT (13.5%) [40].

3.6. Crowdfunding Model in Ireland

A review into renewable energy project case studies [7] determined that for financing solar projects, lending, debenture and equity-based crowdfunding models are the most common while donation and reward-based crowdfunding have seldom been used [46]. The validity of the system dynamic model was tested [46] to confirm: the causal loop diagram contained all important factors; model was dimensionally consistent (unit check function of Vensim PLE software used to confirm measurement units of the variables) and was tested under extreme conditions (sensitivity check of important variables which provided logical behavior of the system) [46]. The testing confirmed the validity of the developed model for simulating solar crowd funding in Ireland.

The review conclusions showed that to be successful project creators and/or campaigners should convey credibility and as well as create project demand [53] by: setting the lowest possible funding amount (as investors/participants were attracted to campaigns with higher percentage funded rather than higher amount funded); decreasing profit margin associated with rewards to encourage more backers; providing tangible reward options rather than gimmicky products e.g., t-shirts stickers etc; and including a short video outlining project, the development timelines, business plans, usage of funds and the motivation and inspiration for the project [53]. It was found that Kickstarter, IndieGoGo, The Funding Circle, Seedrs, Crowdcube etc. which are focused on Lending/Equity based crowdfunding are more suitable for solar crowdfunding of the community-based projects.

4. Discussion

The effect of the project on electricity demand, on primary substation, the MV feeder circuit, the distribution transformer & low voltage (LV) network, the network losses and voltage profile is discussed; the modelled energy and GHG pay-back time and potential carbon credits from the project are reviewed along with the crowdfunding model as a means of funding in Ireland.

4.1. Impact on Electricity Demand of the Project

The annual electricity demand of the project on electricity grid was approximately 210,800 kWh prior to the installation of 1.8 kWp solar PV, following installation, the demand reduced by 21% to 166,514 kWh. The model estimated the solar PV system generated 105,604 kWh electricity annually with 61,250 kWh of excess electricity fed into the electricity grid. These results would allow the unit owner to be compensated by net export rate where this applies and/or become more self-sufficient and less dependent on utility companies, protecting against higher electricity costs and contribute to increasing the security of electricity supply.

4.2. Impact on Primary Substation, MV Feeder Circuit, Distribution Transformer & LV Network, Network Losses and Voltage Profile

The impact of the contribution of the community project and solar PV system the distribution transformer & LV network, the network losses and voltage profile are discussed.

4.2.1. Impact on the Primary Substation

Total demand on the primary substation during a winter peak was 7.19 MVA and the addition of the project increased the demand to 7.58 MVA (a 5.4% increase in substation capacity). The solar PV system contribution at the time of the winter peak was zero as the peak occurred at night [32]. During a summer minimum load demand was 4.93 MVA and the addition of the project increased the demand on the primary substation to 5 MVA (1.4% increase), while the addition of the solar PV covered the increase in demand from the project by reducing the demand on the primary substation back to 4.93 MVA [32]. The impact at the primary substation level is very small. The small contribution of the solar PV system would maintain the demand marginally within the continuous rating of single primary transformer by 0.07 MVA which is around 1.4% [32].

The primary substation is equipped with two 5 MVA, 38/10.47 kV transformers giving a total continuous supply capacity of 10 MVA, one unit is used to meet the electricity demand of customers and another kept in standby mode and brought into operation in the event of failure. By incorporating microgeneration, the demand on the primary substation transformers can be reduced which reduces the temperature hot spot within the winding of the transformer avoiding the most severe electric power outages and increasing the electrical power system security standards [32].

4.2.2. Impact on the MV Feeder Circuit

The total demand from all the 10.47 kV feeders, was 2.34 MVA and 1.91 MVA for the winter and summer peak load reading respectively. The additional electricity demand for the project increased the demand on the all the 10.47 kV feeders to 2.71 MVA and 1.98 MVA for the winter and summer peak load reading respectively (representing a 16.2% and 3.5% increase in feeder capacity respectively). The contribution of the solar PV system at the time of the winter peak is zero (peak occurred at night-time) while in summer it decreased feeder capacity to 1.91 MVA i.e., 3.5% [32]. Overall there is no significant relief to the MV overhead lines and underground cables in the feeder [32].

4.2.3. Impact on the Distribution Transformer and LV Network

The electricity demand on the distribution transformer for the project during a winter peak load reading was 374 kVA. The contribution from the solar PV system at the time of the peak is zero because the peak occurred at night-time [32]. The demand on the distribution transformer during a summer minimum load reading was 68 kVA. The addition of the solar PV increased the demand on the distribution transformer by 56% to 121 kVA. This contribution will increase the demand within the continuous rating of single primary transformer by 53 kVA which is around 8.4% [32]. To allow expansion room for additional loads, the distribution substation should be equipped with one distribution transformer of 630 kVA, 10.47/0.400 kV designed for an emergency rating of 110% of the continuous rating i.e., 693 kVA for certain time period [32].

Incorporating microgeneration of solar PV system in the low voltage network, would increase the demand on the continuous rating of the distribution transformer, which in turn would increase the temperature hot spot within the winding of the transformer, however, the loading of the distribution transformer is significantly below the specified design limits contributing to increasing the electrical power system security [32].

The LV network is equipped 185 mm^2 cross sectional area conductor underground cable which consists of six feeders to supply the project. Each feeder is designed for a rated value of 246.64 kVA (total of 1480 kVA). ESB Network advise LV feeders should be loaded around 30% (74 kVA on each LV feeder or in total 444 kVA). In this study, the loading of the LV feeders is significantly below the specified design limits and further contributes to increasing the electrical power system security standards: avoiding short bursts of higher network losses in the LV and MV network; and voltage fluctuations [32].

4.2.4. Impact on the Network Losses

Total losses in the existing electrical network covering the primary substation and the MV network supplied from the substation during a winter peak load reading was 85 kW (1.18% of the total demand on the substation). The technical losses relate mainly to the primary transformers (40%) and the MV feeder circuit C15, which connected the residential units (44.7%) [32]. The addition of the project increased the technical losses to 106 kW (52.8%). There is no contribution to the time of peak losses from the solar PV system as the peak occurred outside the sunlight hours [32]. During a summer minimum load reading the total losses in the existing electrical network was 61 kW (1.23% of the total demand on the substation). The addition of the project increased technical losses to 64 kW (46.9%) [32]. The addition of the solar PV system reduced total losses to 62 kW which equates to 1.24% of the total demand on the substation. This contribution will maintain the demand marginally by decreasing the proportion of the losses occurring in MV feeder circuit C15 from 46.9% to 45.4% [32]. This suggests that wider deployment of solar PV system can have a significant impact on loss performance at distribution level, with potentially significant cost savings [32].

4.2.5. Impact on the Voltage Profile

The voltage profile at primary substation ranges from maximum 100% to minimum 98.1% at 16 km from the primary substation and at the distribution substation range from maximum 97.8% to minimum 96.2%. The voltage at the LV network entry point to further 170 m towards the project declined by another 2% [32]. The solar PV systems are typically located closer to the consumer load, which provides an opportunity for the electrical network to offset some of the reactive power requirements at the distribution system and provides benefits to the distribution system namely capital expenditure on the reinforcement due to power factor improvements, savings on voltage control equipment and reduced consumption of reactive power and is subject to the microgenerator connection point which is further depended upon the nature and topology of the connection method into the existing electrical distribution network [32].

4.3. Energy Payback Time

The EPBT for the solar installation for the project was 4.59 years i.e., it will take 4.59 years of operation of solar plant to generate the energy used to produce the system itself [4]. The EPBT of the solar PV system decreases as module efficiency increases, with an example of a rooftop mono-crystalline silicon PV system in Southern Europe (solar irradiation of 1700 kWh m^{-2} $year^{-1}$) which has and EPBT of 2.5 to 3 years [39].

4.4. GHG Payback Time and Carbon Credits

The GPBT for the solar installation for the project is 1.5 years. The life cycle GHG emissions from solar electricity production in Ireland are significantly lower than from electricity production on national grid. The net CO_2 mitigated due to incorporation of solar PV electricity of the project over the lifespan of 30 years is 1315 t of CO_2 eq. Carbon credits earned from the proposed project amounts to €27,615 which is approximately €920 per year.

The GHG emissions generally for wind and hydro power amounts to 6.2 to 46 g CO_2 eq kWh^{-1} and 2.2 to 74.8 g CO_2 eq kWh^{-1} respectively, wind power has the lower energy consumption and GHG emissions compared to solar PV system even though solar PV power has larger impact values due to module manufacturing process which has an general emission range of 2.89 to 671 g CO_2 eq kWh^{-1} (quantum dot to mono-si solar cells), but compares favourably to hard coal plant which has a general emission range of 750 to 1050 g CO2 eq kWh^{-1} [39].

4.5. Financial Assessment of the Solar PV System

The total cost of installing the solar PV system in an individual property was €4672 and for the project totaled €317,696. The NPV of the investment per property was calculated at –€1918.52, the negative NPV value means that the present value of the costs exceeds the present value of the returns at the current discount rate. In this scenario the investment cost would be repaid using the savings in electricity costs and payments for the excess electricity sent to the grid after 12 years.

4.6. Crowdfunding Model in Ireland

A review of eight renewable projects illustrates the use of different types of crowdfunding rewards and returns [7]. The choice of platform depends on the business model, lending, equity, reward and donation and the funding/investment amount and needs to be appropriate for the level of crowdfunding risk [46]. Not all the crowdfunding business models are applicable for solar projects. Lending, debenture and equity- based crowdfunding are the most common approaches for financing solar projects in the crowdfunding platforms [46].

This study provides a reference for policymakers in the country and industry practitioners to understand the approaches and processes involved in solar crowdfunding [46]. The system dynamics model outlines the combination of three stage solar crowdfunding process including the identification of potential SCF adopters, factors affecting the adoption and success of SCF, each stage involves numerous factors that shapes the feedback loops impacting the SCF market and provides a perspective to understand the mechanisms and complexity involved in solar crowdfunding which complements the qualitative methods [46].

The successful funding of the community is dependent on investor's motivation, confidence factors as well as restrictions of the government [46]. Feedback from experienced crowdfunding participants suggested the campaign needs to convey credibility and create demand by setting the lowest possible funding amounts, decreasing the profit margin with rewards which are tangible to backers and create a short video outlining project [53]. A summary of findings is presented in Table 5.

Table 5. Summary of findings.

Key Findings	Brief Conclusion
Electricity demand and supply match	There is a low annual match between the project electricity demand profile and the solar electricity generated. As system is configured without battery storage the excess electricity is sent back to the grid.
Solar contribution to reduction of electricity demand	The annual electricity demand of the project without solar PV system was 210,800 kWh, installation of solar PV system reduced the project annual electricity demand to 166,514 kWh
Solar Electricity usage	Results show of 105,604 kWh solar electricity generated, 44,354 kWh was self-consumed, and 61,250 kWh was available for the grid
Impact of project on electricity infrastructure: Primary substation, MV Feeder circuits, Technical Losses, and Distribution Transformer	Power flow analysis showed the addition of the community project to the network increased demand on the primary substation and on the MV feeder circuits, by over 5% and 16% at winter peak and by 1% and 3% at summer minimum respectively. In both cases the addition of the solar PV system had no effect on winter peak demand but the reduction in summer minimum load covered the increase in electricity demand from the project installation. Technical losses increased in both winter (85 kW to 106 kW) and summer (61 to 64 kW) when project was added, solar PV did not reduce losses at winter peak, while in summer losses reduced to 62 kW following addition of solar PV. Total demand on distribution transformer was not reduced by solar PV system at winter peak, while at summer minimum, demand on distribution transformer increased by 8.4% following addition of solar PV.
Energy payback time	4.59 years (of operation to generate same energy used to produce the system).
GHG payback time	1.5 years (of operation to pay back embodied emissions through solar PV generation).
Yearly carbon credit earned	€920 per year, €27,615 over the 30-year lifetime of the project. Carbon credits are awarded for the reduction in GHG emissions, one per ton of CO_2 produced equals one carbon credit.
Total costs (assumed)	€4672 per unit, €317,696 for 68 units.
NPV of project	The financial payback based on NPV of the investment (for a single unit will take 12 years based on energy savings and payments for energy sent to the grid. This payback period may be well above what homeowners might require.
Crowdfunding	Lending, debenture and equity-based crowdfunding are the most common approaches for financing solar projects in the crowdfunding platforms.

5. Conclusions

This study demonstrates the impact of the based solar PV microgeneration project on the electricity grid in Ireland and outlines the most suitable crowdfunding mechanisms for the development of based solar PV microgeneration projects and details the financial costs associated with the project.

The solar PV system (122.4 kWp) designed for the project comprised an array containing 6 modules (300 watts each) on each unit (roof area 10.042 m^2 and 408 modules). The electricity network model was modelled on a typical network supply. The addition of the project to the electrical network resulted in peak demand increase on the MV feeders and increased voltage on MV and LV feeders. Coordination of the contribution from solar PV systems is required to avoid formation of new peak demand on distribution transformer and LV feeders. Controlled operation of larger scale solar PV projects could reduce the peak demand and smooth the load profile and assist in the voltage regulation on the electrical distribution system network. Technical losses on the MV and LV feeders were reduced which shows the potential to reduce the electrical network losses and lower operation costs and savings could be passed on to the consumer. The modelled energy payback time, of the project was 4.59 years and GHG payback time was 1.5 years. The model showed the project mitigated 1315 t of CO_2 and at current market price of €21 per tonne of CO_2 eq. and carbon credits earned was €27,615. The financial payback (NPV) of the investment will take 12 years based on energy savings and payments for energy sent to the grid. This payback period may be well above what homeowners might require.

Crowdfunding categories are based on project risk e.g., Low-risk crowdfunding models involved non-material returns, Low to medium risk crowdfunding models involved material returns and high-risk crowdfunding models involve financial returns. The project creators and/or crowdfunding campaigners need to convey credibility and as well as create demand for the project. Developing a short video was suggested to outline project as well as the motivation & inspiration to help secure crowd funding. The most suitable crowdfunding platform for community solar PV microgeneration projects include Kickstarter, IndieGoGo, The Funding Circle, Seedrs, Crowdcube or similar platforms which are focused on Lending and/or equity-based crowdfunding.

The modelled solar PV microgeneration system in the project would require total investment of €317,696 plus some variable amounts for campaign cost and profit of the project. The model calculated a displacement of 272 t of oil eq. (Appendix B), significant GHG emissions savings and further greening of the national grids. To encourage the accelerated uptake of the solar PV microgeneration projects in Ireland it is essential to extend REFIT to include solar PV systems which will improve sustainability of electricity supply by self-generation, consumption and feeding the excess electricity to the grid.

Author Contributions: V.V.—Conducted the original research work, drafted and corrected manuscript draft based on feedback and provided contributions to journal formatted manuscript. M.H.—Review and correction of original manuscript and preparation and finalizing manuscript for journal. K.M.—Review and correction of original manuscript and project supervisor.

Funding: This research received no external funding.

Conflicts of Interest: The authors declare no conflict of interest.

Appendix A

Table A1. Break-down of embodied energy for a solar PV system [36,37].

Process/Items	Embodied Energy kWh m^{-2}
Material Production Energy (E_{mpe})	
(A) Silicon purification and processing	670
(i) Metallurgical grade silicon production	
(ii) Electronic grade silicon production	
(iii) Silicon crystal growth	
(B) Solar Cell Production	120
(C) PV Module lamination and assembly	190
(i) Steel infrastructure	
(ii) Ethyl vinyl acetate	
(iii) Tedlar production	

Table A1. *Cont.*

Process/Items	Embodied Energy kWh m^{-2}
(iv) Glass Sheet production	
(v) Aluminium frame production	
(vi) Other material	
PV System Installation (E_{inst})	
(A) Support Structure	277.50
(B) Balance of System	
(i) Inverters	33
(ii) Electronic components, cables and miscellaneous items	45
Operation and Maintenance of Equipment ($E_{O\&M}$)	
(A) Instruments	59.5
(i) Tong Meter	
(ii) Solarimeter	
(iii) Temperature sensor	
(iv) Anemometer	
(B) Paints	10
(C) Miscellaneous human labour, wires	12.84
Salvage operation	0
Transportation	53.5
Land Energy Required for Disposal	0

Appendix B

Total fossil resource displaced is calculated by

Annual electricity generated by residential units of the community project * Life span of the solar PV system.

= (105,604 kWh)·(30 years)

= 3168,120 kWh

Factor 1 ktoe = 11,630,000 kWh

Therefore, total fossil resource displaced = 272.40 t of oil equivalent.

Appendix C

To generate simulations using PV*SOL software, run software using the following link http://pvsol-online.valentin-software.com/#/.

Table A2. Parameters and values used for the PV*SOL simulation.

No.	Parameter	Value
1.	Address Search	Belfield, Dublin
2.	Load profile	2 Person Household with 2 children
3.	Annual consumption	3100
4.	PV Modules	Hanwha Q cells; Q peak G4.4 300 Rev1
5.	No of modules	6
6.	Inclination	30°
7.	Orientation	180°
8.	Installation Type	Roof parallel
9.	Albedo	20%
10.	Soil	0%
11.	Shadow	0%
12.	Inverter manufacturer	ABB

To run simulation, supply above parameters, Select "Get best configuration" button. Confirm "not a robot" by checking tick box. Select "Simulate PV System" button. Note: For simulation, software automatically picks up 951.3 kWh m^{-2} and performance ratio as 86.6% based on input parameters. However, in theoretical calculation it is considered 1074 kWh m^{-2} and performance ratio as 80%. To match the theoretical calculations the solar energy value simulated was reduced by approx. 4% in each month.

References

1. Intergovernental Panel on Climate Change (IPCC). Summary for Policymakers. In *Climate Change 2013: The Physical Science Basis. Contribution of Working Group I to the Fifth Assessment Report of the Intergovernmental Panel on Climate Change*; Stocker, T.F., Qin, D., Plattner, G.-K., Tignor, M., Allen, S.K., Boschung, J., Nauels, A., Xia, Y., Bex, V., Midgley, P.M., Eds.; Cambridge University Press: Cambridge, UK; New York, NY, USA, 2013.
2. Available online: https://www.seai.ie/publications/Energy-in-Ireland-2018.pdf (accessed on 16 October 2019).
3. Available online: https://eur-lex.europa.eu/legal-content/EN/ALL/?uri=CELEX:32009L0028 (accessed on 16 October 2019).
4. Murphy, F.; McDonnell, K. A Feasibility Assessment of Photovoltaic Power Systems in Ireland; a Case Study for the Dublin Region. *Sustainability* **2017**, *9*, 302. [CrossRef]
5. SEAI. Solar Electricity Grants—Reduce Home Energy Costs. 2019. Available online: https://www.seai.ie/grants/home-energy-grants/solar-electricity-grant/ (accessed on 25 November 2019).
6. Government of Ireland. Renewable Electricity Support Scheme (RESS) High Level Design. 2018. Available online: https://www.dccae.gov.ie/en-ie/energy/topics/Renewable-Energy/electricity/renewable-electricity-supports/ress/Pages/default.aspx (accessed on 16 October 2019).
7. Lam, P.T.I.; Law, A.O.K. Crowdfunding for renewable and sustainable energy projects: An exploratory case study approach. *Renew. Sustain. Energ. Rev.* **2016**, *60*, 11–20. [CrossRef]
8. Hanna, R.; Leach, M.; Torriti, J. Microgeneration: The installer perspective. *Renew. Energy* **2017**, *116*, 458–469. [CrossRef]
9. Joint Oireachtas Briefing SEAI. Available online: https://data.oireachtas.ie/ie/oireachtas/committee/dail/32/joint_committee_on_communications_climate_action_and_environment/submissions/2019/2019-03-05_opening-statement-jim-gannon-ceo-sustainable-energy-authority-of-ireland-seai_en.pdf (accessed on 16 October 2019).
10. Your Guide to Connecting Micro-Generation to the Electricity Network, Dublin. Available online: https://www.seai.ie/publications/Guide-to-Connecting-Micro-generation-to-the-Electricity-Network.pdf (accessed on 16 October 2019).
11. How do Photovoltaics Work? Available online: https://science.nasa.gov/science-news/science-at-nasa/2002/solarcells (accessed on 19 November 2019).
12. Best Practice Guide Photovoltaics. Available online: https://www.seai.ie/publications/Best_Practice_Guide_for_PV.pdf (accessed on 16 October 2019).
13. Islam, T.; Meade, N. The impact of attribute preferences on adoption timing: The case of photo-voltaic (PV) solar cells for household electricity generation. *Energy Policy* **2013**, *55*, 521–530. [CrossRef]
14. Available online: https://www.dccae.gov.ie/documents/Climate%20Action%20Plan%202019.pdf (accessed on 16 October 2019).
15. Da Rosa, A.V. *Fundamentals of Renewable Energy Processes*, 3rd ed.; Academic Press: Waltham, MA, USA; Oxford, UK, 2013; pp. 485–532.
16. Kannan, N.; Vakeesan, D. Solar energy for future world: A review. *Renew. Sustain. Energy Rev.* **2016**, *62*, 1092–1105. [CrossRef]
17. Honrubia-Escribano, A.; Ramirez, F.J.; Gómez-Lázaro, E.; Garcia-Villaverde, P.M.; Ruiz-Ortega, M.J.; Parra-Requena, G. Influence of solar technology in the economic performance of PV power plants in Europe. A comprehensive analysis. *Renew. Sust Energy Rev.* **2018**, *82*, 488–501. [CrossRef]
18. REN21 Renewables 2018 Global Status Report. 2018. Available online: http://www.ren21.net/gsr-2019/ (accessed on 16 October 2019).
19. Miglietta, M.M.; Huld, T.; Monforti-Ferrario, F. Local Complementarity of Wind and Solar Energy Resources over Europe: An Assessment Study from a Meteorological Perspective. *J. Appl. Meteorol. Clim.* **2016**, *56*, 217–234. [CrossRef]
20. Ayompe, L.M.; Duffy, A.; McCormack, S.J.; Conlon, M. Measured performance of a 1.72 kW rooftop grid connected photovoltaic system in Ireland. *Energy Convers. Manag.* **2011**, *52*, 816–825. [CrossRef]
21. Flood, E.; McDonnell, K.; Murphy, F.; Devlin, G. A Feasibility Analysis of Photovoltaic Solar Power for Small Communities in Ireland. *Renew. Energy* **2011**, *4*, 78–92.

22. Claudy, M.C.; Michelsen, C.; O'Driscoll, A. The diffusion of microgeneration technologies—assessing the influence of perceived product characteristics on home-owners' willingness to pay. *Energy Policy* **2011**, *39*, 1459–1469. [CrossRef]
23. Scarpa, R.; Willis, K. Willingness-to-pay for renewable energy: Primary and discretionary choice of British households' for micro-generation technologies. *Energy Econ.* **2010**, *32*, 129–136. [CrossRef]
24. Balcombe, P.; Rigby, D.; Azapagic, A. Motivations and barriers associated with adopting microgeneration energy technologies in the UK. *Renew. Sustain. Energy. Rev.* **2013**, *22*, 655–666. [CrossRef]
25. Dusonchet, L.; Telaretti, E. Comparative economic analysis of support policies for solar PV in the most representative EU countries. *Renew. Sustain. Energy. Rev.* **2015**, *42*, 986–998. [CrossRef]
26. REFIT Schemes and Supports. Available online: https://www.dccae.gov.ie/en-ie/energy/topics/Renewable-Energy/electricity/renewable-electricity-supports/refit/Pages/REFIT-Schemes-and-Supports.aspx (accessed on 16 October 2019).
27. Statement of Strategy 2019–2021. Available online: https://assets.gov.ie/17523/4d92fe2c91b44767b0e458f8e34eb0fa.pdf (accessed on 16 October 2019).
28. Crowdfunding. Available online: https://ec.europa.eu/info/business-economy-euro/growth-and-investment/financing-investment/crowdfunding_en (accessed on 16 October 2019).
29. The Regulation of Crowdfunding in Ireland: An Update. Available online: https://www.algoodbody.com/insights-publications/the-regulation-of-crowdfunding-in-ireland-an-update (accessed on 16 October 2019).
30. IFS 2020: A Strategy for Ireland's International Financial Services Sector 2015–2020: Action Plan 2019. Available online: https://assets.gov.ie/5949/230119134451-7323c03dfe3f4ca4a2b55e4737eb498c.pdf (accessed on 16 October 2019).
31. How to Calculate Output Energy of PV Solar Systems? Available online: https://photovoltaic-software.com/principle-ressources/how-calculate-solar-energy-power-pv-systems (accessed on 16 October 2019).
32. Ellis, F.; Breathnach, L. FlexiGrid—The Impact of a Storage Network on Existing and Future Grid Infrastructure Using Power Systems Analysis Software, Dublin. 2017. Available online: https://www.seai.ie/publications/SEAI-RDD-101-Solo-Energy-FlexiGrid.pdf (accessed on 16 October 2019).
33. PV SOL Software. Valentin Software GmbH, Stralauer Platz 33-34, 10243 Berlin, Germany, www.valentin.de. PV*Sol copyright ©2005 by Dr. Ing. Gerhard Valentin. Available online: www.pvsol-online.valentin-software.com (accessed on 15 October 2019).
34. Haegermark, M.; Kovacs, P.; Dalenbäck, J.O. Economic feasibility of solar photovoltaic rooftop systems in a complex setting: A Swedish case study. *Energy* **2017**, *127*, 18–29. [CrossRef]
35. La Monaca, S.; Ryan, L. Solar PV where the sun doesn't shine: Estimating the economic impacts of support schemes for residential PV with detailed net demand profiling. *Energy Policy* **2017**, *108*, 731–741. [CrossRef]
36. Khatri, R. Design and assessment of solar PV plant for a girls hostel (GARGI) of MNIT University, Jaipur city: A case study. *Energy Rep.* **2016**, *2*, 89–98. [CrossRef]
37. Tiwari, A.; Barnwal, P.; Sandhu, G.S.; Sodha, M.S. Energy metrics analysis of hybrid photovoltaic (PV) modules. *Appl. Energy* **2009**, *86*, 2615–2625. [CrossRef]
38. Nawaz, I.; Tiwari, G.N. Embodied energy analysis of photovoltaic (PV) system basedon macro- and micro-level. *Energy Policy* **2006**, *34*, 3144–3152. [CrossRef]
39. Ludin, N.A.; Mustafa, N.I.; Hanafiah, M.M.; Ibrahim, M.A.; Asri Mat Teridi, M.; Sepeai, S.; Zaharim, A.; Sopian, K. Prospects of life cycle assessment of renewable energy from solar photovoltaic technologies: A review. *Renew. Sustain. Energy. Rev.* **2018**, *96*, 11–28. [CrossRef]
40. Energy in Ireland 2018 Report, Dublin. Available online: https://www.seai.ie/grants/home-energy-grants/solar-electricity-grant/ (accessed on 16 October 2019).
41. Kristjansdottir, T.F.; Good, C.S.; Inman, M.R.; Schlanbusch, R.D.; Andresen, I. Embodied greenhouse gas emissions from PV systems in Norwegian residential Zero Emission Pilot Buildings. *Sol. Energy* **2016**, *133*, 155–171. [CrossRef]
42. European Energy Exchange. Available online: https://www.eex.com/en/ (accessed on 16 October 2019).
43. Rajput, P.; Singh, Y.K.; Tiwari, G.N.; Sastry, O.S.; Dubey, S.; Pandey, K. Life cycle assessment of the 3.2 kW cadmium telluride (CdTe) photovoltaic system in composite climate of India. *Sol. Energy* **2018**, *159*, 415–422. [CrossRef]
44. Li, Z.; Boyle, F.; Reynolds, A. Domestic application of solar PV systems in Ireland: The reality of their economic viability. *Energy* **2011**, *36*, 5865–5876. [CrossRef]

45. Domestic Solar PV Payback Calculator. Available online: https://www.seai.ie/tools/ (accessed on 14 October 2019).
46. Lu, Y.; Chang, R.; Lim, S. Crowdfunding for solar photovoltaics development: A review and forecast. *Renew. Sustain. Energy. Rev.* **2018**, *93*, 439–450. [CrossRef]
47. Osahenvemwen, O.; Omorogiuwa, O. Parametric Modeling of Voltage Drop in Power Distribution Networks. *IJASTR* **2015**, *3*, 356–359.
48. Zhai, P.; Williams, E.D. Dynamic hybrid life cycle assessment of energy and carbon of multicrystalline silicon photovoltaic systems. *Environ. Sci. Technol.* **2010**, *44*, 7950–7955. [CrossRef]
49. PvXchange Solar PV Price Index. 2019. Available online: https://www.pvxchange.com/en/price-index. (accessed on 16 October 2019).
50. VAT Rates: Electricity Energy Products and Supplies. Available online: https://www.revenue.ie/en/vat/vat-rates/search-vat-rates/E/electricity-energy-products-and-supplies-.aspx (accessed on 25 November 2019).
51. Inverter ABB Range. Available online: https://solartricity.ie/inverter-abb-range/ (accessed on 16 October 2019).
52. PSO Levy Explained. Available online: https://switcher.ie/gas-electricity/guides/energy-bills/pso-levy-explained/ (accessed on 16 October 2019).
53. Was 47 Forbes, H.; Schaefer, D. Guidelines for Successful Crowdfunding. *Procedia CIRP* **2017**, *60*, 398–403. [CrossRef]

© 2019 by the authors. Licensee MDPI, Basel, Switzerland. This article is an open access article distributed under the terms and conditions of the Creative Commons Attribution (CC BY) license (http://creativecommons.org/licenses/by/4.0/).

Article

Design and Testing of a Low Voltage Solid-State Circuit Breaker for a DC Distribution System

Leslie Tracy and Praveen Kumar Sekhar *

Electrical Engineering Program, School of Engineering and Computer Science, Washington State University, Vancouver, WA 98686, USA; leslie.tracy@wsu.edu
* Correspondence: praveen.sekhar@wsu.edu

Received: 17 December 2019; Accepted: 6 January 2020; Published: 10 January 2020

Abstract: In this study, a low voltage solid-state circuit breaker (SSCB) was implemented for a DC distribution system using commercially available components. The design process of the high-side static switch was enabled through a voltage bias. Detailed functional testing of the current sensor, high-side switch, thermal ratings, analog to digital conversion (ADC) techniques, and response times of the SSCB was evaluated. The designed SSCB was capable of low-end lighting protection applications and tested at 50 V. A 15 A continuous current rating was obtained, and the minimum response time of the SSCB was nearly 290 times faster than that of conventional AC protection methods. The SSCB was implemented to fill the gap where traditional AC protection schemes have failed. DC distribution systems are capable of extreme faults that can destroy sensitive power electronic equipment. However, continued research and development of the SSCB is helping to revolutionize the power industry and change the current power distribution methods to better utilize clean renewable energy systems.

Keywords: solid-state circuit breaker; microgrid protection; DC protection; SSCB

1. Introduction

Clean renewable energies have laid a platform for practical DC distribution systems in applications such as commercial buildings, data centers, microgrids, and shipboard power systems [1–5]. Solid-state circuit breakers (SSCBs) have been deemed the most ideal form of protection in developing DC systems due to inadequacies in standard AC protection methods. AC breakers work in connection with the zero crossing of the alternating voltage and current to mechanically extinguish a fault through arcing. Arcing in DC systems is indefinitely sustained unless larger airgaps are used for the same AC rating. Attempts have been made to reduce an AC breaker's voltage and current ratings by as much as half for use in DC systems, but their overall effectiveness remains scrutinized due to the speed at which they operate [6–10]. DC microgrids rely on power electronic converters to operate and are often unable to sustain prolonged faults, making it necessary to reduce instantaneous trip ratings. The fastest SSCBs have reported clearing times of just a few microseconds, where the mechanical AC breaker operates in tens of milliseconds. The comparison of an SSCB and thermomagnetic circuit breaker can be found in Figure 1 [11].

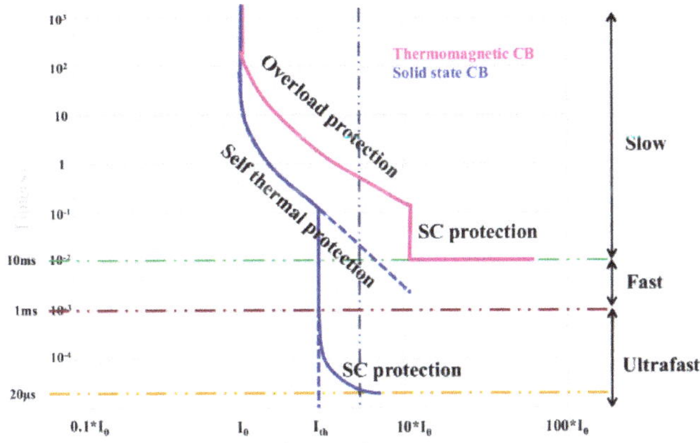

Figure 1. Trip curve comparison of an solid-state circuit breaker (SSCB) and thermomagnetic circuit breaker [11].

Several methods of SSCB design have been researched using different types of semiconductor devices for various applications. Silicon insolated-gate bipolar transistor (IGBT) devices are the most commonly reported [12–14]. More expensive wide-bandgap semiconductors with higher voltage ratings and lower losses have also been investigated [15–17]. Figure 2 depicts the basic structure of a DC distribution system, illustrating a general protection scheme. The SSCB is built to operate in a unidirectional or bidirectional mode, depending on its placement in the distribution system due to multiple generation sources. The placement of the devices determines the system voltage level, which can range from 28 V for low-level systems, such as lighting, to 1 kV for general distributions [18].

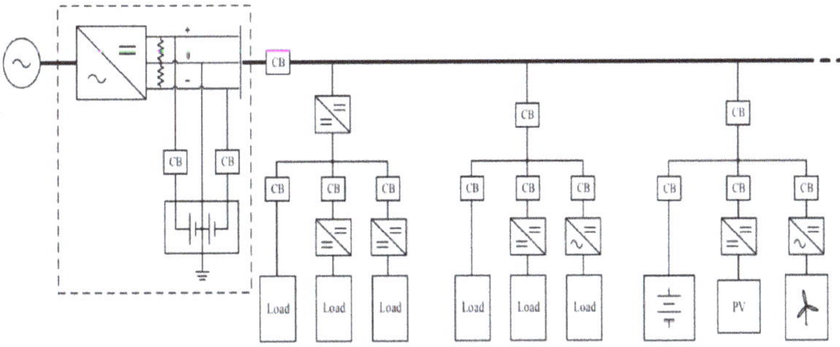

Figure 2. Generalized block diagram for a DC microgrid protection scheme with multiple generation sources [18].

At least two commercial SSCBs have been designed for use in a conventional three-phase AC distribution system, but, currently, the AC breaker remains the most widely used protection device in experimental DC systems. The development of new SSCB technology shows promise for the future of integrated power systems, but the cost associated with new technologies remains a hindrance in the growth of commercial DC systems. Continued research into affordable and reliable protection schemes is essential to the development of efficient clean energy systems. Expanding interests in the DC protection market provide a unique opportunity for the research and growth of future DC distribution systems. In this context, this article investigates the viability of a low voltage SSCB using commercially available low-cost off-the-shelf components.

The SSCB proposed in this article was used to provide the building blocks necessary to anyone interested in the subject, provide insight into possible improvements in the design, and further the research into effective low-cost solutions of SSCB implementation while highlighting the most troublesome areas of the design process, such as switching, thermal requirements, and speed.

The following section details the design process for a unidirectional low voltage SSCB using a silicon IGBT. The SSCB was designed for the low-end protection of lighting and/or other sensitive low current electronics. The device was tested in a laboratory setting, and the results were used to verify the SSCB voltage and current ratings. All external experiments were conducted using randomly generated test conditions to minimize error and improve the output reliability. The materials used in the creation of the device were purchased through an electronics vendor, and information and data sheets are available online.

2. Current Sensor Output

The ACS770ECB current sensor was selected for its continuous current rating of up to 200 A. The device features a high overcurrent rating and a rise time of 4.2 μs (0–60 A). The sensor is of the hall effect-type, and uses the magnetic field generated by the current to produce a linear output voltage. A test of the input current versus output voltage was conducted on the sensor using an omicron CMC 356 relay test set. The input current was varied from 0 to 45 A in 1 A increments and the output voltage was recorded (as shown in Figure 3).

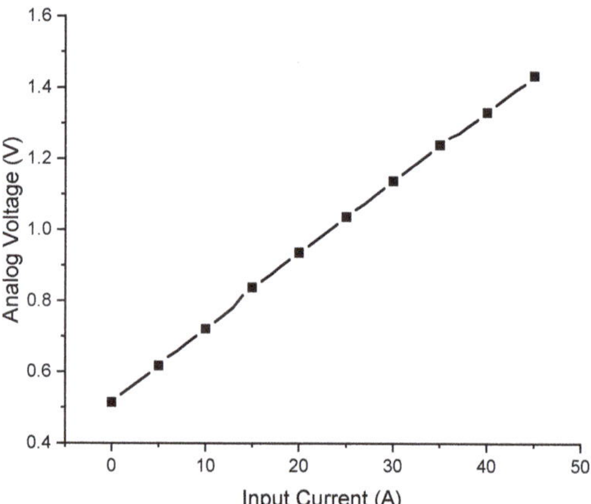

Figure 3. Experimental results of current vs. output voltage for the ACS770ECB current sensor.

A linear trendline was used to fit the collected data. The corresponding linear equation was found to be

$$V_a = 0.0204 * I_{in} + 0.5211, \quad (1)$$

where, V_a is the analog voltage output from the sensor and I_{in} is the input current. The output voltage showed a 20.4 mV increase per amp passing through the sensor. The linear output is essential for the easy implementation of analog to digital conversion (ADC) settings of the SSCB and is further discussed in Section 5.

3. Static Switch Design

The static nature of an SSCB requires the use of a high-side switch capable of a continuous on-time. The most common high-side drive configuration is the bootstrap circuit as it has a low propagation delay. Furthermore, it is simple and inexpensive. However, without modifications, the bootstrap topology is incapable of the continuous on-time needed for SSCB applications [19].

The introduction of an isolated voltage bias to the bootstrap topology provides the continuous operation needed, improves isolation, and simplifies the overall circuit [20]. Several considerations for the voltage bias must be met for successful implementation, and a careful selection process should be adhered to. The bias must be isolated as the secondary voltage will swing with the voltage at the emitter of the IGBT. The converter must have high dv/dt tolerances to keep voltage spikes across the bootstrap capacitor below those that would damage the high-side driver. As the IGBT is a voltage-controlled device, the converter must also have a low voltage drift under no-load conditions, preferably under 10%, in order to maintain the desired output of the converter.

In essence, the converter's main purpose is to maintain the charge lost due to the gate operation and leakage currents on the bootstrap capacitor, and the following equations found in [19] have been modified to reflect the voltage bias and the static nature of the SSCB. The capacitor is selected based on the gate charge of the transistor and is governed by the following equations:

$$C_g = Q_g/V_{bias}, \quad (2)$$

$$C_{boot} \geq 10 * C_g, \quad (3)$$

where, C_g is the total gate capacitance, Q_g is the gate charge of the transistor, V_{bias} is the voltage supplied by the converter, and C_{boot} is the capacitance value of the bootstrap capacitor. Likewise, the high-side gate driver source and sink requirements are dependent on Q_g and are expressed as

$$I_{source} = 1.5 * Q_g/t_{sw\_on}, \quad (4)$$

$$I_{sink} = 1.5 * Q_g/t_{sw\_off}, \quad (5)$$

where I_{source} and I_{sink} are the source and sink requirements of the high-side gate drive, respectively; 1.5 is an empirical constant; and t_{sw_on}/t_{sw_off} are the transition times experienced when switching the transistor on or off, respectively. Finally, a minimum gate resistance must be implemented to ensure that the source and sink currents do not exceed that of the maximum rating of the high-side gate drive and is governed by the following:

$$Rg\_min \geq Vbias/Isource/sink, \quad (6)$$

where R_{g_min} is the minimum gate resistance.

The simplified circuit diagram utilizing an isolated 15 V 1:1 REPCOM DC/DC converter, single output FAN7371MX high-side driver, and IXXH110N65C4 IGBT is shown in Figure 4.

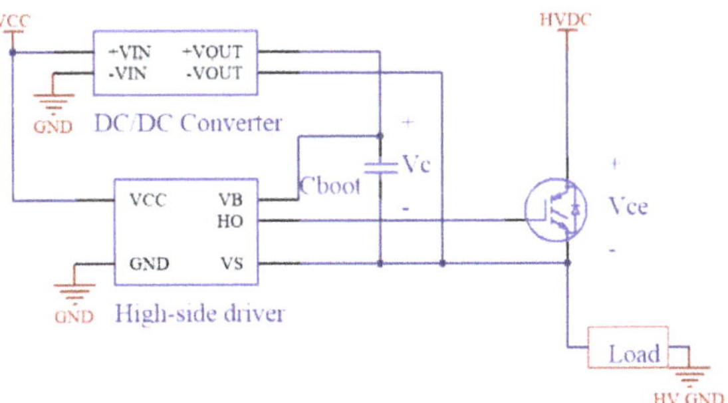

Figure 4. Simplified high-side switch design circuit capable of continuous on-time.

The designed static switch circuit is capable of operating with voltages up to a maximum of 600 V, and the gate driver can supply up to 4 A for source and sink operations. The voltage bias runs at a 1:1 conversion and operates at 15 V and 1 W. The gate charge, Q_g, of the IXXH110N65C4 IGBT is 167 nC. A 0.1 uF ceramic capacitor was selected for C_{boot}, and a gate resistance of 4.7 ohms was implemented.

Static Switch Experimental Results

The static switch was tested to ensure the safe operation of the voltage bias under any foreseeable condition. The recommended maximum for the safe operating range for the voltage across V_c is 20 V, while the absolute maximum rating is listed at 25 V. It was found that the switch was fully capable of continuous on-time, and two experiments were conducted for the turning on and turning off of the switch to obtain the maximum safe operating region for the voltage bias.

The turn-off experiment shown in Figure 5 was conducted at 30 V/1 A, without a snubber. The voltage across V_c reached a maximum and minimum voltage of 15.9 and 15.1 V, respectively, and poses no threat to the safe operation of the device. The turn-on experiment was run under an open circuit condition to increase the dV/dt of the emitter to its maximum value.

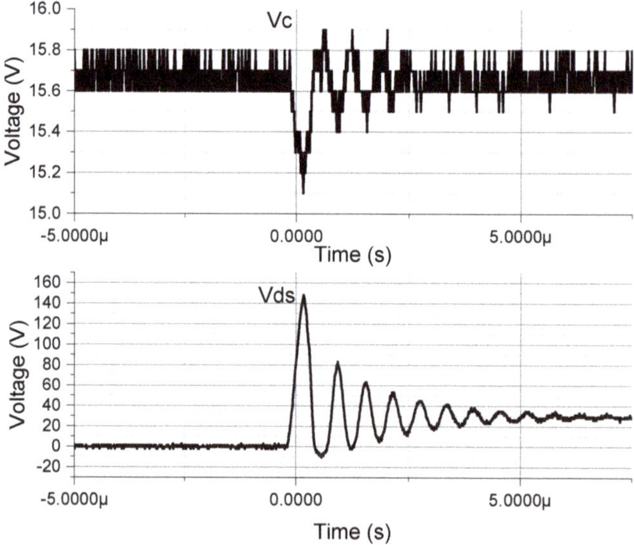

Figure 5. Voltage at turn-off across V_c (**above**) and V_{ce} (**below**).

The test voltage was increased until the maximum safe operating area of 20 V across V_c was reached, as shown in Figure 6. The switching period was completed in 50 ns, creating 6 kV/µs dV/dt at the emitter of the transistor. This effect, coupled with the need for voltage bias to recharge the capacitor, created a transient event with a maximum voltage of 19.7 V that lasted roughly 20 µs, before returning to a steady state operation. The test shows that the maximum safe operating area of the device was 300 V as the SSCB must be able to operate under any load, without fear of damage to the switch.

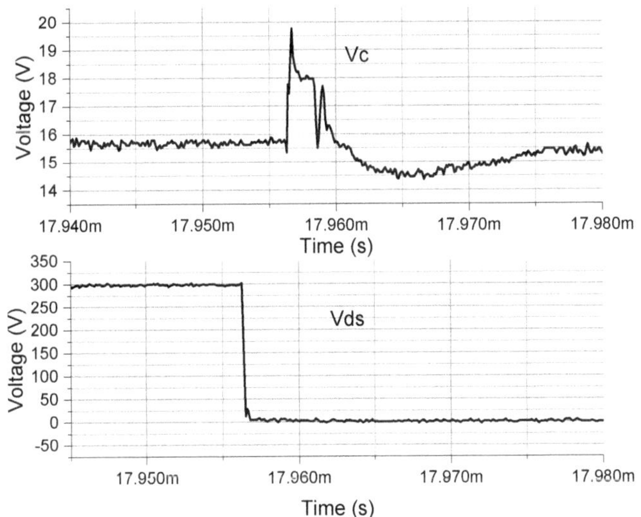

Figure 6. Voltage at turn-on across V_c (**above**) and V_{ce} (**below**).

4. Heat and Power Loss

The IXXH110N65C4 IGBT has a high-power density capable of 235 to 110 A of continuous current at 25 and 110 °C, respectively. The short circuit safe operating areas are listed at 600 A for 1 ms and 220 A for 10 µs at 25 and 150 °C, respectively. The switch was heat-tested in open air using a R2A-CT4–38E heatsink to determine the maximum continuous current rating of the SSCB. The experiment was conducted using a DC power source supplying 50.9 V DC to the NHR 4760 DC load bank. The current level was varied from 0 to 10 A, and each level was maintained for a period of no less than 4 min. The maximum surface temperature of the heatsink was recorded via an infrared thermometer using an emissivity of 0.94, and the forward voltage drop was recorded at every step. The power loss of the switch was then calculated, and the results are shown in Table 1.

Table 1. Experimental results of heat and power loss.

Amperage	Temperature	Forward Voltage Drop	Power Loss
0	23.8	0	0
1	27.2	0.76	0.76
2	32.5	0.825	1.65
3	38.3	0.862	2.586
4	43.3	0.895	3.58
5	47.9	0.925	4.625
6	52.9	0.935	5.61
7	57.6	0.955	6.685
8	64	0.973	7.784
9	69.9	0.99	8.91
10	74.5	1.005	10.05

The following equations, derived from Figure 7, were used to predict the on-time losses and temperature rise of the SSCB during a steady state operation:

$$P_d = 1.0347 * I_c - 0.4667, \text{ where } I_c \geq 1 \text{ A}, \tag{7}$$

$$T_{sw} = 5.2358 * I_c + 22.013, \text{ where } I_c \geq 1 \text{ A}, \tag{8}$$

where I_c is the continuous current of the SSCB, P_d is the power dissipated in the switch, and T_{sw} is the temperature of the switch. The slope of T_{sw} indicates that the temperature will increase 5.24 °C above the ambient per amp of continuous current. Combining Equations (7) and (8) reveals a temperature increase of 5.06 °C per watt consumed. The temperature analysis indicates that, under the current open-air conditions, the IGBT's optimum safe current rating is between 8 and 15 A and a 41.92 °C to 78.6 °C rise above the ambient temperature is to be expected. However, alternate cooling techniques that could be used to improve the thermal performance are discussed in [21] and should be considered when additional current is needed or ambient temperatures are unpredictable.

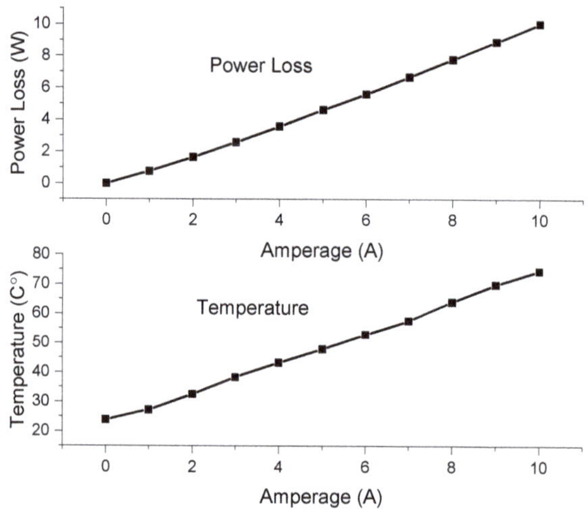

Figure 7. Power dissipated in the switch (**above**) and heat generated (**below**).

5. Analog to Digital Conversion

The SSCB was tested using the MEGA 2560 R3 microcontroller. The controller has a 10-bit resolution ADC and operates with a 16 MHz clock. The ADC conversion rate was increased from 9600 Hz to 50 kHz, resulting in one read every 20 µs. The equation governing the output of the ADC converter is

$$V_a = R_v * 5/1023, \tag{9}$$

where V_a is the analog voltage read by the converter and R_v is the integer return value seen by the controller. Equations (1) and (9) were combined to determine the current passing through the SSCB, resulting in the following equation:

$$I_{in} = 0.239 * R_v - 25.544. \tag{10}$$

Equation (10) was multiplied by 1000 to avoid the use of floating-point numbers and coded as

$$I_{input} = 239 * R_v - 25544, \tag{11}$$

where $I_{input} = I_{in} * 1000$. The current will be incremented up or down by a minimum of 0.239 A due to the ADC integer return values from 0 to 1023. Therefore, it should be noted that a slight range in variation is to be expected at any given setpoint.

6. ADC-Based Pickup and Clearing Times

The block diagram in Figure 8 shows the SSCB configuration that was implemented in this experiment. The snubbing circuit used in this experiment was a simple RC snubber optimized for 10 A at 1.19 µH of line inductance, as described in [22]. Advanced snubbing circuits are a continuing point of research in SSCBs, as inductances are not always known and can be found in [23,24].

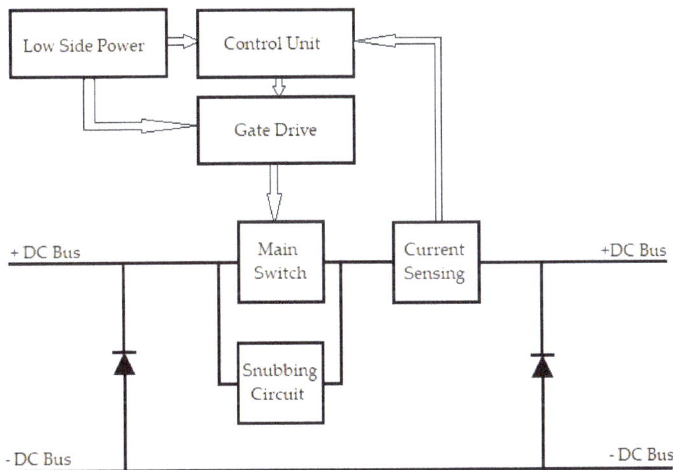

Figure 8. Generalized block diagram for an auxiliary-powered unidirectional SSCB.

6.1. Experimental Setup

The experiment was conducted at a trip setting of 4 A, where I_{input} = 4000. The pickup (PU) time was defined as the time taken after the current reached the defined trip setting to the point that the gate drive control signal went from high to low. The total clearing time (TCT) was defined as the time taken for the current to stabilize within two percent of the zero value after reaching the rated trip setting, and the switching time (SWT) was the difference between the two measurements. The test bench and sensing equipment consisted of a Feedback 67–113 200-ohm variable-load resistor, KEYSIGHT AC6804A power supply, KEYSIGHT InfiniVision MSOX4104A oscilloscope, KEYSIGHT N7026A current probe, KEYSIGHT N2790A differential probe, KEYSIGHT N2790A passive probe, and Tektronix PWS2323 DC Power Supply. The AC6804A was used to provide the high-side voltage and current through the switch to the 200-ohm variable-load resistor. The PWS2323 DC power supply was used to power the low-side electronics. The KEYSIGHT oscilloscope and probes were used in all measurements. In total, 25 test inputs were formulated and randomized with varying current levels from 4 to 8 A, and the N7026A current probe was zeroed out at the start of each session. The MSOX4104A oscilloscope was then set to the plus or minus 100 µs range, and each waveform to be measured was set within the expected output levels. The trigger function was set to a negative slope, and the trigger point was set to 500 mA. The MEGA microcontroller was programmed to the designated trip setting, and the 67–113 variable resistor was set to the needed resistance determined by the current that would be delivered to the circuit. The oscilloscope was set to single-waveform capture and the voltage from the AC6804A power source was turned on. The switch was then activated, allowing current to flow through the circuit and the resulting trip was documented as follows.

Cursor Y1 was set to zero and Y2 was set to the trip setting amperage of 4 A, as shown in Figure 9. X1 was then set at the crossing point of Y2, and X2 was set at which the point microcontroller responded. The resulting PU measurement between X1 and X2 was recorded. X2 was then moved to the zero-crossing, as shown in Figure 10, and the resulting TCT measurement was recorded.

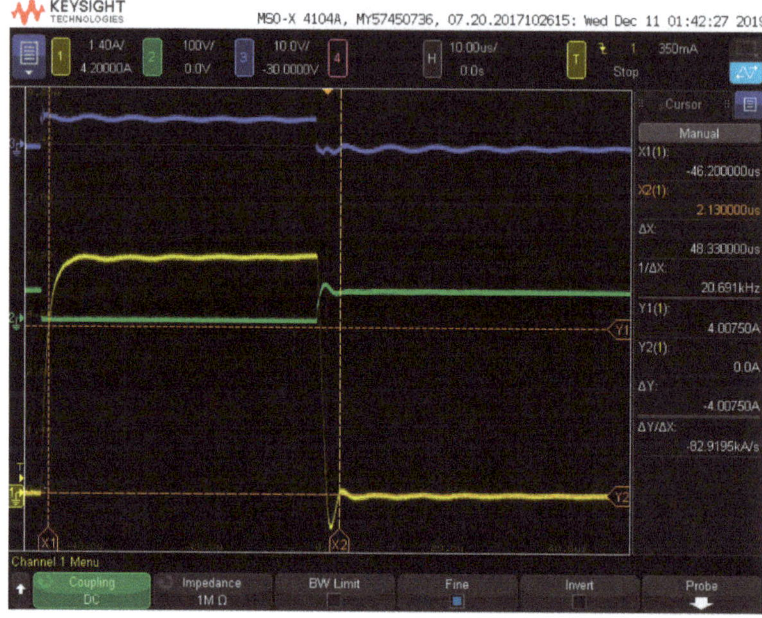

Figure 9. Trip setting measurement setup. Crossing point of cursors X1 and Y1.

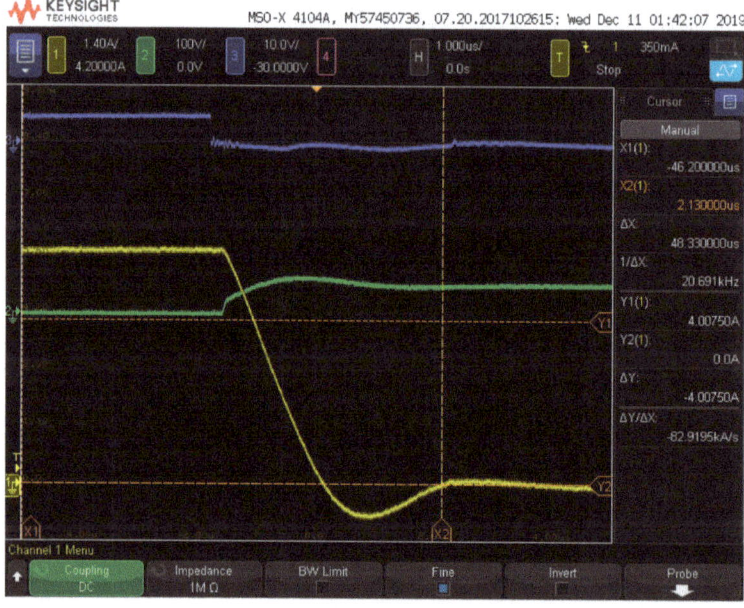

Figure 10. Total clearing time measurement setup showing time between X1 and X2.

The SWT measurement was taken as the difference between the PU and TCT measuring the propagation delay of the high-side driver and the switching action of the circuit. The process was repeated until all 25 inputs were completed. Channel 1 depicts the current (yellow), channel 2 depicts V_{CE} as shown in Figure 4 (green), and channel 3 shows the logic control signal for the high-side switch from the microcontroller (blue).

6.2. Switching Results

The experimental results are shown in Tables 2–4.

Table 2. Total clearing time results.

Amperage (A)	Total Clearing Time (µs)					Average (µs)
4	48.33	46.34	46.34	45.75	45.35	46.422
5	47.26	47.3	47.29	47.5	47.31	47.332
6	47.86	47.88	47.85	47.88	47.68	47.83
7	48.44	47.84	48.03	48.02	48.02	48.07
8	48.03	48.04	48.02	48.03	48.03	48.03

Table 3. Pickup time results.

Amperage (A)	Pickup Time (µs)					Average (µs)
4	44.43	42.22	42.21	41.59	41.18	42.326
5	43.36	43.36	43.35	43.6	43.34	43.402
6	44.14	44.14	44.12	44.1	43.9	44.08
7	44.78	44.16	44.43	44.4	44.41	44.436
8	44.46	44.5	44.5	44.45	44.46	44.474

Table 4. Switching time.

Amperage (A)	Switching Time (µs)					Average (µs)
4	3.9	4.12	4.13	4.16	4.17	4.096
5	3.9	3.94	3.94	3.9	3.97	3.93
6	3.72	3.74	3.73	3.78	3.78	3.75
7	3.66	3.68	3.6	3.62	3.61	3.634
8	3.57	3.54	3.52	3.58	3.57	3.556

The overall average response and clearings times for PU, TCT, and SWT were 43.74, 47.53, and 3.79 µs, respectively. As the amperage increases, the TCT and PU have an increasing average, while the SWT shows a decreasing trend of nearly 100 ns/A due to the snubbing circuit's overdamped response across V_{ce} at low current levels. The SSCB's response time, when compared to the generalized curve of its mechanical circuit breaker (MCB) counterpart depicted in Figure 11, shows that the average TCT during an instantaneous fault is greatly reduced by a factor of 1000. Furthermore, slower MCBs often have an instantaneous trip time between 40 and 55 ms, faster units will trip within 16 ms or one 60 Hz cycle [25,26]. The resulting SSCB's average TCT was 294.7 times faster than that of the single-cycle MCB using purely ADC techniques.

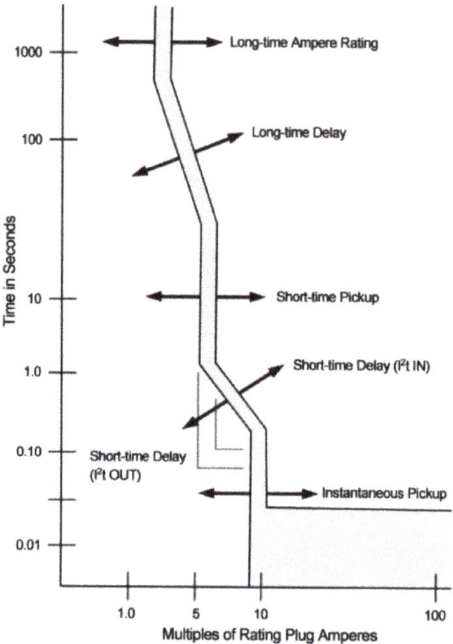

Figure 11. Generalized trip curve for mechanical circuit breakers [27].

The PU time is the main contributing factor in the speed of the SSCB under test, resulting in 92% of the overall average TCT delay. The PU speed is governed by the speed of the ADC capabilities of the microcontroller used in the experiment. Further reduction of the SSCB response time is accomplished via improved external sensing or faster ADC equipment, but may come at an increased cost to the design. The TCT of the SSCB is important for the success of the overall design due to the lack of large inductances from transformers in DC systems. In many cases, the wire inductance will be the largest limiting factor in the DC system. The change in current over time from a DC source, di/dt, is governed by Equation (12).

$$di/dt = V_{dc}/L, \tag{12}$$

where V_{dc} is the system voltage and L is the system inductance. Therefore, a 10-m 8 AWG wire with roughly 13 μH of inductance and a bus voltage of 50 volts would result in a di/dt of 3.85 A/μs, producing a 180.8 A increase in just 47 μs if limited by wire inductance alone. The same fault sustained for 16 ms would reach an excess of 61 kA, causing extreme thermal and physical stress on the system. This problem is further compounded as voltages and wire diameters increase in higher-rated systems [11,18]. Fast, easily programmable SSCBs can be used in tandem with fault-limiting converters and advanced protection algorithms to minimize the effects of extreme faults in the DC system.

7. Conclusions

A unidirectional low voltage SSCB design process was framed, and then constructed and tested, in this study. The assembled SSCB is a viable option for the protection of low-level sensitive equipment in a DC distribution system as the circuit can operate at 50 V/15 A of continuous current, without failure.

The fault clearing speed of the design is limited to an average of 47.53 μs by the speed of the ADC, which is 294.7 times faster than a fast operating AC counterpart.

Author Contributions: Conceptualization, methodology, and formal analysis: L.T. Resources, review, editing, and supervision: P.K.S. All authors have contributed substantially to the work reported. All authors have read and agreed to the published version of the manuscript.

Funding: This research received no external funding.

Conflicts of Interest: The authors declare no conflict of interest.

References

1. Salomonsson, D.; Sannino, A. Low-Voltage dc distribution system for commercial power systems with sensitive electronic loads. *IEEE Trans. Power Deliv.* **2007**, *22*, 1620–1627. [CrossRef]
2. Salomonsson, D.; Soder, L.; Sannino, A. Protection of low-voltage dc microgrids. *IEEE Trans. Power Deliv.* **2009**, *24*, 1045–1053. [CrossRef]
3. Salato, M.; Zolj, A.; Becker, D.J.; Sonnenberg, B.J. Power system architectures for 380V dc distribution in telecom datacenters. In Proceedings of the IEEE Intelec, Scottsdale, AZ, USA, 30 September 2012.
4. ABB Delivers First Onboard DC Grid System. Available online: http://www.abb.com/cawp/seitp202/7199db5e8cd3924e85257b3b00491f07.aspx (accessed on 16 December 2019).
5. Ton, M.; Fortenbery, B.; Tschudi, W. DC Power for Improved Datacenter Efficiency. Available online: http://www.chip2grid.com/docs/DCDemoFinalReport.pdf (accessed on 16 December 2019).
6. ABB Circuit Breakers for Direct Current Applications. Available online: http://www04.abb.com/global/seitp/seitp202.nsf/0/6b16aa3f34983211c125761f004fd7f9/$file/vol.5.pdf (accessed on 16 December 2019).
7. Gregory, G.D. Applying low-voltage circuit breakers in direct current systems. *IEEE Trans. Ind. Appl.* **1995**, *31*, 650–657. [CrossRef]
8. Geary, D.E.; Mohr, D.P.; Owen, D. 380V DC eco-system development present status and future challenges. In Proceedings of the IEEE Intelec, Hamburg, Germany, 13–17 October 2013.
9. Pugliese, H.; Von Kannewurff, M. Discovering DC: A primer on dc circuit breakers, their advantages, and design. *IEEE Ind. Mag.* **2013**, *19*, 22–28. [CrossRef]
10. CX-series Circuit Breakers. Available online: https://www.carlingtech.com/sites/default/files/documents/CX-Series_Details_%26_COS_010714.pdf (accessed on 16 December 2019).
11. Li, L.Q.; Antonello, A.; Luca, R. Design of solid-state circuit breaker-based protection for DC shipboard power systems. *IEEE J. Emerg. Sel. Top. Power Electron.* **2017**, *5*, 260–268.
12. Cuzner, R.; Venkataramanan, G. The status of DC micro-grid protection. In Proceedings of the IEEE Industrial Applications Society Annual Meeting, Edmonton, AI, Canada, 5–9 October 2008.
13. Slobodan, K. Circuit breaker technologies for advanced ship power systems. In Proceedings of the Electric Ship Technologies Symposium, Arlington, VA, USA, 21–23 May 2007.
14. Schmerda, R.F.; Krstic, S.; Wellner, E.L. IGCTs vs. IGBTs for circuit breakers in advanced ship electrical systems. In Proceedings of the Electric Ship Technologies Symposium, Baltimore, MD, USA, 20–22 April 2009.
15. Veliadis, V.; Urciuoli, D.; Hearne, H. 600-V/2-A symmetrical bi-directional power flow using vertical-channel JFETs connected in common source configuration. *Mater. Sci. Forum* **2010**, *645–648*, 1147–1150. [CrossRef]
16. Sato, Y.; Tanaka, Y.; Fukui, A. SiC-SIT circuit breakers with controllable interruption voltage for 400-V DC distribution systems. *IEEE Trans. Power Electron.* **2014**, *29*, 2597–2605. [CrossRef]
17. Urciuoli, D.P.; Veliadis, V.; Ha, H.C. Demonstration of a 600-V, 60-A, bidirectional silicon carbide solid-state circuit breaker. In Proceedings of the Twenty-Sixth Annual IEEE Applied Power Electronics Conference and Exposition, Fort Worth, TX, USA, 6–11 May 2011.
18. Shen, J.Z.; Miao, Z.; Roshandeh, A.M. Solid state circuit breakers for DC micrgrids: Current status and future trends. In Proceedings of the IEEE First International Conference on DC Microgrids, Atlanta, GA, USA, 7–10 June 2015.
19. AN-6076 Design and Application Guide of Bootstrap Circuit for High-Voltage Gate-Drive IC. Available online: https://www.onsemi.com/pub/Collateral/AN-6076.pdf.pdf (accessed on 16 December 2019).
20. Using a Single-Output Gate-Driver for High-Side or Low-Side Drive. Available online: http://www.ti.com/lit/an/slua669a/slua669a.pdf (accessed on 16 December 2019).
21. Qian, C.; Gheitaghy, A.M.; Fan, J. Thermal Management on IGBT Power Electronic Devices and Modules. *IEEE Access* **2018**, *6*, 12868–12884. [CrossRef]

22. Application Guide Snubber Capacitors. Available online: https://www.cde.com/resources/catalogs/igbtAPPguide.pdf (accessed on 16 December 2019).
23. Liu, F.; Liu, W.; Zha, X. Solid-state circuit breaker snubber design for transient overvoltage suppression at bus fault interruption in low-voltage DC microgrid. *IEEE Trans. Power Electron.* **2017**, *32*, 3007–3021. [CrossRef]
24. Liu, W.; Yang, H.; Liu, F. An improved RCD snubber for solid-state circuit breaker protection against bus fault in low-voltage DC microgrid. In Proceedings of the IEEE 2nd International Future Energy Electronics, Taipei, Taiwan, 1–4 November 2015.
25. Low Voltage Selectivity with ABB Circuit-Breakers. Available online: https://library.e.abb.com/public/65ddf36f7c3bd0fec1257ac500377a37/1SDC007100G0204.pdf (accessed on 28 December 2019).
26. Working with the Trip Characteristic Curves of ABB SACE Low Voltage Circuit-Breakers. Available online: http://www04.abb.com/global/seitp/seitp202.nsf/0/440613170f6c8628c125761f00506afe/%24file/White%2BPaper%2BVolume%2B1.pdf (accessed on 28 December 2019).
27. Characteristics of Circuit Breaker Trip Curves and Coordination. Available online: https://testguy.net/content/197-Characteristics-of-Circuit-Breaker-Trip-Curves-and-Coordination (accessed on 28 December 2019).

© 2020 by the authors. Licensee MDPI, Basel, Switzerland. This article is an open access article distributed under the terms and conditions of the Creative Commons Attribution (CC BY) license (http://creativecommons.org/licenses/by/4.0/).

Article

A Two-Stage Industrial Load Forecasting Scheme for Day-Ahead Combined Cooling, Heating and Power Scheduling †

Sungwoo Park [1], Jihoon Moon [1], Seungwon Jung [1], Seungmin Rho [2], Sung Wook Baik [2] and Eenjun Hwang [1,*]

1 School of Electrical Engineering, Korea University, 145 Anam-ro, Seongbuk-gu, Seoul 02841, Korea; psw5574@korea.ac.kr (S.P.); johnny89@korea.ac.kr (J.M.); jsw161@korea.ac.kr (S.J.)

2 Department of Software, Sejong University, 209 Neungdong-ro, Gwangjin-gu, Seoul 05006, Korea; smrho@sejong.edu (S.R.); sbaik@sejong.ac.kr (S.W.B.)

* Correspondence: ehwang04@korea.ac.kr; Tel.: +82-2-3290-3256

† This paper is an extended version of our paper published in Proceedings of the 2019 IEEE International Conference on Power Electronic and Drive Systems (PEDS), Toulouse, France, 9–12 July 2019.

Received: 7 December 2019; Accepted: 15 January 2020; Published: 16 January 2020

Abstract: Smart grid systems, which have gained much attention due to its ability to reduce operation and management costs of power systems, consist of diverse components including energy storage, renewable energy, and combined cooling, heating and power (CCHP) systems. The CCHP has been investigated to reduce energy costs by using the thermal energy generated during the power generation process. For efficient utilization of CCHP and numerous power generation systems, accurate short-term load forecasting (STLF) is necessary. So far, even though many single algorithm-based STLF models have been proposed, they showed limited success in terms of applicability and coverage. This problem can be alleviated by combining such single algorithm-based models in ways that take advantage of their strengths. In this paper, we propose a novel two-stage STLF scheme; extreme gradient boosting and random forest models are executed in the first stage, and deep neural networks are executed in the second stage to combine them. To show the effectiveness of our proposed scheme, we compare our model with other popular single algorithm-based forecasting models and then show how much electric charges can be saved by operating CCHP based on the schedules made by the economic analysis on the predicted electric loads.

Keywords: short-term load forecasting; two-stage forecasting model; combined cooling heating and power; energy operation plan; economic analysis

---

1. Introduction

Recently, as the amount of resources consumed by one person has increased, there are growing concerns about environmental problems caused by carbon dioxide emitted during energy generation and energy shortage problems [1]. Smart grid technologies have been gaining much attention because they help to solve these problems by enabling more efficient use of energy [2]. A smart grid is an intelligent power grid that combines information and communication technology with the existing power grid and integrates the work of all users in the power network by using computer-based remote control and automation [3]. It allows monitoring, analyzing, controlling, and communication within the supply chain to improve efficiency, reduce energy consumption and costs, and maximize the transparency and reliability of the energy supply chain [4]. In addition, by intelligentizing the power grid, it is possible to construct a bi-directional supply system such as a microgrid and distributed power supply system where suppliers and consumers can exchange information that they need [5].

Based on this information, energy prosumers can be more active in the trade of electricity. For instance, prosumers on the demand side can choose the supplier that can supply electricity at a lower price, and prosumers on the supply side can create opportunities to sell electricity more expensive.

Typical smart grids are closely related to various energy systems such as the energy storage system (ESS), renewable energy system (RES), combined cooling, heating and power (CCHP), and so on [6]. In particular, CCHP is a cogeneration technology that integrates an absorption chiller to produce cooling. Thermal energy produced during the power generation process is collected to meet cooling and heating demands via the absorption chiller and heating unit [7]. Besides, natural gas-based CCHP has the advantage of lower fuel prices and lower carbon dioxide emissions compared to existing fossil fuel-based power generations [8]. For the efficient operation of CCHP, accurate short-term load forecasting (STLF) is required [9]. STLF is the basis of the design and implementation of the control strategy of the CCHP system, and the results of the STLF affect the overall energy efficiency of the system directly [10]. CCHP uses the primary energy to drive the generator to generate electricity and then recycle waste heat using waste heat equipment. Therefore, running CCHP without accurate predictions can increase the unnecessary operation cost of the power generation facility [11].

Electric energy consumption can be affected by diverse factors, which include architectural structures, thermal properties of physical materials, lighting, time zones, climatic conditions, and electric rates [12]. In addition, there are complicated electric load correlations between current and previous times [13]. They should be considered appropriately for accurate electric energy consumption forecasting. For instance, many STLF models have proposed a single machine learning algorithm to consider them [14]. However, such models do not always provide good prediction performance because electric energy consumption patterns are intricate, and uncertain external factors can cause a shift in the demand curve [15]. Besides, the domains that they show good performance could be different. Thus, it is not effective to use a single STLF model for prediction in diverse domains. This limitation can be alleviated by combining multiple models of this type [16].

To address these issues, many previous studies have suggested a two-stage STLF model that uses linear regression in the second stage for improving the accuracy of electric load forecasting [17]. These models performed better than previous studies that use a single algorithm by combining the predicted values obtained in the first stage [18]. However, there still are many deficits in the linearly combined model. For instance, the fixed weights of the linear combination can ignore the importance of potential nonlinear terms, which leads to a reduction in prediction accuracy. Additionally, the linear combination can give poor forecasting results when there is a strong nonlinear relationship between individual predictors and outcomes [19]. South Korea is one of the highest energy consumption countries and is interested in using smart grids to improve energy efficiency [20]. However, although studies on the electric load forecasting model have been sufficiently conducted, there are not many cases of configuring a power system in conjunction with CCHP. We focus on the features of the Korean power system and develop an application for scheduling CCHP operations to provide a bi-directional benefit to power suppliers and users.

In this paper, we propose a novel two-stage STLF scheme based on nonlinear combination of forecasting methods to solve this problem. In the first stage, we build two STLF models based on extreme gradient boosting (XGBoost) and random forest (RF), which are known to be popular tree-based ensemble models in time series prediction. In the second stage, we build a deep neural network (DNN)-based STLF model to combine the predicted values of XGBoost and RF. Further, we propose an economic analysis-based operation scheduling scheme for CCHP to effectively utilize the results of the STLF. For instance, in Korea, electric rates and contract demand, especially for industrial services should be considered in the electric rate system. Contract demand indicates the instantaneous peak load contracted with the power supply company. Based on the contract demands, a power supply company can make a stable power plan. Basically, the lower the contract demand is, the lower the basic electricity bill is. Hence, to derive more accurate contract demands, the following policy is used: If the consumer sets the contract demand too low, a progressive tax penalty will be added to the excess power, which results in higher electricity charges. On the contrary, if the contract demand is set too high, consumers have to pay unnecessarily high electricity bills. The economic analysis shows the electric rate and contract demand that should be made to achieve the lowest electric charges. In order to intuitively display the outcome of the economic analysis, a graphical representation of CCHP scheduling is shown with the amount of economic benefits gained from the schedule. Figure 1 shows the overall architecture of our scheme.

The main contributions of this paper are as follows:

(1) We propose a novel two-stage STLF scheme that can predict electric energy consumption accurately compared to other previous methods.
(2) We propose a method to generate an optimal operation schedule of CCHP based on the predictive values of electric energy consumption and electric/gas charges in South Korea.
(3) We propose a method to minimize the electric charge by calculating optimal contract demand and electric rate.

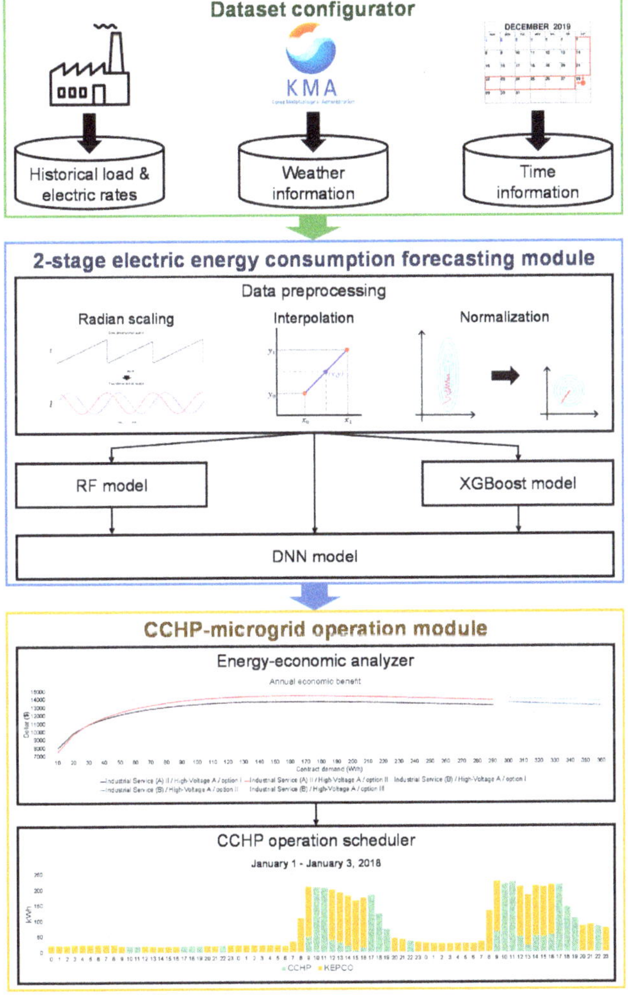

Figure 1. Overall system architecture.

The rest of this paper is organized as follows: In Section 2, related studies on STLF are reviewed. In Section 3, we explain the input variables for constructing the STLF model. In Section 4, we describe the structure of our forecasting model, and then, explain how to make CCHP operational scheduling in Section 5. In Section 6, we describe some of the experiments for performance evaluation of the proposed model and CCHP operation scheduling. Finally, in Section 7 we summarize our study.

2. Related Works

So far, many studies have been done to efficiently perform STLF. In recent years, diverse machine learning algorithms in particular have been tested to build more accurate STLF models [21]. In this section, we first introduce diverse STLF models and then describe our previous works for STLF.

2.1. Short-Term Load Forecasting

Typical approaches for STLF have been applied to statistical and machine learning methods for diverse external information such as time factors, weather conditions, and so on. Table 1 summarizes STLF-related studies using statistical techniques and machine learning. Veghefi et al. [22] proposed an STLF model based on the Cochrane–Orcutt estimation technique that combines the multiple linear regression (MLR) and seasonal auto-regressive integrated moving average (SARIMA) models to predict cooling and electric energy consumption effectively. Bagnasco et al. [23] constructed an artificial neural network (ANN)-based STLF model considering holiday indicators and weather conditions as input variables for forecasting electric energy consumption of Cellini Medical Clinic. Powell et al. [24] constructed an STLF model based on a nonlinear autoregressive model with exogenous inputs (NARX) for heating, cooling, and electric energy consumption of a district energy system. This study was unique because it covered a large-scale district energy system that simultaneously produced combined heat and power (CHP), chilled water thermal energy storage (TES), gas turbines, steam turbines, heat recovery steam generators (HRSGs), and auxiliary boilers for a large campus. Jurado et al. [25] constructed several prediction models using RF, ANN, and fuzzy inductive reasoning (FIR). They then compared the prediction models with an ARIMA model by predicting electric energy consumptions in three different buildings at Catalonia Technical University, Catalonia, Spain. They confirmed that FIR showed the best prediction performance. Sandels et al. [26] presented a data analysis framework for identifying and generating models that can predict energy consumption on load level in North European office building floors. The models were based on a simplified statistical approach that did not require detailed knowledge about the office building floor. Grolinger et al. [27] constructed two STLF models based on support vector regression (SVR) and ANN. They considered time data, historical electric load data, and event information and compared their prediction performances with other methods for a large entertainment building in Canada. With daily data, the ANN model achieved better accuracy than the SVR. Gerossier et al. [28] presented a forecasting model for hourly household electric load based on quantile smoothing spline regression using the previous day's hourly load, last week's mid-load, and temperature. They computed the mean of the predicted quantile distribution and used it as a single-point forecast. These statistical approaches exhibited excellent performance for simple demand patterns but inaccurate prediction performance for intricate demand patterns. Chen et al. [29] developed a combination of a hybrid SVR model and multiresolution wavelet decomposition (MWD) to predict the hourly electric energy consumption of a hotel and mall. Dong et al. [30] proposed a seasonal SVR with a chaotic cuckoo search (CCS) named SSVRCCS to predict electric energy consumptions in the National Electricity Market and New York Independent System Operator. By using the CCS model, their proposed model can enlarge the population in cuckoo search (CS) to prevent the local optimal problem and increased the search space. By using the seasonal SVR model, it can deal with the seasonal cyclic nature of electric load for accurate and better prediction. However, the computational time is increased due to a large number of iterations. Fan et al. [31] proposed a novel electricity load forecasting model by hybridizing the phase space reconstruction (PSR) algorithm with the bi-square kernel (BSK) regression model, namely the PSR-BSK model. The authors investigated the performance of the model using an hourly dataset of NYISO, USA, and New South Wales market. Hong et al. [32] proposed an electric load prediction model, namely the H-EMD-SVR-PSO model, which combines the empirical mode decomposition (EMD) method, particle swarm optimization (PSO) algorithm, and SVR, to improve predictive accuracy. Based on electrical load data from the Australian electricity market, experimental results showed that the proposed H-EMD-SVR-PSO model received more satisfactory prediction performance than other comparable models.

These studies suggested the construction of non-generic forecasting models by considering the characteristics of buildings and microgrids. On the other hand, CCHP can be installed and used in various places with possibly different features. Moreover, different types of schedules may be required for CCHP depending on electric energy consumption patterns.

Table 1. Summary of several approaches for short-term load forecasting.

Author (Year)	Target	Types of Input Variables	Forecasting Method	Description
Vaghefi et al. [22] (2014)	A CCHP plant at University of California	Historical load, weather, time	Cochrane–Orcutt estimation	This approach utilized the advantage of both time series and regression methods.
Bagnasco et al. [23] (2014)	Cellini Medical Clinic	Historical load, weather, time	ANN	The model can be easily integrated into a building energy management system or into a real-time monitoring system.
Powell et al. [24] (2014)	College campus building	Weather, time	Nonlinear autoregressive model	95% confidence limits are used to quantify the uncertainty of the predictions.
Jurado et al. [25] (2015)	Three buildings of the UPC	Historical load, time	RF, ANN, FIR, ARIMA	The approaches discussed generate fast and reliable models, with low computational costs.
Sandels et al. [26] (2015)	Office building	Weather, electricity price, occupancy	Regression model	Occupancy is correlated with appliance load, and outdoor temperature and a temporal variable defining work hours are connected with ventilation and cooling load.
Grolinger et al. [27] (2016)	A large event-organizing venue	Historical load, time, event	ANN, SVR	Daily data intervals resulted in higher consumption prediction accuracy than hourly or 15 min readings.
Gerossier et al. [28] (2017)	A neighborhood comprising 226 individual buildings	Historical load, weather	Quantile smoothing splines regression	Providing probabilistic forecasts by computing a list of quantiles.
Chen et al. [29] (2017)	Mall and hotel	Historical load, weather	SVR	Multi-resolution wavelet decomposition can always improve the predicting accuracy for the hotel, while it is not necessary for the mall.
Dong et al. [30] (2018)	National electricity market, New York Independent system operator	Historical load	SSVRCCS	SSVRCCS model is employed to improve the forecasting accuracy level by sufficiently capturing the non-linear and cyclic tendency of electric load changes.
Fan et al. [31] (2018)	New South Wales market, New York Independent system operator	Historical load	PSR-BSK	Their proposed model can extract some valuable features embedded in the time series to demonstrate the relationships of the nonlinearity.
Hong et al. [32] (2019)	New South Wales market	Historical load	H-EMD-SVR-PSO	Decomposed intrinsic mode functions could be defined as three items and there three items would be modeled separately by the SVR-PSO model.

2.2. Our Previous Works

In this section, we briefly describe several STLF models that we proposed in our previous studies and their differences from the proposed model.

In [33], we built two STLF models using the ANN and SVR for four building clusters of a private university in South Korea. For the prediction, we considered not only weather information and time data but also university events, office hours, and class hours. Subsequently, we evaluated the prediction performance of each model by using 5-fold cross-validation. The comparison showed that the ANN-based forecasting model had better performance than the SVR-based model. In [34], we proposed another STLF model based on an auto-encoder (AE) and RF. The AE was used to extract weather information features and time factors effectively. We constructed an RF-based forecasting model using feature extraction values and historical electric loads for day-ahead electric load forecasting. The model was evaluated using the electric energy consumption data of university campuses and the results showed that it gave a better performance than the proposed model in [33]. In [35], we proposed a recurrent inception convolution neural network (RICNN) that combines recurrent neural networks (RNN) and 1-dimensional convolutional neural networks (CNN) to forecast multiple short-term electric loads (48 time steps with an interval of 30 min). A 1-D convolution inception module was used to calibrate the prediction time and hidden state vector values calculated from nearby time steps. By doing so, the inception module could generate an optimized network via the prediction time generated in the RNN and nearby hidden state vectors. The proposed RICNN model was verified using the electric energy consumption data of three large distribution complexes in South Korea. In [36], we constructed diverse ANN models using different numbers of hidden layers and diverse activation functions and compared their performances in a 30 min STLF resolution. To compare the prediction performance, we considered electric load data collected for two years from five different types of buildings (including the dataset used in this study). The comparison showed that a scaled exponential linear unit (SELU)-based ANN model with five hidden layers had a better average performance than other ANN-based STLF models. In [37], we proposed a two-stage electric load forecasting model that combined XGBoost and RF using MLR for the efficient operation of CCHP. To construct this model, an hourly load forecasting was performed using XGBoost and RF. The forecasting results were then combined using a sliding-window based MLR to reflect the energy consumption pattern. The model had a better prediction performance compared with several popular single algorithm-based forecasting models.

The difference between the papers mentioned above and our paper is as follows.

The models in [33] were tailored for university campuses; they were challenging to apply for other types of buildings. The model in [34] used AE to extract features. However, since the performance of AE heavily depends on the size of the training set, it is challenging to show excellent performance if there is not enough quantity of data. In [35], we proposed the RICNN model. However, the RICNN, which purposed a probabilistic approach, is a different purpose because we focus on day-ahead point load forecasting. In [36], the SELU-based ANN model with five hidden layers showed that the dataset we used in this study exhibited insufficient prediction accuracy compared to the other building types because its electric loads are close to zero. In [37], we proposed a two-stage electric load forecasting model to combine XGBoost and RF using MLR. However, to use the forecasted values from one-stage more efficiently, we have to consider existing input variables. Eventually, we further develop our research and propose integrated applications with CCHP operation scheduling and electric rate recommendations, not just ending with forecasts.

3. Input Variable Configuration

In this study, we use hourly electric energy consumption data collected from 1 January 2015 to 31 December 2018 from an industrial building in Incheon, South Korea. Table 2 summarizes some statistics of the collected data. To construct our STLF model, we consider time factors, weather information, historical electric energy consumption, and electric rates for input variable configuration. The details are described in the following subsections.

Table 2. Statistics of energy consumption data from industrial building.

Statistics	Energy Consumption (kWh)
Mean	46.26
Standard error	0.26
Median	22.94
Mode	11.88
Standard deviation	49.59
Sample variance	2459.36
Kurtosis	2.94
Skewness	1.80
Range	287.52
Minimum	6.63
Maximum	294.15
Sum	1,622,214.17
Count	35,064

3.1. Time Data

As time is a very critical factor in the trends of electric energy consumption, we consider all variables that express time such as month, day, hour, day of the week, and holiday. Table 3 shows a list of the time factors we considered as input variables. Herein, month, day, and hour have a sequence form. It is difficult to reflect periodic information in machine learning algorithms when data are in a sequential format. Therefore, we enhanced the data to 2-dimensional data through the periodic function [36]. Table 4 summarizes some regression statistics of 1-dimensional, 2-dimensional, and 1-dimensional + 2-dimensional time factors. The table shows that 1-dimensional + 2-dimensional space data can represent the time factor most effectively. Therefore, we use both 1-dimensional data and continuous 2-dimensional data to represent time factor.

Table 3. Input variables of time factors.

No.	Input Variables	Type of Variables
1	Month	Continuous on [1, 12]
2	Day	Continuous on [1, 31]
3	Hour	Continuous on [0, 23]
4	$month_x$	Continuous on [−1, 1]
5	$month_y$	Continuous on [−1, 1]
6	day_x	Continuous on [−1, 1]
7	day_y	Continuous on [−1, 1]
8	$hour_x$	Continuous on [−1, 1]
9	$hour_y$	Continuous on [−1, 1]
10	$day\_of\_the\_week_x$	Continuous on [−1, 1]
11	$day\_of\_the\_week_y$	Continuous on [−1, 1]
12	Monday	Binary
13	Tuesday	Binary
14	Wednesday	Binary
15	Thursday	Binary
16	Friday	Binary
17	Saturday	Binary
18	Sunday	Binary
19	Holiday	Binary

Table 4. Regression statistics of time data representation.

Regression Statistics	1-Dimensional	2-Dimensional	1-Dimensional + 2-Dimensional
Multiple R	0.190	0.609	0.614
R-squared	0.036	0.371	0.377
Adjusted R-squared	0.036	0.370	0.377
Standard error	48.978	39.579	39.365

3.2. Weather Data

Because the frequency of use of high-power consumption products such as air conditioners and radiators is closely related to weather [38], weather conditions have generally been used for constructing STLF models in many studies [39]. In South Korea, various weather forecast information including temperature, humidity, wind speed, and so on are provided by the Korea Meteorological Administration (KMA). However, KMA provides weather data using two different time resolutions depending on the type of forecast. Very short-term weather forecast provides weather data up to 4 h by 30 min resolution, and short-term weather forecast provides weather data resolution up to 67 h by 3 h resolution. Since our goal is to predict day-ahead electric energy consumption, we need weather data for up to 24 h. Thus, we used the short-term weather forecast data that have 3 h resolution and used linear interpolation to calculate 1 h weather forecast data from them. The short-term weather forecast data consists of values such as daily minimum temperature, daily average temperature, daily maximum temperature, temperature, humidity, wind speed, and precipitation, as shown in Figure 2.

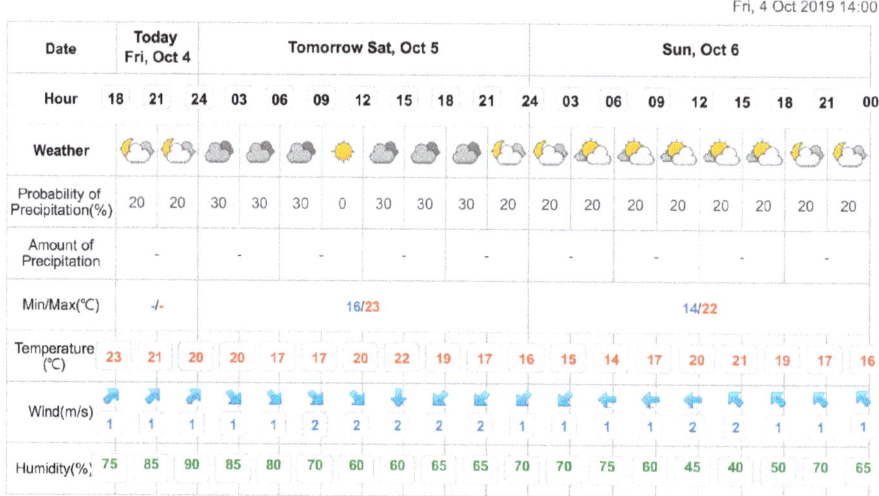
Fri, 4 Oct 2019 14:00

Date	Today Fri, Oct 4		Tomorrow Sat, Oct 5									Sun, Oct 6								
Hour	18	21	24	03	06	09	12	15	18	21	24	03	06	09	12	15	18	21	00
Weather																			
Probability of Precipitation(%)	20	20	30	30	30	0	30	30	30	20	20	20	20	20	20	20	20	20	
Amount of Precipitation	-		-		-		-		-		-		-		-		-		
Min/Max(°C)	-/-		16/23									14/22							
Temperature (°C)	23	21	20	20	17	17	20	22	19	17	16	15	14	17	20	21	19	17	16
Wind(m/s)	1	1	1	1	1	2	2	2	2	2	1	1	1	1	2	2	1	1	1
Humidity(%)	75	85	90	85	80	70	60	60	65	65	70	70	75	60	45	40	50	70	65

Figure 2. Example of short-term weather forecast information provided by KMA.

In addition, to establish a more direct correlation between weather data and electric energy consumption, we considered the discomfort index (DI) [40] and wind chill (WC) [41]. *DI* and *WC* are defined using Equations (1) and (2), respectively. Here, *T*, *H*, and *WS* represent the temperature, humidity, and wind speed, respectively.

$$DI = (1.8 \times T + 32) - [(0.55 - 0.0055 \times H) \times (1.8 \times T - 26)] \quad (1)$$

$$WC = 13.12 + 0.06215 \times T - 11.37 \times WS^{0.16} + 0.3965 \times T \times WS^{0.16} \quad (2)$$

As a result, we use nine types of weather data (i.e., daily maximum temperature, daily average temperature, daily minimum temperature, temperature, humidity, wind speed, precipitation, discomfort index, and wind chill) for the STLF model construction. Table 5 summarizes an example of weather conditions considered for the input variables.

Table 5. An example of weather conditions on December 25, 2018.

Hour	Avg. Temp.	Min. Temp.	Max. Temp.	Prec.	Temp.	Wind speed	Humi.	*DI*	*WC*
1	1.1	−2.3	4.7	0	−1.8	1.6	40.4	38.34	−1.03
2	1.1	−2.3	4.7	0	−2.0	1.3	42.5	37.76	−0.81
3	1.1	−2.3	4.7	0	−1.6	2.9	43.1	38.16	−2.11
4	1.1	−2.3	4.7	0	−1.9	3.9	43.0	37.80	−3.13
5	1.1	−2.3	4.7	0	−2.0	2.8	41.4	37.94	−2.46
6	1.1	−2.3	4.7	0	−2.3	3.3	41.4	37.57	−3.18
7	1.1	−2.3	4.7	0	−2.3	2.9	41.4	37.57	−2.87
8	1.1	−2.3	4.7	0	−2.2	2.4	41.2	37.73	−2.33
9	1.1	−2.3	4.7	0	2.1	2.4	43.2	37.52	−2.22
10	1.1	−2.3	4.7	0	−1.1	3.3	40.1	39.24	−1.85
11	1.1	−2.3	4.7	0	−0.1	4.1	42.0	40.17	−1.24
12	1.1	−2.3	4.7	0	1.1	3.9	44.3	41.34	0.21
13	1.1	−2.3	4.7	0	2.7	2.8	46.4	43.09	2.65
14	1.1	−2.3	4.7	0	4.7	2.5	46.7	45.60	5.03
15	1.1	−2.3	4.7	0	3.2	4.0	57.1	42.54	2.50
16	1.1	−2.3	4.7	0	2.7	3.6	76.1	39.64	2.16
17	1.1	−2.3	4.7	0	2.4	2.8	85.4	38.06	2.33
18	1.1	−2.3	4.7	0	2.6	3.7	86.0	38.32	1.99
19	1.1	−2.3	4.7	0	2.5	2.8	83.1	38.50	2.44
20	1.1	−2.3	4.7	0	2.8	3.1	81.3	39.20	2.56
21	1.1	−2.3	4.7	0	2.8	2.8	82.8	39.02	2.76
22	1.1	−2.3	4.7	0	2.8	3.4	83.6	38.93	2.38
23	1.1	−2.3	4.7	0	2.9	1.4	82.9	39.17	4.14
24	1.1	−2.3	4.7	0	3.1	2.1	81.1	39.70	3.63

3.3. Historical Electric Energy Consumption

Historical electric energy consumption is a good indicator of electric usage forecasts as it shows electricity usage patterns and trends [42]. We consider a specific time for 10 different days to reflect historical electric energy consumption. Herein, the historical electric energy consumption we considered included the last six days and the same day of the previous four weeks. For instance, if the forecast time is Saturday, 29 June 2019, 4 p.m., we use historical load data measured at 4 p.m. from 1, 8, 15, 22, 23, 24, 25, 26, 27, and 28 June. However, the weekday and weekend load patterns could be different. To reflect trends through historical load data more accurately, we use additional data that indicate whether the ten days used as input variables were holidays or not.

3.4. Electric Rates

Because one of the operational goals of the smart grid is to reduce electric charges, many STLF studies have used electric rate information as one of the input variables [43]. Thus, we also consider information on electric rates as input variables [44]. In South Korea, three different sections are used by Korea Electric Power Corporation (KEPCO) for electric rate: off-peak, mid-peak, and on-peak loads [45]. Electric rates are determined by the amount of electric energy consumption, the intended use of the building (i.e., residential, general, educational, industrial, etc.) and the season or month. As in the day of the week and holiday data, one-hot encoding method is used to represent the electric rate information. Hence, depending on the time and rate section, the input variable is set to 1; otherwise, it is set to zero.

4. Two-Stage STLF Model Construction

So far, various single algorithm-based STLF models have been proposed [46]. Even though they showed good performance in the domains that were focused on, their performance was limited in the other domains or electric energy consumption patterns were intricate. To alleviate this limitation, we propose a two-stage day-ahead STLF model that combines two single algorithm-based STLF models using DNNs.

4.1. The First Stage: Constructing Two STLF Models

In the first stage, we build two STLF models based on XGBoost and RF, which are well-known tree-based ensemble models in time series prediction [47], by using various input variables. They are based on boosting and bootstrap aggregating (bagging) algorithms, respectively. Compared to other boosting algorithms and bagging algorithms, the XGBoost and RF models show better predictive accuracy and have the highest correlation with actual power consumption. In addition, as XGBoost supports various loss functions, we can choose an appropriate loss function depending on the characteristics of the data. On the contrary, it suffers from overfitting during training [48]. RF can handle high dimensional data well, but it cannot give precise value for the regression model because the final prediction is the average of all the predictions from the subset trees [49]. By using the predicted values of the XGBoost model and the RF model together with other input variables, it is possible to prevent overfitting and to make more accurate prediction.

4.1.1. Extreme Boosting Machine

XGBoost, which was proposed by Chen and Guestrin [50], is a scalable machine learning model used in tree-boosting. It has been widely used for forecasting purposes such as STLF and store sales forecasting [51]. The basic principle of XGBoost is boosting, which combines a weak basic learning model with an active learner in an iterative fashion [52]. At each iteration of boosting, the residuals can modify the previous predictor to optimize the specified loss function. XGBoost provides faster learning and expandability based on parallel and distributed computing by further developing the existing boosting technique. It establishes an objective function to measure model performance by adding regularization to loss functions to improve performance. In addition, missing values can be handled easily because they are recognized and automatically supplemented to perform boosting. XGBoost gradually increases the depth of the tree at the beginning of learning. If the gain information obtained in the tree with increased depth is smaller than the of Gamma value, the depth stops increasing.

4.1.2. Random Forest

RF is a flexible machine learning algorithm that produces excellent results even without hyper parameter tuning. It has become one of the most commonly used machine learning algorithms because it can be easily used for classification and regression. Moreover, it can work efficiently on a large amount of data and handle thousands of input variables without deleting them, which is why it performs well. The basic principle of RF is called the bagging algorithm [53]. Bagging is an ensemble algorithm designed to improve the stability and accuracy of individual forecasting models such as decision trees. It selects a random sample of size n from the training set, fits it in the individual forecasting models, and produces a result that is averaged or voted on all individual forecasting models. The bagging algorithm in RF helps reduce the variance and influence of overfitting of decision trees.

4.2. The Second Stage: Combining STLF Models Using DNN

An ANN, which is also known as a multilayer perceptron (MLP), is a type of machine learning algorithm that is a feed-forward neural network architecture with an input layer, hidden layer, and output layer [54]. It aims to learn the nonlinear and complex structure of data by duplicating human brain functions [55]. Each layer in the neural network consists of several nodes. Each node receives values from the nodes in the previous layer to determine the output and provide values for the nodes in the next layer. As this process repeats, the nodes in the output layer provide the required values [56]. The number of hidden layers determines whether the network is deep or shallow. For instance, when the number of hidden layers is two or more, then the network is called a deep neural network [57]. Recently, various DNN models have shown excellent prediction performances due to the remarkably improved computing performance [58].

In the second stage, we construct an STLF model by combining the results of the two STLF models built in the first stage using a DNN. For training the DNN model, we used the predicted values of XGBoost and RF as input variables to reflect the characteristics of bagging and boosting algorithms. We also considered time factors, weather data, historical electric energy consumption data, and electric rate as input variables to further improve the forecasting performance. In our DNN model, we use the SELU function as an activation function and the number of hidden layers is set as five [36]. Additionally, the number of neurons in the hidden layer is set by two thirds of the number of input variables [59].

5. Economic Analysis Based CCHP Operation Scheduling

CCHP is known to improve energy utilization, reduce energy costs, and respond to peak loads by using thermal energy generated from the power generation process for heating and cooling. In addition, by using natural gas, CCHP can be a solution to environmental pollution [60]. Natural gas is a relatively clean-burning fossil fuel [61]. Burning natural gas for energy gives fewer emissions of nearly all types of air pollutants and carbon dioxide than burning coal or petroleum products to produce an equal amount of energy [62]. In this section, to see the applicability of our proposed scheme, we describe how daily CCHP operation scheduling can be made based on the forecasted daily electric energy consumption of 1 h resolution. In particular, we consider natural gas as the primary energy source of CCHP, and the economic benefit of CCHP operation is changed according to power generation efficiency. Hence, for its economic analysis, the cost of natural gas consumed in power generation must be determined by the power generation efficiency. Table 6 summarizes the gas charge sections of the industrial service in South Korea.

Table 6. Gas charge sections for industrial buildings.

Classification	Charge (Won/MJ)
Winter (Dec., Jan., Feb., Mar.)	14.627
Summer (Jun., Jul., Aug., Sep.)	13.988
Spring/Fall (Apr., May, Oct., Nov.)	14.061

The electric rate system should be considered for a more accurate economic analysis. There are several considerations in the electric rate system of South Korea.

- The electric rate system divides the types of contracts according to the purpose of electricity usage, and applies the corresponding charges. The contract types are divided into six classes, namely, residential, general, industrial, educational, agricultural, and streetlights service. Some contract types have more granular rates, depending on the size of the voltage or the contract demand.
- The electric rate consists of the demand and energy charges. The demand charge recovers the fixed costs related to the electric energy supply equipment. It is determined based on the contract demand or peak load. On the other hand, the energy charge recovers the variable costs in proportion to usage.
- Seasonal and hourly differential electric rates are applied to some contract types, including industrial and general service. To reflect the differences in supply costs by time zone according to seasonal demand, high rates are charged in seasons and time zones with high electric energy consumption, and low rates are applied in seasons and time zones with low electric energy consumption.

- The electric rate system offers three options depending on the relative amount of the demand and energy charges: Option I, Option II, and Option III. For the demand charge, Option I > Option II > Option III and for the energy charges, Option I < Option II < Option III. These options are for reducing energy consumption, inducing voluntary peak time load management, and ultimately reducing the cost of power equipment by enabling consumers to select an electric rate depending on their load pattern.

As we focus on the industrial building in this study, we have more refined electric rates depending on the supply voltage and the contract demand. First, depending on whether the contract demand exceeds 300 kW or not, there are two electric rates: Type A and Type B. For each rate, there are four groups depending on the size of the supply voltage: low voltage, high voltage A, high voltage B, and high voltage C. Each group then offers three options: Option I, Option II, and Option III. Table 7 summarizes the electric rates for Industrial Service (B), High Voltage A, and Option I. Industrial service (B) is an electric rate that can be used when contract demand is more than 300 kW.

Table 7. An example of electric rates table for an industrial building.

Classification		Demand Charge (Won/kW)	Energy charge (Won/kWh)			
			Time Period	Summer (1 June~31 August)	Spring/Fall (1 March~31 May/1 September~31 October)	Winter (1 November~28 February)
High-Voltage A	Option I	7220	Off-peak load	61.6	61.6	68.6
			Mid-load	114.5	84.1	114.7
			Peak-load	196.6	114.8	172.2

Operation scheduling is created to maximize annual economic benefits. Equations (3)–(6) represent detailed formulas for calculating annual economic benefits. Economic benefits are composed of two parts: (i) reduced electric charges, which are the direct economic benefits of using CCHP, and (ii) reduced heating/cooling charges by using thermal energy generated by CCHP. In the experiment, we assume CCHP can make 1.43 Mcal of thermal energy while generating 1 kWh [63]. We calculate how much it would cost to obtain this 1.43 Mcal of thermal energy using electric energy and reflect it in the formulas. Algorithm 1 shows the generation of an operational schedule for maximizing annual economic benefits. Basically, the schedule tells how much energy should be generated by CCHP and how much energy should be supplied by the public power system for each scheduling hour.

$$Contract\ Demand \geq max\left(Electric_{m,d,h} - CCHP_{m,d,h}\right) \tag{3}$$

$$\begin{aligned} Annual\ Economic\quad Benefit \\ &= Reduced\ Annual\ Electric\ Charges \\ &+ Reduced\ Annual\ Heating/Cooling\ Charges \end{aligned} \tag{4}$$

$$\begin{aligned} Reduced\ Annual\quad Electric\ Charges \\ &= Demand\ Rate1 \times max\left(Electric_{m,d,h}\right) \times 12 \\ &- Demand\ Rate2 \times Contract\ Demand \times 12 \\ &+ \sum_{m=1}^{12} \sum_{d=1}^{EoM} \sum_{h=1}^{24} ((Load_{m,d,h} \times Energy\ Rate1_{m,h}) \\ &- \sum_{m=1}^{12} \sum_{d=1}^{EoM} \sum_{h=1}^{24} (\left(CCHP_{m,d,h}\right) \times \left(\frac{Gas\ Rate_{m,h} \times \frac{42.377}{12.2}}{E_{CCHP}}\right) \\ &- \left(Electric_{m,d,h} - CCHP_{m,d,h}\right) \times Energy\ Rate2_{m,h}) \end{aligned} \tag{5}$$

$$\begin{aligned} Reduced\ Annual\quad Heating/Cooling\ Charges \\ &= \sum_{m=1}^{12} \sum_{d=1}^{EoM} \sum_{h=1}^{24} ((\tfrac{1.43}{2.3} \times CCHP_{m,d,h} \times Energy\ Rate2_{m,h}) \end{aligned} \tag{6}$$

Algorithm 1. CCHP operation scheduling

Input: month m, day d, hour h

Output: generation amount of CCHP $CCHP_{m,d,h}$, supply amount of public system $Public\ System_{m,d,h}$

```
if Contract Demand > 300
    Energy Rate_{m,h} = Energy Rate1_{m,h}
else
    Energy Rate_{m,h} = Energy Rate2_{m,h}
if Energy Rate_{m,h} − Gas Rate_{m,h} × (42.377 / (12.2×E_CCHP) − 5.9871) > 0
    CCHP_{m,d,h} = Electric_{m,d,h}
    Public System_{m,d,h} = 0
else
    if Load_{m,d,h} > Contract Demand
        CCHP_{m,d,h} = Electric_{m,d,h} − Contract Demand
        Public System_{m,d,h} = Contract Demand
    else
        CCHP_{m,d,h} = 0
        Public System_{m,d,h} = Electric_{m,d,h}
return CCHP_{m,d,h}, Public System_{m,d,h}
```

6. Experimental Results

6.1. Comparison of Prediction Performance with Various STLF Models

In this paper, we compare popular machine learning algorithms such as decision tree (DT), gradient boosting machine (GBM), bagging algorithm, and so on, to explain why we chose XGBoost and RF models in the first stage. Besides, we compare the performance with the prediction model (Persistence) which is actually using in the data collection environment. Persistence model uses the previous day (or the corresponding day in the previous week) as a prediction. Persistence implies that future values of the time series are calculated on the assumption that conditions remain unchanged between the current time and future time. As the second stage of our proposed model uses the predicted values of these two models, we divide the dataset into training set 1 (training the first-stage model), training set 2 (training the second-stage model), and test set (forecasting electric energy consumption and economic analysis), at a ratio of 50:25:25. Specifically, data collected from January 2015 to December 2016 was used as training set 1, data collected from January 2017 to December 2017 was used as training set 2, and data collected from January 2018 to December 2018 was used as test set. The performance of each machine learning algorithm was measured using the training set 2. Figure 3 shows monthly energy consumption and divided dataset. In addition, we compare our proposed model with various STLF models composed of different machine learning algorithms in the first-stage, and several forecasting models from our previous studies in the second-stage. To do this, we divided the dataset into training and test sets, at a ratio of 75:25.

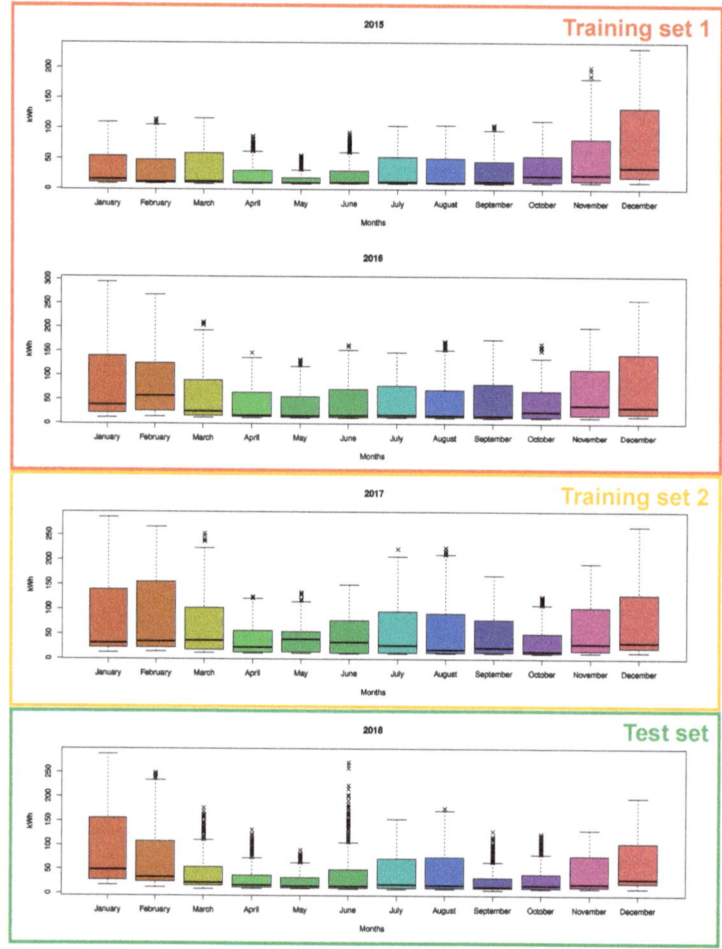

Figure 3. Box plots by the monthly electric load for separation of training and test sets.

Additionally, we selected a coefficient of variation of the root mean square error (CVRMSE) and mean absolute percentage error (MAPE) because they are easier to understand than other performance indicators such as the root mean square error (RMSE) or mean squared error (MSE) [64]. They were then used to evaluate the prediction performance of the proposed model. The CVRMSE and MAPE equations are shown in (7) and (8), respectively. Here, n is the number of time observed, $\overline{Y}$ is an average of the actual values. Y_i and $\hat{Y}_i$ are the actual and predicted values, respectively. Figure 4 exhibits the comparison of CVRMSE and MAPE results for each machine learning algorithm.

$$CVRMSE = \frac{100}{\overline{Y}}\sqrt{\frac{\sum_{i=1}^{n}\left(Y_i - \hat{Y}_i\right)^2}{n}} \tag{7}$$

$$MAPE = \frac{100}{n}\sum_{i=1}^{n}\left|\frac{Y_i - \hat{Y}_i}{Y_i}\right| \tag{8}$$

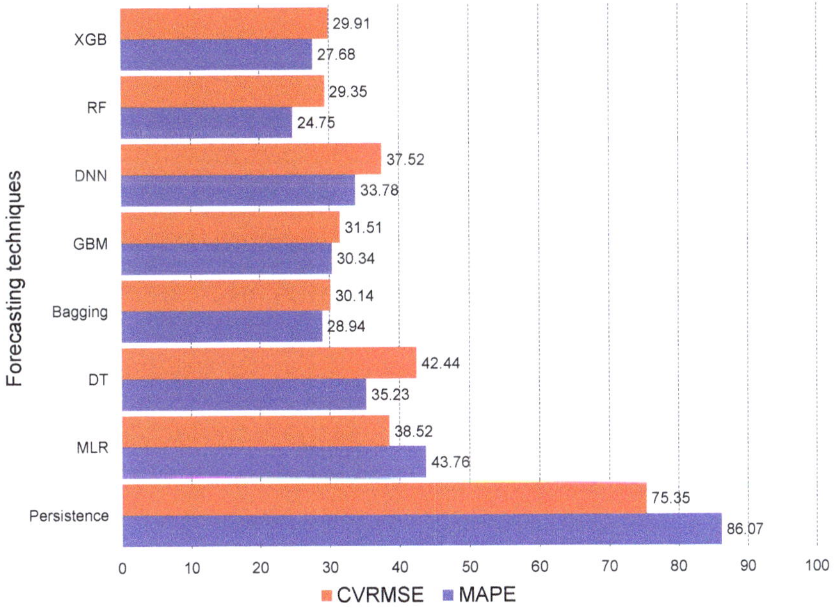

Figure 4. CVRMSE and MAPE comparison of machine learning algorithms.

As shown in Figure 4, XGBoost and RF models show better prediction performance in training set 2 compared with other machine learning algorithms. The performance of machine learning techniques is better than the persistence model which is a statistical model. In addition, the performance of the XGBoost and RF models was better than the other machine learning algorithms. XGBoost performed well because it allows users to choose an appropriate loss function depending on the characteristics of the data. RF performed well because it can handle high-dimensional data well. Table 8 summarizes the Pearson correlation coefficients between the forecasted values of machine learning algorithms and actual electric energy consumptions. We found that the forecasted values of XGBoost and RF present higher correlation coefficients than those of other machine learning algorithms. Therefore, we used the forecasted values of XGBoost and RF as new input variables for the second stage.

Table 8. Comparison of the Pearson correlation coefficients.

Machine Learning Algorithms	Correlation Coefficient
Extreme gradient boosting	0.941
Random forest	0.942
Deep neural networks	0.916
Gradient boosting machine	0.916
Bagging	0.924
Decision tree	0.890
Multiple linear regression	0.913
Persistence	0.612

Tables 9–11 summarize the comparison of our proposed model with other 2-stage models and several forecasting models of our previous studies in terms of CVRMSE and MAPE. As summarized in Tables 9–11, our proposed model showed an almost better prediction performance than other forecasting models.

Table 9. MAPE comparison for each month (The best in bold).

Forecasting Model	Jan.	Feb.	Mar.	Apr.	May	Jun.	Jul.	Aug.	Sep.	Oct.	Nov.	Dec.	Avg.
Moon et al. (2018) [33]	28.58	37.49	39.13	39.55	45.00	42.30	36.39	39.71	50.19	50.85	45.27	40.77	41.27
Son et al. (2018) [34]	23.11	34.42	33.57	23.31	22.13	33.06	29.28	26.43	29.68	28.49	23.71	27.47	27.84
Moon et al. (2019) [36]	22.35	30.79	37.78	30.89	32.88	33.15	26.51	25.77	30.23	33.11	30.27	32.19	30.48
Park et al. (2019) [37]	15.45	20.92	26.78	18.67	22.96	40.55	28.34	23.48	18.38	32.17	23.89	20.13	24.32
2-stage RF	22.39	37.06	37.08	27.38	40.23	48.22	43.18	41.88	47.28	42.41	25.80	24.24	36.42
2-stage XGBoost	22.36	36.93	33.73	25.06	37.20	35.47	31.31	28.66	34.96	32.67	21.00	22.94	30.15
Proposed Model	**14.62**	**13.98**	**20.35**	**18.58**	**18.95**	**16.50**	**19.42**	**18.45**	**21.09**	**18.33**	**19.81**	**18.24**	**18.22**

Table 10. CVRMSE comparison for each month (The best in bold).

Forecasting Model	Jan.	Feb.	Mar.	Apr.	May	Jun.	Jul.	Aug.	Sep.	Oct.	Nov.	Dec.	Avg.
Moon et al. (2018) [33]	28.87	40.90	57.15	57.55	66.32	74.95	43.67	46.17	75.25	69.58	51.00	44.36	50.59
Son et al. (2018) [34]	20.18	34.54	52.83	43.97	42.65	72.61	42.62	37.77	48.81	44.94	28.84	29.81	39.73
Moon et al. (2019) [36]	20.12	36.95	57.12	51.21	58.31	77.71	44.93	45.45	52.19	48.01	31.60	35.23	43.63
Park et al. (2019) [37]	18.05	31.03	47.38	35.30	41.95	70.52	39.76	35.64	44.25	49.43	29.77	23.77	36.85
2-stage RF	21.96	39.79	53.16	43.04	51.10	74.39	43.84	39.91	59.91	53.16	28.69	27.83	42.26
2-stage XGBoost	20.96	39.92	51.35	42.38	48.29	74.87	42.35	39.27	56.63	51.39	27.78	27.27	41.39
Proposed Model	**14.42**	**16.28**	**29.83**	**36.09**	**38.50**	**31.67**	**31.64**	**28.17**	**41.89**	**35.15**	**27.15**	**21.43**	**30.74**

Table 11. Statistical analysis of APEs by each forecasting model.

Forecasting Model	Min.	1st Qu.	Median	Mean	3rd Qu.	Max.
Moon et al. (2018) [33]	0.002	12.418	27.718	41.266	53.016	1607.788
Son et al. (2018) [34]	0.001	7.452	17.194	27.839	33.725	971.726
Moon et al. (2019) [36]	0.014	8.591	19.149	30.484	37.038	1132.967
Park et al. (2019) [37]	0.005	5.835	14.159	24.326	29.943	1210.511
2-stage RF	0.004	11.401	23.924	36.415	44.974	1378.919
2-stage XGBoost	0.001	9.666	20.742	30.147	38.180	1388.752
Proposed Model	0.001	5.454	12.414	18.221	23.445	1098.146

Finally, to ensure the significant contribution in terms of forecasting accuracy improvement for the proposed model, the Wilcoxon test and the Friedman test are conducted [30]. Wilcoxon test was used to test the null hypothesis by setting the null hypothesis to determine whether there was a significant difference between the two models. If the p-value is less than the significance level, the null hypothesis is rejected and the two models are judged to have significant differences. Friedman test is a multiple comparisons test that aims to detect significant differences between the results of two or more algorithms model. The results of the Wilcoxon test with the significance level set to 0.05 are shown in Table 12. Since the p-value in all cases is below the significance level, it was proven that proposed model is superior to the other models.

Table 12. Results of Wilcoxon test and Friedman test.

Compared Models		Wilcoxon Test (p-Value)	Friedman Test
Proposed Model	Moon et al. (2018) [33]	0	Friedman chi-squared = 4368.7 p-values < 2.2×10^{-16}
	Son et al. (2018) [34]	2.170288×10^{-88}	
	Moon et al. (2019) [36]	$5.954351 \times 10^{-158}$	
	Park et al. (2019) [37]	1.931605×10^{-21}	
	2-stage RF	0	
	2-stage XGBoost	1.37878×10^{-211}	

6.2. Economic Analysis Based CCHP Operation Scheduling

In this section, we describe how CCHP operation scheduling is made based on economic analysis. To maximize the annual economic benefits, it is also essential to determine the electric rate and amount of contract demand at the same time. We perform an experiment to find the optimal electric rate and contract demand to maximize on economic benefits.

A monthly economic analysis using the test set confirms that the economic benefits are similar to the monthly energy consumption, as shown in Figure 5. We can see that high economic benefits can be obtained in summer and winter when energy consumption is high.

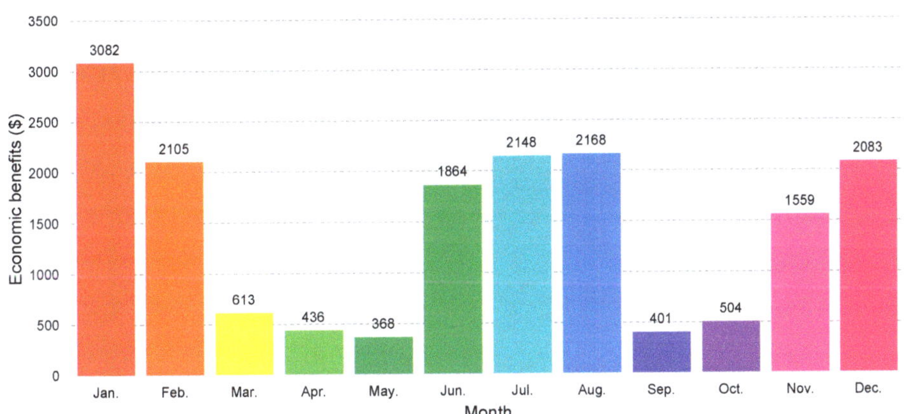

Figure 5. Monthly economic benefits.

Because the industrial building where the electric energy consumption data was collected is equipped with advanced meters, the electric rate of industrial service (A) II and industrial service (B) can be chosen. In addition, since the building's supply voltage is between 3300 V and 66,000 V, we choose the high voltage A as the electric rate of the building. Industrial service (A) II has two options, and industrial service (B) has three options. As a result, five different electric rates are compared in the experiment. Figure 6 shows the annual economic benefit of each electric rate based on contract demand.

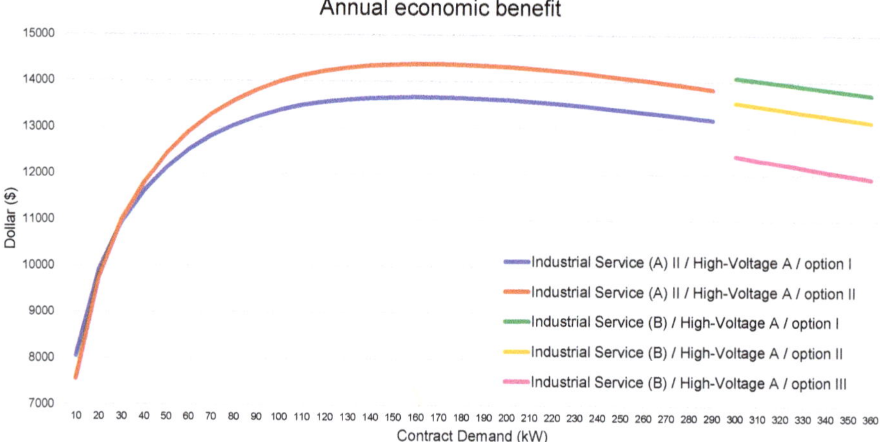

Figure 6. Annual economic benefit of each electric rate based on contract demand.

Figure 6 shows that "industrial service (A) II / high voltage A / Option II" electric rate with 160 kW contract demand can make the highest annual economic benefit and Figures 7–9 show the scheduling result of the CCHP operation according to this electric rate. In the figure, the yellow boxes represent electric energy supplied by the public power system and the green boxes represent electric energy generated by the CCHP system.

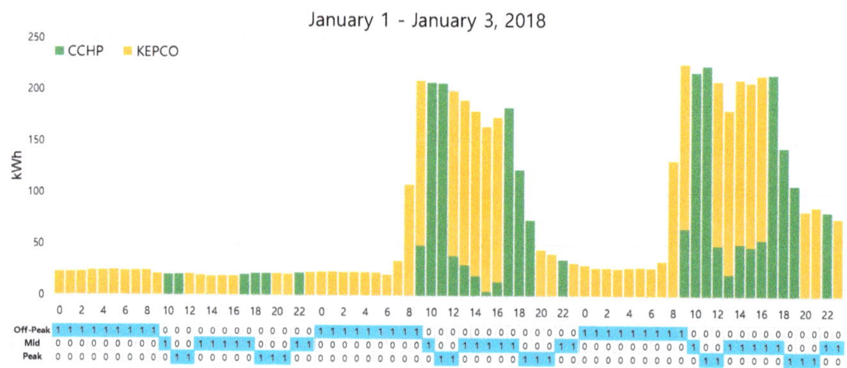

(**a**) CCHP operation scheduling based on predicted electric loads.

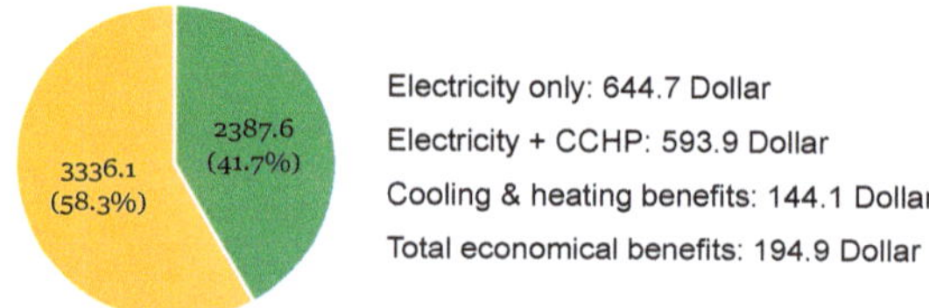

(**b**) Results of economic analysis based on predicted electric loads.

Figure 7. Example of CCHP operation scheduling.

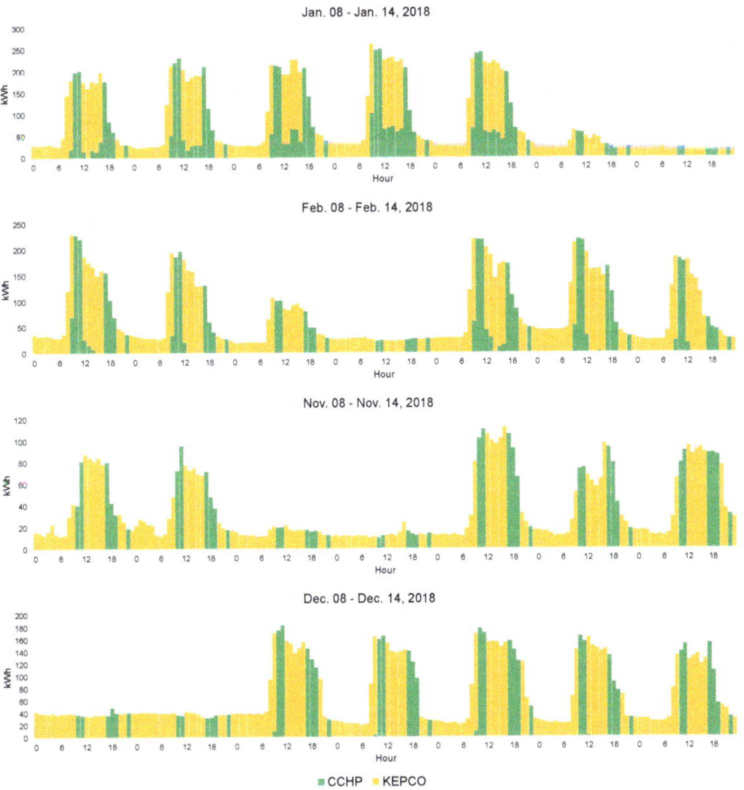

Figure 8. Example of CCHP operation scheduling in winter.

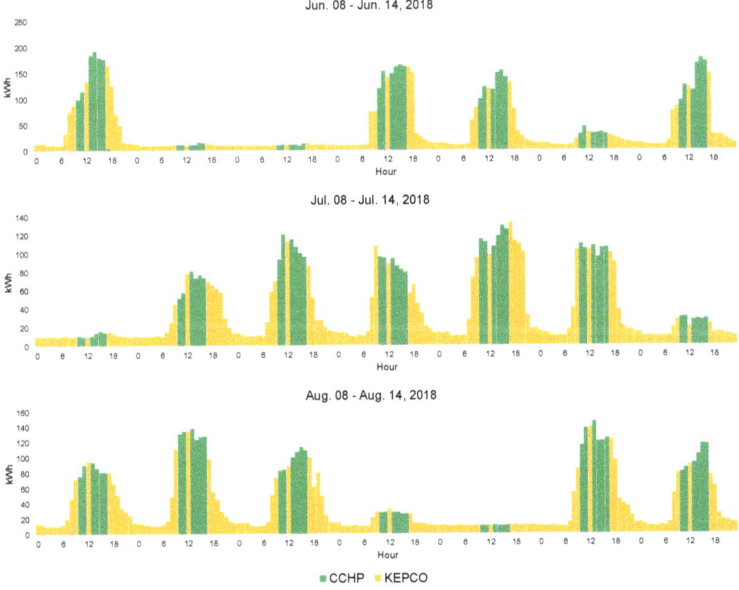

Figure 9. Example of CCHP operation scheduling in summer.

According to the schedule, an economic benefit of USD 195 can be made when using CCHP with a public power system for three days. Moreover, economic benefits of more than USD 14,000 annually can be achieved by using CCHP with the public power system.

7. Conclusions

In this study, we proposed a novel 2-stage STLF model that combines popular STLF models by using a DNN to further expand the domain of applicability. In the first stage, we used XGBoost and RF algorithms to predict day-ahead electric energy consumption. In the second stage, we built a load forecasting model based on DNN by using the forecasted results of XGBoost and RF and other external data as new input variables. To verify the forecasting performance of our proposed model, we performed day-ahead forecasting using actual factory electric energy consumption data and compared its accuracy with several machine learning methods and our previous forecasting models. The comparison showed that our proposed model showed the best prediction performance in terms of CVRMSE and MAPE.

Additionally, to show the applicability of our model, we performed CCHP operation scheduling based on forecasting and economic analysis, decided the best electric rate and contract demand, and showed how much could be saved by the decision. According to the experiment, the electric cost was reduced by 37% annually.

Author Contributions: All authors have read and agreed to the published version of the manuscript. Conceptualization, S.P. and J.M.; methodology, S.P. and J.M.; software, S.J.; validation, S.P., J.M. and S.J.; formal analysis, S.R. and E.H.; data curation, S.R.; writing—original draft preparation, S.P. and E.H.; writing—review and editing, E.H. and S.W.B.; visualization, S.J.; supervision, E.H.; project administration, E.H. and S.W.B.; funding acquisition, S.R.

Funding: This research was supported in part by the Korea Electric Power Corporation (grant number: R18XA05) and in part by Energy Cloud R&D Program (grant number: 2019M3F2A1073179) through the National Research Foundation of Korea (NRF) funded by the Ministry of Science and ICT.

Conflicts of Interest: The authors declare no conflict of interest.

References

1. Tabrizchi, H.; Javidi, M.M.; Amirzadeh, V. Estimates of residential building energy consumption using a multi-verse optimizer-based support vector machine with k-fold cross-validation. *Evol. Syst.* **2019**, 1–13. [CrossRef]
2. Morandin, M.; Bolognani, S.; Faggion, A. Active Torque Damping for an ICE-Based Domestic CHP System with an SPM Machine Drive. *IEEE Trans. Ind. Appl.* **2015**, *51*, 3137–3146. [CrossRef]
3. Lee, S.K.; Mogi, G. Relative efficiency of energy technologies in the Korean mid-term strategic energy technology development plan. *Renew. Sustain. Energy Rev.* **2018**, *91*, 472–482. [CrossRef]
4. Kenner, S.; Thaler, R.; Kucera, M.; Volbert, K.; Waas, T. Smart Grid Architecture for Monitoring and Analyzing, Including Modbus and REST Performance Comparison. In Proceedings of the 2015 12th International Workshop on Intelligent Solutions in Embedded Systems (WISES), Ancona, Italy, 29–30 October 2015; pp. 91–96.
5. Saponara, S.; Saletti, R.; Mihet-Popa, L. Hybrid Micro-Grids Exploiting Renewables Sources, Battery Energy Storages, and Bi-Directional Converters. *Appl. Sci.* **2019**, *9*, 4973. [CrossRef]
6. Ding, Z.; Lee, W.J.; Wang, J. Stochastic Resource Planning Strategy to Improve the Efficiency of Microgrid Operation. In Proceedings of the 2014 IEEE Industry Application Society Annual Meeting, Vancouver, BC, Canada, 5–9 October 2014; IEEE: Piscataway, NJ, USA, 2014.
7. Banerji, A.; Biswas, S.K.; Singh, B. Enhancing Quality of Power to Sensitive Loads with Microgrid. *IEEE Trans. Ind. Appl.* **2015**, *52*, 360–368. [CrossRef]
8. Pantaleo, A.M.; Camporeale, S.; Shah, N. Natural gas-biomass dual fuelled microturbines: Comparison of operating strategies in the Italian residential sector. *Appl. Therm. Eng.* **2014**, *71*, 686–696. [CrossRef]

9. Mortaji, M.; Ow, S.H.; Moghavvemi, M.; Almurib, H.A.F. Load Shedding and Smart-Direct Load Control Using Internet of Things in Smart Grid Demand Response Management. *IEEE Trans. Ind. Appl.* **2017**, *53*, 5155–5163. [CrossRef]
10. Zhu, R.; Guo, W.; Gong, X. Short-Term Load Forecasting for CCHP Systems Considering the Correlation between Heating, Gas and Electrical Loads Based on Deep Learning. *Energies* **2019**, *12*, 3308. [CrossRef]
11. Sun, Z.; Li, L.; Bego, A.; Dababneh, F. Customer-side electricity load management for sustainable manufacturing systems utilizing combined heat and power generation system. *Int. J. Prod. Econ.* **2015**, *165*, 112–119. [CrossRef]
12. Chen, Y.T. The Factors Affecting Electricity Consumption and the Consumption Characteristics in the Residential Sector—A Case Example of Taiwan. *Sustainability* **2017**, *9*, 1484. [CrossRef]
13. Mocanu, E.; Nguyen, P.H.; Gibescu, M.; Kling, W.L. Deep Learning for Estimating Building Energy Consumption. *Sustain. Energy Grids Netw.* **2016**, *6*, 91–99. [CrossRef]
14. Ahmad, A.S.; Hassan, M.Y.; Abdullah, M.P.; Rahman, H.A.; Hussin, F.; Abdullah, H.; Saidur, R. A review on applications of ANN and SVM for building electrical energy consumption forecasting. *Renew. Sustain. Energy Rev.* **2014**, *33*, 102–109. [CrossRef]
15. Hong, T.; Fan, S. Probabilistic electric load forecasting: A tutorial review. *Int. J. Forecast.* **2016**, *32*, 914–938. [CrossRef]
16. Dedinec, A.; Filiposka, S.; Dedinec, A.; Kocarev, L. Deep Belief Network based Electricity Load Forecasting: An Analysis of Macedonian Case. *Energy* **2016**, *115*, 1688–1700. [CrossRef]
17. Guo, Y.; Nazarian, E.; Ko, J.; Rajurkar, K. Hourly cooling load forecasting using time-indexed ARX models with two-stage weighted least squares regression. *Energy Convers. Manag.* **2014**, *80*, 46–53. [CrossRef]
18. Fan, S.; Chen, L. Short-term load forecasting based on an adaptive hybrid method. *IEEE Trans. Power Syst.* **2006**, *21*, 392–401. [CrossRef]
19. Tian, C.; Hao, Y. A novel nonlinear combined forecasting system for short-term load forecasting. *Energies* **2018**, *11*, 712. [CrossRef]
20. Kim, J.C.; Cho, S.M.; Shin, H.S. Advanced power distribution system configuration for smart grid. *IEEE Trans. Smart Grid* **2013**, *4*, 353–358. [CrossRef]
21. Armstrong, J.S. Combining Forecasts: The End of the Beginning or the Beginning of the End? *Int. J. Forecast.* **1989**, *5*, 585–588. [CrossRef]
22. Vaghefi, A.; Jafari, M.A.; Bisse, E.; Lu, Y.; Brouwer, J. Modeling and forecasting of cooling and electricity load demand. *Appl. Energy* **2014**, *136*, 186–196. [CrossRef]
23. Bagnasco, A.; Fresi, F.; Saviozzi, M.; Silvestro, F.; Vinci, A. Electrical consumption forecasting in hospital facilities: An application case. *Energy Build.* **2015**, *103*, 261–270. [CrossRef]
24. Powell, K.M.; Sriprasad, A.; Cole, W.J.; Edgar, T.F. Heating, cooling, and electrical load forecasting for a large-scale district energy system. *Energy* **2014**, *74*, 877–885. [CrossRef]
25. Jurado, S.; Nebot, À.; Mugica, F.; Avellana, N. Hybrid methodologies for electricity load forecasting: Entropy-based feature selection with machine learning and soft computing techniques. *Energy* **2015**, *86*, 276–291. [CrossRef]
26. Sandels, C.; Widén, J.; Nordström, L.; Andersson, E. Day-ahead predictions of electricity consumption in a Swedish office building from weather, occupancy, and temporal data. *Energy Build.* **2015**, *108*, 279–290. [CrossRef]
27. Grolinger, K.; L'Heureux, A.; Capretz, M.A.; Seewald, L. Energy forecasting for event venues: Big data and prediction accuracy. *Energy Build.* **2016**, *112*, 222–233. [CrossRef]
28. Gerossier, A.; Girard, R.; Kariniotakis, G.; Michiorri, A. Probabilistic Day-Ahead Forecasting of Household Electricity Demand. *CIRED-Open Access Proc. J.* **2017**, *2017*, 2500–2504. [CrossRef]
29. Chen, Y.; Tan, H. Short-term prediction of electric demand in building sector via hybrid support vector regression. *Appl. Energy* **2017**, *204*, 1363–1374. [CrossRef]
30. Dong, Y.; Zhang, Z.; Hong, W.C. A hybrid seasonal mechanism with a chaotic cuckoo search algorithm with a support vector regression model for electric load forecasting. *Energies* **2018**, *11*, 1009. [CrossRef]
31. Fan, G.F.; Peng, L.L.; Hong, W.C. Short term load forecasting based on phase space reconstruction algorithm and bi-square kernel regression model. *Appl. Energy* **2018**, *224*, 13–33. [CrossRef]
32. Hong, W.C.; Fan, G.F. Hybrid Empirical Mode Decomposition with Support Vector Regression Model for Short Term Load Forecasting. *Energies* **2019**, *12*, 1093. [CrossRef]

33. Moon, J.; Park, J.; Hwang, E.; Jun, S. Forecasting power consumption for higher educational institutions based on machine learning. *J. Supercomput.* **2018**, *74*, 3778–3800. [CrossRef]
34. Son, M.; Moon, J.; Jung, S.; Hwang, E. A Short-Term Load Forecasting Scheme Based on Auto-Encoder and Random Forest. In Proceedings of the 3rd International Conference on Applied Physics, System Science and Computers (APSAC), Dubrovnik, Croatia, 26–28 September 2018; pp. 138–144.
35. Kim, J.; Moon, J.; Hwang, E.; Kang, P. Recurrent inception convolution neural network for multi short-term load forecasting. *Energy Build.* **2019**, *194*, 328–341. [CrossRef]
36. Moon, J.; Park, S.; Rho, S.; Hwang, E. A comparative analysis of artificial neural network architectures for building energy consumption forecasting. *Int. J. Distrib. Sens. Netw.* **2019**, *15*, 1550147719877616. [CrossRef]
37. Park, S.; Moon, J.; Hwang, E. 2-Stage Electric Load Forecasting Scheme for Day-Ahead CCHP Scheduling. In Proceedings of the 13th IEEE International Conference on Power Electronics and Drive Systems (PEDS), Toulouse, France, 9–12 July 2019.
38. Moon, J.; Kim, Y.; Son, M.; Hwang, E. Hybrid Short-Term Load Forecasting Scheme Using Random Forest and Multilayer Perceptron. *Energies* **2018**, *11*, 3283. [CrossRef]
39. Wang, P.; Liu, B.; Hong, T. Electric load forecasting with recency effect: A big data approach. *Int. J. Forecast.* **2016**, *32*, 585–597. [CrossRef]
40. Xie, J.; Chen, Y.; Hong, T.; Laing, T.D. Relative humidity for load forecasting models. *IEEE Trans. Smart Grid* **2016**, *9*, 191–198. [CrossRef]
41. Xie, J.; Hong, T. Wind speed for load forecasting models. *Sustainability* **2017**, *9*, 795. [CrossRef]
42. Iglesias, F.; Kastner, W. Analysis of similarity measures in times series clustering for the discovery of building energy patterns. *Energies* **2013**, *6*, 579–597. [CrossRef]
43. Lee, W.; Jung, J.; Lee, M. Development of 24-hour optimal scheduling algorithm for energy storage system using load forecasting and renewable energy forecasting. In Proceedings of the 2017 IEEE Power & Energy Society General Meeting (PESGM), Chicago, IL, USA, 16–20 July 2017.
44. Raza, M.Q.; Khosravi, A. A review on artificial intelligence based load demand forecasting techniques for smart grid and buildings. *Renew. Sustain. Energy Rev.* **2015**, *50*, 1352–1372. [CrossRef]
45. Lee, W.; Jung, J.; Kang, B.O. Cost-Benefit Analysis for Industrial Customers-Installed Energy Storage System in South Korea. In Proceedings of the 2018 IEEE Innovative Smart Grid Technologies-Asia (ISGT Asia), Singapore, 22–25 May 2018.
46. Le, T.; Vo, M.T.; Vo, B.; Hwang, E.; Rho, S.; Baik, S.W. Improving electric energy consumption prediction using CNN and Bi-LSTM. *Appl. Sci.* **2019**, *9*, 4237. [CrossRef]
47. Lindner, C.; Bromiley, P.A.; Ionita, M.C.; Cootes, T.F. Robust and accurate shape model matching using random forest regression-voting. *IEEE Trans. Pattern Anal. Mach. Intell.* **2014**, *37*, 1862–1874. [CrossRef] [PubMed]
48. Gómez-Ríos, A.; Luengo, J.; Herrera, F. A study on the noise label influence in boosting algorithms: AdaBoost, GBM and XGBoost. In Proceedings of the 12th International Conference, HAIS 2017, La Rioja, Spain, 21–23 June 2017; Springer: Cham, Switzerland, 2017; pp. 268–280.
49. Qingqing, K.; Xiangqian, D.; Huili, G. Application of improved random forest pruning algorithm in tobacco origin identification of near infrared spectrum. *Laser Optoelectron. Prog.* **2018**, *55*, 446–451.
50. Chen, T.; Guestrin, C. Xgboost: A scalable tree boosting system. In Proceedings of the 22nd ACM SIGKDD International Conference on Knowledge Discovery and Data Mining, San Francisco, CA, USA, 13–17 August 2016.
51. Ruiz-Abellón, M.; Gabaldón, A.; Guillamón, A. Load forecasting for a campus university using ensemble methods based on regression trees. *Energies* **2018**, *11*, 2038. [CrossRef]
52. Schapire, R.E. The boosting approach to machine learning: An overview. In *Nonlinear Estimation and Classification*; Springer: Berlin, Germany, 2003; pp. 149–171.
53. Breiman, L. Bagging predictors. *Mach. Learn.* **1996**, *24*, 123–140. [CrossRef]
54. Liang, Y.; Niu, D.; Hong, W.C. Short term load forecasting based on feature extraction and improved general regression neural network model. *Energy* **2019**, *166*, 653–663. [CrossRef]
55. Baek, S.J.; Yoon, S.G. Short-Term Load Forecasting for Campus Building with Small-Scale Loads by Types Using Artificial Neural Network. In Proceedings of the 2019 IEEE Power & Energy Society Innovative Smart Grid Technologies Conference (ISGT), Washington, DC, USA, 18–21 February 2019.

56. Liu, P.; Zheng, P.; Chen, Z. Deep Learning with Stacked Denoising Auto-Encoder for Short-Term Electric Load Forecasting. *Energies* **2019**, *12*, 2445. [CrossRef]
57. Ryu, S.; Noh, J.; Kim, H. Deep neural network based demand side short term load forecasting. *Energies* **2017**, *10*, 3. [CrossRef]
58. Shi, H.; Xu, M.; Li, R. Deep learning for household load forecasting—A novel pooling deep RNN. *IEEE Trans. Smart Grid* **2017**, *9*, 5271–5280. [CrossRef]
59. Heaton, J. *Introduction to Neural Networks with Java*; Heaton Research, Inc.: Chesterfield, MO, USA, 2008; ISBN 1-60439-008-5.
60. Li, H.; Fu, L.; Geng, K.; Jiang, Y. Energy utilization evaluation of CCHP systems. *Energy Build.* **2006**, *38*, 253–257. [CrossRef]
61. Allen, D.T.; Torres, V.M.; Thomas, J.; Sullivan, D.W.; Harrison, M.; Hendler, A.; Herndon, S.C.; Kolb, C.E.; Fraser, M.P.; Hill, A.D.; et al. Measurements of methane emissions at natural gas production sites in the United States. *Proc. Natl. Acad. Sci. USA* **2013**, *110*, 17768–17773. [CrossRef]
62. Van Ruijven, B.; Van Vuuren, D.P. Oil and natural gas prices and greenhouse gas emission mitigation. *Energy Policy* **2009**, *37*, 4797–4808. [CrossRef]
63. Yun, R. Economic analysis of CHP system for building by CHP capacity optimizer. *Korean J. Air-Cond. Refrig. Eng.* **2008**, *20*, 321–326.
64. Kuo, P.H.; Huang, C.J. A high precision artificial neural networks model for short-term energy load forecasting. *Energies* **2018**, *11*, 213. [CrossRef]

© 2020 by the authors. Licensee MDPI, Basel, Switzerland. This article is an open access article distributed under the terms and conditions of the Creative Commons Attribution (CC BY) license (http://creativecommons.org/licenses/by/4.0/).

MDPI
St. Alban-Anlage 66
4052 Basel
Switzerland
Tel. +41 61 683 77 34
Fax +41 61 302 89 18
www.mdpi.com

Energies Editorial Office
E-mail: energies@mdpi.com
www.mdpi.com/journal/energies

www.ingramcontent.com/pod-product-compliance
Lightning Source LLC
LaVergne TN
LVHW070127170726
843515LV00007B/1761
* 9 7 8 3 0 3 9 2 8 7 0 0 0 *